A-Level Year 2

Physics

The Complete Course for OCR A

Ever stared up at a beautiful twinkling star in the night sky and thought, "I wish I knew how to calculate how far away that was, making full use of the angle of parallax due to the Earth's orbit around the Sun"?

If you've made it to Year 2 of A-Level Physics, your answer might be yes. And this fantastic CGP book might be your new favourite thing — it'll help you understand the universe *and* score top marks in your exams.

What more could you ask for — a free Online Edition? Oh, go on then.

CGP

How to get your free Online Edition

Go to **cgpbooks.co.uk/extras** and enter this code...

4229 7136 0103 1012

This code will only work once. If someone has used this book before you, they may have already claimed the Online Edition.

Contents

Introduction

How Science Works

Practical Skills

Practical Skills for the Practical Endorsement

Module 5

Section 1: Thermal Physics

Section 2: Circular Motion and Oscillations

Section 3: Gravitational Fields

Section 4: Astrophysics and Cosmology

Module 6

Section 1: Capacitors

Section 2: Electric Fields

Section 3: Electromagnetism

Section 4: Nuclear and Particle Physics

Section 5: Medical Imaging

Maths Skills

MATHS SKILLS

Exam Help

EXAM HELP

Reference

How to use this book

Learning Objectives

- These tell you exactly what you need to learn, or be able to do, for the exam.
- There's a specification reference at the bottom that links to the OCR A specification.

Examples

These are here to help you understand the theory.

Tips and Exam Tips

There are tips throughout the book to help with all sorts of things, including exam tips to do with answering exam questions.

Learning Objectives:

- Understand what is meant by induced nuclear fission.
- Understand what is meant by nuclear fusion.
- Know why fusion reactions require very high temperatures.
- Be able to balance nuclear transformation equations.
- Be able to use the graph of binding energy per nucleon against nucleon number to calculate energy changes in reactions.
- Be able to calculate the energy released (or absorbed) in simple nuclear reactions.

Specification Reference 6.4.4

Tip: Heavy nuclei don't always fission to form the same daughter nuclei — ^{235}U can split into lots of different pairs of nuclei (with nucleon numbers around 90 and 140).

Tip: Just like decay equations, equations showing nuclear fission and fusion need to be balanced — see page 196.

10. Nuclear Fission and Fusion

Radioactive decay isn't the only way that nuclei can change — they can also split into two smaller nuclei, or fuse with other nuclei to form larger ones.

Nuclear fission

Heavy nuclei (e.g. uranium) are unstable, and some can randomly split into two smaller nuclei — this is called **nuclear fission**. This process is called spontaneous if it just happens by itself, or induced if we encourage it to happen. When nuclear fission occurs, in addition to the two smaller nuclei, a large amount of energy is released, along with a number of free neutrons.

Example

Fission can be induced by making a neutron enter a ^{235}U nucleus, causing it to become very unstable. Only low-energy neutrons can be captured in this way. A low-energy neutron is called a **thermal neutron**.

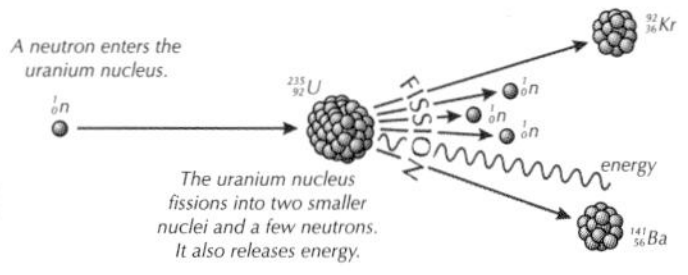

Figure 1: A possible fission of a uranium-235 nucleus.

Energy is released during nuclear fission because the new, smaller nuclei have a higher average binding energy per nucleon (see page 208). The larger the nucleus, the more unstable it will be — so large nuclei are more likely to spontaneously fission. This means that spontaneous fission limits the number of nucleons that a nucleus can contain — in other words, it limits the number of possible elements.

Nuclear fusion

Two light nuclei can combine to create a larger nucleus. This is called **nuclear fusion**.

Example

In the Sun, hydrogen nuclei fuse in a series of reactions to form helium. One of the reactions is: $^2_1H + ^1_1H \longrightarrow ^3_2He$ + energy.

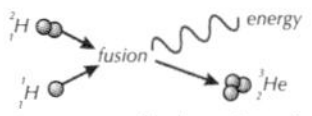

Figure 2: Two isotopes of hydrogen fuse to form helium.

210 Module 6: Section 4 Nuclear and Particle Physics

Example — Maths Skills

The graph on the right shows the activity-time graph of a sample of a radioisotope. Calculate the half-life of the isotope.

The starting activity of the sample is 120 Bq. So to find the half-life, read off the time at the point when the activity has dropped to half this value.

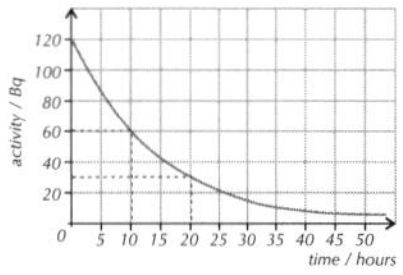

120 ÷ 2 = 60. When the activity is 60 Bq, time = 10 hours.

Check this by finding the time taken for the activity to fall to a quarter of its initial value and halving it. When the activity is 30 Bq, time = 20 hours. 20 ÷ 2 = 10 hours. So, the half-life is 10 hours.

Investigating half-life

PRACTICAL ACTIVITY GROUP 7

You can determine the half-life of an isotope for yourself by measuring how the count rate detected from a sample decreases with time. You're most likely to do this using the isotope protactinium-234. Protactinium-234 is formed when uranium decays (via another isotope). You can measure protactinium-234's decay rate using a protactinium generator — a bottle containing a uranium salt and two solvents, which separate out into layers, as shown in Figure 3. The uranium salt is only soluble in the bottom layer. The uranium salt decays into protactinium-234 (as well as some other isotopes, but you don't need to worry about those).

When the generator is shaken, the solvents are mixed. Only the protactinium-234 can form a solution with the top layer solvent, so when the layers in the bottle are allowed to separate, the top layer will only contain the protactinium-234 in solution, while the uranium salt and other decay products are still left in the bottom layer.

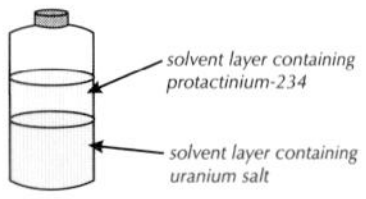

Figure 3: A protactinium generator.

So if a Geiger-Müller tube is directed towards the top layer only, the activity of protactinium-234 can be monitored without confusing its counts with those from other decaying isotopes.

1. Start by setting up the equipment shown in Figure 5. Don't shake the protactinium generator yet. Find the background counts by measuring the number of counts detected over a time period (at least 30 s). Repeat this two more times, find the mean of your three results, and divide the mean counts by the number of seconds. This is the background count rate you will need to subtract from your future results.

Tip: Make sure you do a risk assessment before carrying out this experiment.

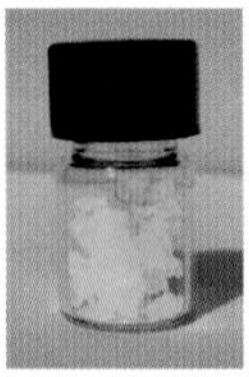

Figure 4: Uranyl nitrate, a uranium salt. It can be dissolved in water and used in protactinium generators.

Tip: The uranium in the salt is always decaying into protactinium-234, so you can reuse the protactinium generator by shaking the bottle again to dissolve more protactinium-234 in the top layer.

Module 6: Section 4 Nuclear and Particle Physics 203

Maths Skills

There's a range of maths skills you could be expected to apply in your exams. Examples that show these maths skills in action are marked up like this. There's also a maths skills section at the back of the book.

Practical Activity Groups

You need to show you've mastered some key practical skills in your Practical Endorsement. Information on the skills you need and opportunities to apply them are marked up throughout the book.

There's also a section on the Practical Endorsement near the beginning of the book.

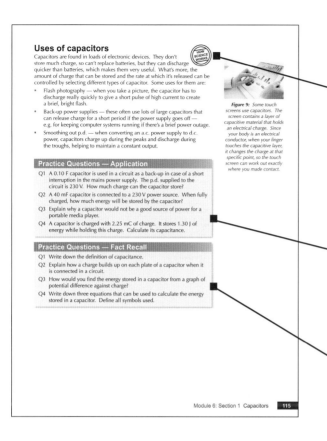

Uses of capacitors

Capacitors are found in loads of electronic devices. They don't store much charge, so can't replace batteries, but they can discharge quicker than batteries, which makes them very useful. What's more, the amount of charge that can be stored and the rate at which it's released can be controlled by selecting different types of capacitor. Some uses for them are:

- Flash photography — when you take a picture, the capacitor has to discharge really quickly to give a short pulse of high current to create a brief, bright flash.
- Back-up power supplies — these often use lots of large capacitors that can release charge for a short period if the power supply goes off — e.g. for keeping computer systems running if there's a brief power outage.
- Smoothing out p.d. — when converting an a.c. power supply to d.c. power, capacitors charge up during the peaks and discharge during the troughs, helping to maintain a constant output.

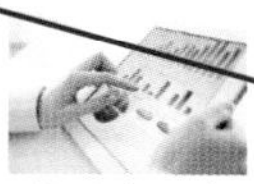

Figure 9: Some touch screens use capacitors. The screen contains a layer of capacitive material that holds an electrical charge. Since your body is an electrical conductor, when your finger touches the capacitive layer, it changes the charge at that specific point, so the touch screen can work out exactly where you made contact.

Practice Questions — Application

Q1 A 0.10 F capacitor is used in a circuit as a back-up in case of a short interruption in the mains power supply. The p.d. supplied to the circuit is 230 V. How much charge can the capacitor store?

Q2 A 40 mF capacitor is connected to a 230 V power source. When fully charged, how much energy will be stored by the capacitor?

Q3 Explain why a capacitor would not be a good source of power for a portable media player.

Q4 A capacitor is charged with 2.25 mC of charge. It stores 1.30 J of energy while holding this charge. Calculate its capacitance.

Practice Questions — Fact Recall

Q1 Write down the definition of capacitance.

Q2 Explain how a charge builds up on each plate of a capacitor when it is connected in a circuit.

Q3 How would you find the energy stored in a capacitor from a graph of potential difference against charge?

Q4 Write down three equations that can be used to calculate the energy stored in a capacitor. Define all symbols used.

Module 6: Section 1 Capacitors 115

How Science Works

- You need to know about How Science Works. There's a section on it at the front of the book.
- How Science Works is also covered throughout the book wherever you see this symbol.

HOW SCIENCE WORKS

Practice Questions — Application

- Annoyingly, the examiners expect you to be able to apply your knowledge to new situations — these questions are here to give you plenty of practice at doing this.
- All the answers are in the back of the book (including any calculation workings).

Practice Questions — Fact Recall

- There are a lot of facts you need to learn — these questions are here to test that you know them.
- All the answers are in the back of the book.

Exam-style Questions

- Practising exam-style questions is really important — you'll find some at the end of each section.
- They're the same style as the ones you'll get in the real exams — some will test your knowledge and understanding and some will test that you can apply your knowledge.
- All the answers are in the back of the book, along with a mark scheme to show you how you get the marks.

Exam-style Questions

1 A 33 cm long conducting rod moves perpendicular to a magnetic field of magnetic flux density 21 mT. An e.m.f. of 4.5 mV is generated across its length. At what velocity is the bar moving?

A 0.071 ms^{-1}

B 14.1 ms^{-1}

C 0.65 ms^{-1}

D 1.5 ms^{-1}

(1 mark)

2 In a velocity selector, electrons are travelling perpendicularly through an electric field and a magnetic field, which are at right angles to each other. The velocity selector is used to select electrons moving with a velocity of 42 kms^{-1} by varying the magnetic flux density, so only electrons travelling at this speed will travel in a straight line and pass through a hole in a collimator. The field strength of the electric field is $37 \times 10^3 \text{ NC}^{-1}$. What is the magnetic flux density of the magnetic field?

A 0.88 T

B 1.1 T

C 8.8×10^2 T

D 1.1×10^{-3} T

(1 mark)

3 An electron is fired into a uniform magnetic field with a flux density of 0.93 T at a speed of 8.1×10^7 m, as **Fig 3.1** shows.

(a) What shape will the electron's path take? You can assume that the electron's velocity is perpendicular to the magnetic field.

(1 mark)

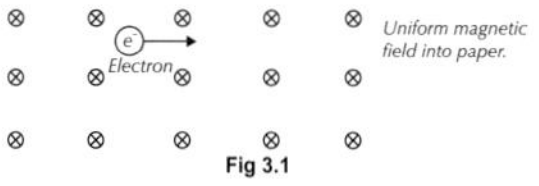

Fig 3.1

(b) Calculate the magnitude of the force the electron will experience, and state its direction.

(2 marks)

(c) An alpha particle of charge +2e is fired into the same magnetic field as the electron at the same speed. State the magnitude and direction of the force experienced by the alpha particle.

(2 marks)

174 Module 6: Section 3 Electromagnetism

Exam Help

There's a section at the back of the book stuffed full of things to help with your exams.

Glossary

There's a glossary at the back of the book full of useful words — perfect for looking up key words and their meanings.

Published by CGP

Editors:
Sarah Armstrong, Emily Garrett, Duncan Lindsay, Andy Park, Frances Rooney and Charlotte Whiteley.

Contributors:
Peter Cecil, Mark Edwards, Barbara Mascetti, John Myers, Zoe Nye, Moira Stevens and Andy Williams.

ISBN: 978 1 78294 791 2

With thanks to Mark Edwards for the proofreading.
With thanks to Ana Pungartnik for the copyright research.

Printed by Elanders Ltd, Newcastle upon Tyne.
Clipart from Corel®

The Scientific Process

Science tries to explain how and why things happen. It's all about gaining knowledge about the world. Scientists do this by asking questions, suggesting answers and then doing tests to see if they're correct — the scientific process.

Developing and testing theories

A **theory** is a possible explanation for something. Theories usually come about when scientists observe something and wonder why or how it happens. Scientists also sometimes form a **model** too — a simplified picture or representation of a real physical situation. Scientific theories and models are developed and tested in the following way:

- Ask a question — make an observation and ask why or how whatever you've observed happens.
- Suggest an answer, or part of an answer, by forming a theory or a model.
- Make a **prediction** or **hypothesis** — a specific testable statement, based on the theory, about what will happen in a test situation.
- Carry out tests — to provide evidence that will support the hypothesis or refute it (help to disprove it).

Tip: A theory is only scientific if it can be tested.

Tip: The results of one test can't prove that a theory is true — they can only suggest that it's true. They can however disprove a theory — show that it's wrong.

Example

Question: What stops beta radiation?

Theory: Beta radiation will be stopped by a 5 mm thick sheet of aluminium.

Hypothesis: If beta radiation is stopped by a 5 mm thick sheet of aluminium a Geiger-Müller tube and counter will not detect radiation from a beta source if the aluminium sheet is placed between the source and the tube.

Test: Measure the background radiation count rate using the Geiger-Müller tube and counter without a source present. Then, place a beta-emitting source in front of the tube. Record the new count rate. Then, place a 5 mm thick sheet of aluminium between the source and the tube and once more record the count rate. If the count rate has dropped to a level similar to the background count rate, then the beta radiation emitted by the source has been absorbed by the aluminium sheet, and this evidence supports the hypothesis.

Communicating results

The results of testing a scientific theory are published — scientists need to let others know about their work. Scientists publish their results in scientific journals. These are just like normal magazines, only they contain scientific reports (called papers) that use scientific terminology, instead of the latest celebrity gossip.

Scientific reports are similar to the lab write-ups you do in school. And just as a lab write-up is reviewed (marked) by your teacher, reports in scientific journals undergo **peer review** before they're published. The report is sent out to peers — other scientists who are experts in the same area.

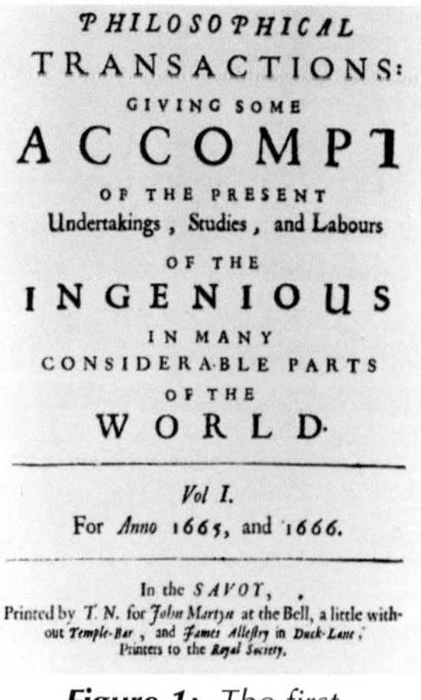
PHILOSOPHICAL
TRANSACTIONS:
GIVING SOME
ACCOMPT
OF THE PRESENT
Undertakings, Studies, and Labours
OF THE
INGENIOUS
IN MANY
CONSIDERABLE PARTS
OF THE
WORLD.

Vol I.
For Anno 1665, and 1666.

In the SAVOY,
Printed by T. N. for John Martyn at the Bell, a little without Temple-Bar, and James Allestry in Duck-Lane, Printers to the Royal Society.

***Figure 1:** The first British scientific journal, 'Philosophical Transactions of the Royal Society', published in 1665.*

Tip: Scientific research is often funded by companies who have a vested interest in its outcomes. Scientists are ethically obliged to make sure that this does not bias their results.

The other scientists go through it bit by bit, examining the methods and data, and checking it's all clear and logical. Thorough evaluation allows decisions to be made about what makes a good methodology or experimental technique. Individual scientists may have their own ethical codes (based on their humanistic, moral and religious beliefs), but having their work scrutinised by other scientists helps to reduce the effect of personal bias (either deliberate or accidental) on the conclusions drawn from the results.

When the report is approved, it's published. This makes sure that work published in scientific journals is of a good standard. But peer review can't guarantee the science is correct — other scientists still need to reproduce it. Sometimes mistakes are made and bad work is published. Peer review isn't perfect but it's probably the best way for scientists to self-regulate their work and to publish quality reports.

Validating theories

Other scientists read the published theories and results, and try to test the theory themselves in order to validate it (back it up). This involves:

- Repeating the exact same experiments.
- Using the theory to make new predictions and then testing them with new experiments.

Tip: Once an experimental method is found to give good evidence it becomes a <u>protocol</u> — an accepted method to test that particular thing that all scientists can use.

Examples

- In 1989, two scientists claimed that they'd produced 'cold fusion' (the energy source of the Sun but without the high temperatures). If it was true, it would have meant cheap energy for the world forever. However, other scientists just couldn't reproduce the results, so the theory of 'cold fusion' couldn't be validated.
- In the 1960s it was proposed that particles like protons were made up of smaller particles called quarks — the quark model (p.189). After this, many experiments were conducted that supported this idea, and so validated the quark model.

How do theories evolve?

Tip: Sometimes data from one experiment can be the starting point for developing a new theory.

If multiple experiments show a theory to be incorrect then scientists either have to modify the theory or develop a new one, and start the testing again. If all the experiments in all the world provide good evidence to back a theory up, the theory is thought of as scientific 'fact' (for now) — see Figure 2. But it will never become totally indisputable fact. Scientific breakthroughs or advances could provide new ways to question and test the theory, which could lead to new evidence that conflicts with the current evidence. Then the testing starts all over again... And this, my friend, is the tentative nature of scientific knowledge — it's always changing and evolving.

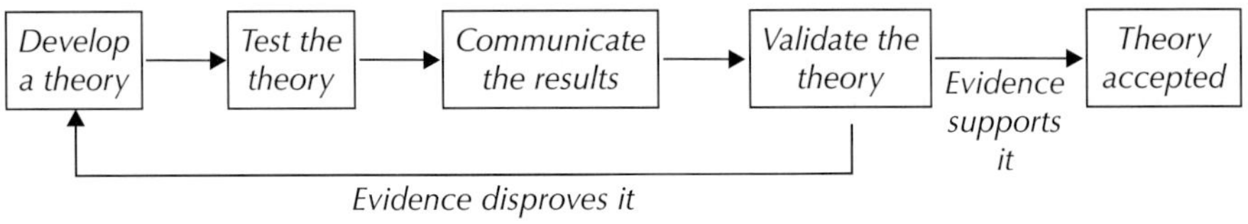

Figure 2: *Flow diagram summarising the scientific process.*

Example

The structure of the atom

It took years and years for the current model of the atom to be developed and accepted.

Dalton's theory in the early 1800s, that atoms were solid spheres, was disputed by the results of Thomson's experiments at the end of that century. As a result, Thomson developed the 'plum pudding' model of the atom, which was later proven wrong by Rutherford's alpha-scattering experiments in the early 1900s. Rutherford's 'nuclear model' has since been developed and modified further to create the currently accepted model of the atom we use today — but scientists are still searching for more accurate models.

Tip: See p.176 for more on the development of the model of the atom.

Collecting evidence

1. Evidence from lab experiments

Results from controlled experiments in laboratories are great. A lab is the easiest place to control **variables** so that they're all kept constant (except for the one you're investigating). This means you can draw meaningful conclusions.

Tip: Pages 5-12 are all about carrying out practicals. You should also remember all the skills covered in Module 1 in year 1.

Example

The pressure of a gas

If you're investigating how the volume of a gas affects its pressure, you need to keep all other factors constant. This means controlling things like the temperature of the gas. Otherwise there's no way of knowing that the volume change is what is causing the pressure to change.

2. Investigations outside the lab

There are things you can't study in a lab. And outside the lab, controlling the variables is tricky, if not impossible.

Example

Does living near power lines increase the risk of developing certain cancers?

You could compare the number of cancer cases in a group of people who live near power lines to a group of people who don't. But there are always differences between groups of people. The best you can do is to have a well-designed study using matched groups — choose two groups of people (those who live near power lines and those who don't) that are as similar as possible (same mix of ages, same mix of diets etc.). But you still can't rule out every possibility.

Figure 3: *Studies are ongoing to determine if there is a link between proximity to power lines and the risk of cancer.*

Science and decision making

Scientific knowledge is used by society (that's you, me and everyone else) to make decisions — about the way we live, what we eat, what we drive, etc. All sections of society use scientific evidence to make decisions, e.g. politicians use it to devise policies and individuals use science to make decisions about their own lives.

Tip: Don't get mixed up — it's not the scientists who make the decisions, it's society. Scientists just produce evidence to help society make the decisions.

Example

X-rays are used in dentistry to see images of the teeth inside the human body. They're used to monitor dental health, including seeing when adult teeth or wisdom teeth are coming through.

However, science has found a link between exposure to X-rays and increased risk of cancer. To minimise this risk, there are restrictions on when X-ray images of a patient's mouth can be taken (e.g. not during pregnancy), and how often. As a result, patients don't tend to have X-ray images taken as part of their standard check-ups, unless there is a strong medical reason to do so.

Factors affecting decision making

Tip: People's perception of risk also affects how they make decisions. For example, people may underestimate the risk of things with long-term harmful effects that aren't immediately seen.

The scientific evidence we have can be overshadowed by other influences such as personal bias and beliefs, public opinion, and the media.

Economic factors

Tip: Economic factors just mean anything to do with money, and social factors just mean anything to do with people.

Society has to consider the cost of implementing changes based on scientific conclusions. Sometimes it decides the cost outweighs the benefits.

Example

Building new power plants that use renewable energy resources helps to reduce our contribution to global warming but it costs money. Sometimes the cost of building a new plant is just too much for the government to justify it, especially when government money could be put to more immediate use in, for example, the NHS or schools.

Social factors

Decisions affect people's lives — sometimes people don't want to follow advice, or are strongly against some recommendations.

Examples

- Exposure to UV radiation from tanning beds can lead to cancer. Scientists recommend that people don't use tanning beds, but shouldn't we be able to choose whether we want to use them or not?
- People may not want new wind farms to be built in certain locations, as they believe that the turbines spoil the view.

Figure 4: *Wind farms are often built at sea to reduce the impact on people.*

Environmental factors

Some scientific research and breakthroughs might affect the environment. Not everyone thinks the benefits are worth the possible environmental damage.

Example

Hydroelectricity requires the building of a dam. This often destroys the habitats of plants and animals, so many think that other renewable energy resources should be used instead.

Practical Skills for the Practical Endorsement

What is the Practical Endorsement?

The Practical Endorsement is assessed slightly differently to the rest of your course. Unlike the exams, you don't get a mark for the Practical Endorsement — you just have to get a pass grade. The Practical Endorsement is split into twelve categories, called Practical Activity Groups (PAGs). Each PAG covers a variety of practical techniques, for example, using an oscilloscope or constructing circuits. All the PAGs are listed on page 11 and the practical techniques are listed on page 12. In order to pass the Practical Endorsement, you'll have to carry out at least twelve experiments, and demonstrate that you can carry out each of the required techniques. You'll do the experiments in class, and your teacher will assess you as you're doing them.

You'll need to keep a record of all your assessed practical activities. You may have already done some experiments that count towards the Practical Endorsement in Year 1 of the course. For example, investigating motion (PAG1) and using an oscilloscope to investigate waves (PAG5) fit in with the material you learned in Year 1, so it's quite likely you carried out these practical activities then. You'll also meet some of the PAGs in this book.

Tip: Throughout this book, experiments and skills that you could use for your Practical Endorsement are marked with a big PAG stamp, like this one:

PRACTICAL ACTIVITY GROUP 7

1. General Practical Skills

PRACTICAL ACTIVITY GROUP 11

The way you do an experiment is important. You may be given a method, or you may have to plan it yourself. It's important that you follow all the steps in a method — this ensures that you work safely, and also makes your results more likely to be precise.

Solving problems in a practical context

HOW SCIENCE WORKS

Practical experiments are used to solve problems or test whether a theoretical model works in a practical setting. If you're given a method to follow for an experiment, you should carry out each step, as described, in the correct order.

It's possible you'll be given a problem and asked to solve it using your own knowledge. In Module 1, there's loads of information about how to plan and carry out experiments correctly. Have a look back at your Year 1 book if it's all a bit hazy. Here's a quick round-up of some of the things you'll need to think about when you plan an experiment:

- First, identify the aim of your experiment.
- Next, work out how to achieve the aim. You'll often need to identify a **variable** that you'll change (the **independent variable**) and the variable you'll measure (the **dependent variable**) in order to gather data that meets the aim of your experiment.
- Identify all the variables that will need to be controlled during your experiment, and how to control them.
- Think about how to make your data as **precise** and **accurate** as possible. This could include repeating your experiment a number of times, or choosing equipment with a scale that has the right sensitivity for the data you're trying to collect.

Tip: Experiments are often used to test scientific theories to see if they are true in a practical context.

Tip: Variables include things like temperature, time, mass and volume.

Tip: Precise results are results that don't vary much from the mean. Accurate results are close to the true value.

Implementing experiments

Once your experiment is planned, you can carry it out, working carefully to make sure your results are precise. Relevant practical techniques will be covered in this book as they come up during the course. You'll have already come across some techniques in Year 1. All of these techniques should be carried out safely (see pages 7-8) and correctly.

Tip: If you need more information about analysing data, including the maths you'll need to know, have a look back at your Year 1 notes — you should have covered it in Module 1.

Recording and analysing data

As you carry out your experiment, you should record your results in a well laid-out table, leaving space for any data analysis you might want to do later. This may include calculating quantities or finding averages of repeat results.

Your table should include a heading for each column. The units for any measurements should be included in the heading, not in the table itself (see Figure 1).

heading *column* *units* *row*

P.d. / V	Current / A Run 1	Current / A Run 2	Current / A Run 3	Mean current / A (to 3 s.f.)
0.0	0.000	0.000	0.000	**0.000**
1.0	0.104	0.105	0.102	**0.104**
2.0	0.150	0.151	0.149	**0.150**

Figure 1: *Table showing the effect of varying p.d. on current through a resistor.*

Tip: Any recorded results that don't fit in with the rest of the data are called anomalous results. If you can work out that they were due to a mistake when carrying out your experiment, you can leave them out of your data analysis.

Presenting data

Presenting your data can make it easier for you to understand your results and spot any trends. How you present your data will depend on the type of data and what you want to find out. You'll have covered all this in Module 1, but you'll mostly want to present your results on a scatter graph.

Drawing a scatter graph lets you show how two variables are related (or correlated). You can draw a line (or curve) of best fit on the scatter graph to help show the trend in your results. The trend is called the **correlation** — see Figure 2.

Tip: A correlation between two variables can be positive (as one variable increases, the other increases), negative (as one variable increases, the other decreases), or there may be no correlation present.

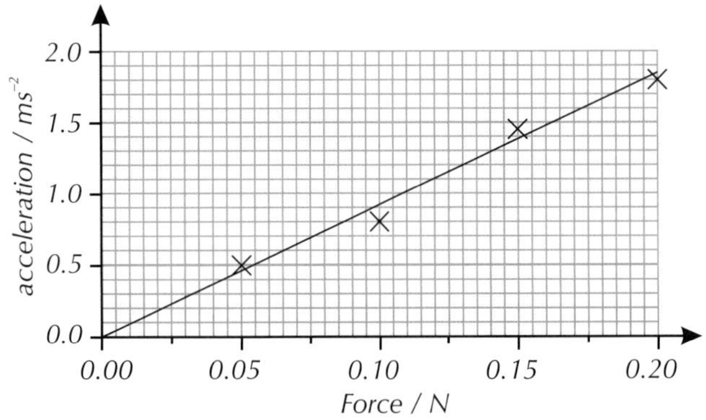

Figure 2: *Scatter graph showing a positive correlation between the force acting on a trolley and its acceleration.*

You may even be able to use software to process your data and generate graphs for you, especially if you've used a data logger in your experiment.

Tip: Take a look back at your Year 1 notes for more information about how to draw and interpret graphs.

Conclusions and evaluations

When you've collected and analysed your data, it's time to wrap up your experiment with some conclusions and an evaluation. Conclusions explain what your results showed. They need to be supported by the data you collected, and shouldn't make sweeping generalisations. If you've found a correlation between two variables, you should be cautious about claiming that the change in one has caused the change in the other — there may be another factor that is causing both variables to change.

An evaluation is a chance for you to look at what you did and think about how you could have improved your method to improve the **validity**, accuracy and precision of your results. You should consider how well your results address the original aim of your experiment, how you could reduce errors in your results, and whether you need to repeat your experiment further to show your results can be reproduced.

Tip: When evaluating your experiment, you could comment on the uncertainty of your measurements and work out the absolute uncertainty or percentage error. The smaller these values are, the more precise your results will be. Look back at your Module 2 notes for how to work them out.

Minimising risks

Many physics experiments have risks associated with them. These can include risks associated with the equipment you're using, as well as risks associated with, say, using radioactive materials.

When you're planning an experiment, you need to identify all the hazards and what the risk is from each hazard — this is a risk assessment. A risk assessment includes working out how likely it is that something could go wrong, and how serious it would be if it did go wrong. You then need to think of ways to reduce these risks.

One precaution that's worth taking in most experiments is to wear goggles. These will protect the eyes from moving objects, snapping wires, chemicals and other dangers.

Other precautions you need to take might depend on specific risk factors associated with particular experiments or materials.

Tip: A hazard is anything that has the potential to cause harm or damage. The risk associated with that hazard is the probability of someone (or something) being harmed if they are exposed to the hazard.

Electricity

Some of the experiments that you are expected to do during your course involve electrical circuits. When working with electricity, always take the following steps to avoid shocks, overheating or damage to equipment:

- Make sure you turn the power supply off while you're adjusting your equipment, and then only turn the power on when you're ready to actually carry out the experiment.
- Never get electrical equipment wet. Wet electrical equipment can lead to short circuits, creating a risk of fire or electric shock.
- Always set up your equipment carefully, following all the instructions you've been given. And pay attention to any safety warnings on the equipment itself.
- Be very careful when working with capacitors. Even when not connected to a circuit, they can store enough charge to give a shock. Make sure you connect them the right way round in the circuit too, or you could damage the capacitor.

Figure 3: *Safety warning on a piece of electrical equipment.*

Tip: Some capacitors can catch fire or explode if connected the wrong way round.

Lasers

Lasers can be very dangerous because laser light is focused into a very direct, powerful beam of monochromatic light which can easily damage someone's eyesight.

When carrying out any experiments involving lasers, it's especially important to wear special laser goggles. To make sure you don't cause damage while using lasers, make sure you:

- Never shine the laser towards a person.
- Remember to turn it off when it's not in use.
- Wear laser safety goggles.
- Avoid shining the laser beam at a reflective surface.
- Have a warning sign on display (see Figure 4).

Figure 4: *Laser warning signs should be used to alert people of the danger of laser beams.*

Radioactive materials

Ionising radiation (e.g. alpha, beta or gamma radiation) can damage living cells. If a cell's DNA is damaged, there's a risk it may become cancerous. So when you're carrying out experiments involving radiative materials, remember how dangerous they can be and follow these precautions:

- Radioactive sources should be kept in a lead-lined box when they're not being used.
- Always wear gloves when working with radioactive materials.
- Radioactive materials should only be picked up using long-handled tongs or forceps.
- You must take care not to point them at anyone, and always keep a safe distance from them.

Figure 5: *Hazard symbol to indicate the presence of radioactive substances.*

Tip: Your teacher should closely supervise any use of radioactive sources.

Glassware

Many pieces of equipment that you may use during your course are made of glass. Broken glass can cause serious injury, so you should take care to transport glass items safely and check them for cracks and flaws before you use them. Any broken pieces of glass should be moved away from the work area immediately and disposed of in an appropriate container (not the normal waste bin).

Tip: The CLEAPSS® website has student safety sheets, and your school or college may have CLEAPSS® Hazcards® you can use. These are all good sources of information if you're writing a risk assessment.

Appropriate clothing

When working in the lab, you should make sure that you are wearing sensible clothing to reduce the risk of injury, e.g. open shoes or sandals won't protect your feet against things that have been accidentally dropped. You should also wear a lab coat to protect your skin and clothing. It's all about using your common sense really.

2. Keeping Scientific Records

When you carry out experiments, it's important to keep records of everything you do. The records should be detailed and clear enough that a complete stranger would be able to read them and understand what you did.

Records of scientific experiments

Throughout your A-Level Physics course, you should keep a record of all the experiments you carry out, the results you obtain and the solutions to any data analysis you do. This could be done in a physical lab book, or kept in folders on a computer. However you choose to keep your records, the information for each experiment should include:

- The aim of the experiment.
- A detailed method for how you carried out the experiment, including any safety precautions you had to take.
- The results of your experiment, clearly set out in a table. The results may be hand-written, or a print-out of data collected by a data logger.
- Any other important observations you made whilst carrying out your experiment, for example, anything that went wrong or anything you did slightly differently from how it was described in the method.
- The solutions to any analysis you did on your results, or any graphs drawn using your results. These should be clearly labelled to show what analysis has been done or what graph has been drawn.
- Citations of any references you used.

Tip: Try to keep all your lab reports in the same place — write them in the same book, or keep them in the same folder on your computer. That way you'll know where everything is.

Tip: A clear and detailed method is important, as it could be used by another scientist who is trying to reproduce your results.

Tip: There's loads more detail about making observations, recording data and analysing your results in Module 1.

Sources of information

PRACTICAL ACTIVITY GROUP 12

It's possible you'll have to do some research to find out information before you get started with an experiment. Useful sources of information include:

Websites

Using the Internet for research is really convenient, but you have to be slightly wary as not all the information you find will be true. It's hard to know where information comes from on forums, blogs and websites that can be edited by the general public, so you should avoid using these. Websites of universities or other respected institutions (e.g. the Institute of Physics), provide lots of information based on reliable scientific sources. To decide whether a website gives reliable information, think about the following things:

- Who has written the information — was it a scientist, a teacher, or just a member of the public?
- Whether or not anyone will have checked the source — articles on websites for scientific organisations will have people reading through the information and checking all the facts. Information on forums or blogs is likely to have been written by an individual, and won't necessarily have been thoroughly checked.
- What the purpose of the website is — if it's a website all about physics, then it's likely whoever has written it will know quite a lot. If it's a website where you can also find out how to make a laser gun from objects you'd find in a typical garden shed, then the depth and quality of the information may not be enough.

Tip: If you're unsure whether the information on a website is true or not, try and find the same piece of information in a different place. The more sources you can find for the information, the more likely it is to be correct.

Textbooks

Your school or public library is likely to have textbooks covering specific areas of physics in a lot of detail.

Scientific papers

You can find papers in online catalogues (e.g. arxiv.org), as well as in journals that are often available in public libraries.

The source you do your research from needs to give the right level of information. It's no good trawling through a scientific paper if you're just looking for the speed of light — you'll probably end up wading through lots of complicated information that you don't need to understand. Equally, if you're researching the theory behind an experiment, you want a source that gives enough detail. A GCSE textbook will probably be too simplistic — you're better off finding a book that deals specifically with the subject in a library instead.

Tip: Scientific papers are checked by other scientists who are experts in the subject of the paper. This is called peer review (see page 1).

Using references and making citations

PRACTICAL ACTIVITY GROUP 12

It sounds obvious, but when you're using the information that you've found during your research, you can't just copy it down word for word. Any data you're looking up should be copied accurately, but you should rewrite everything else in your own words.

When you've used information from a source, you need to cite the reference properly. **Citations** allow someone else to go back and find the source of your information. This means they can check your information and see you're not making things up out of thin air. Citations also mean you've properly credited other people's data that you've used in your work.

Citations are included in the main text of a report and are usually written in brackets after the relevant piece of information. They can either include the entire reference or link the information to a list of references at the end of the report (e.g. using a number — see Figure 1). **References** for each piece of information may include the title of the book, paper or website where you found the information, the author and/or the publisher of the document and the date the document was published.

Tip: There are lots of slightly different ways of referencing sources, but the important thing is that it's clear where you found the information.

Tip: You should include page numbers with your citation if you quote directly from the text or copy a diagram.

Tip: Scientific papers are often written by large groups of people. References often include the name of just a small number of these authors, followed by 'et al', which means 'and others'.

Referencing a website: include the author(s), year, title [online], date accessed, URL.

Referencing a book: include the author(s), publication year, book title, edition, publisher's location and publisher.

Referencing a paper: include the author(s), publication year, title of the paper, the journal it was published in, the volume number and page numbers.

<u>Report</u>

The mass of an electron is approximately 9.11×10^{-31} kg (2, p.28)...

<u>References</u>

1. McNaught, A.D. and Wilkinson, A. (1997). *IUPAC. Compendium of Chemical Terminology, 2nd ed. (the "Gold Book")* [online], Accessed 16 March 2017: https://goldbook.iupac.org/E02008.html
2. Young, H and Freedman, R (2006). *University Physics with modern physics*, 12th Edition, Boston, Addison Wesley
3. Sturm, S, Köhler, F. et al (2014), *'High-precision measurement of the atomic mass of the electron'*, Nature vol. 506: p. 467-470

Figure 1: *Example of a citation in the main text of a report and the corresponding references document.*

3. Practical Activity Groups

This section tells you all the Practical Activity Groups you'll be expected to have carried out for A-Level Physics, as well as the techniques included in them. You'll have met many of them before, and others will be covered in more detail as they crop up throughout the book.

PAGs

There are 12 Practical Activity Groups (PAGs) that you should have covered by the end of your A-Level course. These are shown in the table below, along with an example of the type of activity you could carry out for each one.

PAG		Example activity
1	Investigating motion	Determining the acceleration of free fall.
2	Investigating properties of materials	Determining Young's Modulus for a metal.
3	Investigating electrical properties	Determining the resistivity/conductivity of a metal.
4	Investigating electrical circuits	Investigation of potential divider circuits.
5	Investigating waves	Determination of the wavelength of light with a diffraction grating (see page 93).
6	Investigating quantum effects	Determination of Planck's constant using LEDs.
7	Investigating ionising radiation	Absorption of α or β or γ radiation (see pages 194-195).
8	Investigating gases	Determining an estimate of absolute zero using variation of gas temperature with pressure (see pages 24-25).
9	Investigating capacitors	Determining how the current through a circuit containing a capacitor varies as the capacitor charges (see page 121).
10	Investigating simple harmonic motion	Investigating the factors affecting the period of a simple harmonic oscillator (see pages 52-54).
11	Investigation	Apply investigative approaches and methods to your practical work (see page 5).
12	Research skills	Researching online for further information on a topic (e.g. see page 9).

Tip: You'll probably have carried out some of these experiments in Year 1.

Tip: PAG11 and PAG12 are more general than the others — there are lots of areas in the course where these might be covered.

Tip: Research skills include being able to cite sources of information. There's more about citations on p.10.

Practical techniques

As part of each PAG you'll be expected to show that you can carry out certain techniques, such as:

- use of appropriate analogue apparatus to record a range of measurements (to include length/distance, temperature, pressure, force, angles and volume) and to interpolate between scale markings
- use of appropriate digital instruments, including electrical multimeters, to obtain a range of measurements (to include time, current, voltage, resistance and mass)
- use of methods to increase accuracy of measurements, such as timing over multiple oscillations, or use of fiducial markers, set square or plumb line
- use of a stopwatch or light gates for timing
- use of calipers and micrometers for small distances, using digital or vernier scales
- correctly constructing circuits from circuit diagrams using d.c. power supplies, cells, and a range of circuit components, including those where polarity is important
- designing, constructing and checking circuits using d.c. power supplies, cells, and a range of circuit components
- use of a signal generator and oscilloscope, including volts/division and time-base
- generating and measuring waves, using microphone and loudspeaker, or ripple tank, or vibration transducer, or microwave/radio wave source
- use of a laser or light source to investigate characteristics of light, including interference and diffraction
- use of ICT such as computer modelling, or data logger with a variety of sensors to collect data, or use of software to process data
- use of ionising radiation, including detectors
- applying investigative approaches and methods to practical work
- using online and offline research skills
- correctly citing sources of information

All of these techniques should be covered across all of the PAGs.

Tip: Each individual PAG won't cover every single one of these techniques, but you should have covered all the techniques you need to know by the end of your course.

Module 5 Section 1: Thermal Physics

1. Phases of Matter and Temperature

You should remember the three phases of matter from GCSE. Solids, liquids and gases are described by the kinetic model of matter.

Three phases of matter

Solids, liquids and gases are three different phases that matter can exist in. Particles behave differently in each phase (see Figure 1):

- Particles in solids vibrate about fixed positions in a regular lattice. They're close together, and are held in position by strong forces of attraction.
- Particles in liquids are constantly moving around and are free to move past one another. The attraction between particles in liquids is weaker than for particles in solids. They're fairly close together but have an irregular arrangement.
- Particles in gases are far apart and free to move around with constant random motion, and so are not in any particular order. There are no forces of attraction between particles in an **ideal gas** (p.29).

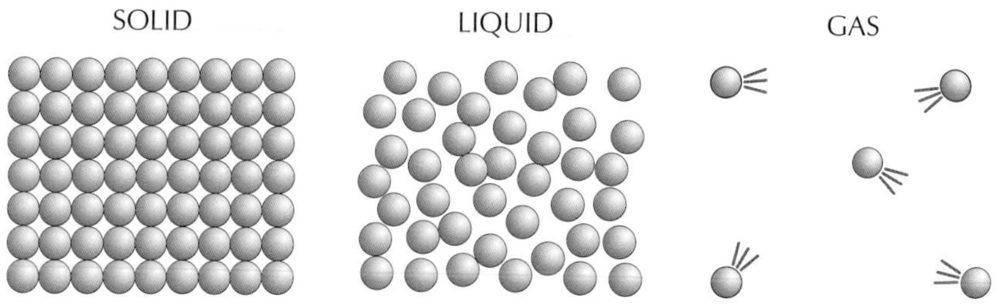

***Figure 1**: The solid, liquid and gas phases of matter.*

The idea that solids, liquids and gases are made up of tiny moving or vibrating particles is called the **kinetic model of matter** (or **kinetic theory**). It seems obvious now, but this wasn't always accepted by the scientific community. It took several scientists and hundreds of years to develop a controversial idea into an accepted theory.

Brownian motion

HOW SCIENCE WORKS

In 1827, botanist Robert Brown noticed that tiny particles of pollen suspended in water moved with a zigzag, random motion. This type of movement of any particles suspended in a fluid is known as **Brownian motion**.

Brown couldn't explain this, but nearly 80 years later Einstein showed that this provided evidence for the existence of atoms or molecules in the water (the kinetic model of matter). The randomly moving water particles were hitting the pollen particles unevenly, causing this motion.

Learning Objectives:

- Know the simple kinetic model that describes solids, liquids and gases in terms of the spacing, ordering and motion of atoms or molecules.
- Be able to explain Brownian motion in terms of the kinetic model and describe a simple demonstration using smoke particles suspended in air.
- Know the meaning of 'internal energy'.
- Know that the internal energy of a body increases as its temperature rises.
- Know that during a change of phase, the internal energy of a substance changes but the temperature is constant.
- Know that the absolute scale of temperature does not depend on the property of any particular substance.
- Know that absolute zero (0 K) is the lowest limit for temperature and is the temperature at which a substance has minimum internal energy.
- Know that temperature measurements can be recorded in both degrees Celsius (°C) and kelvin (K), where $T(\text{K}) \approx \theta\,(°\text{C}) + 273$.
- Know the meaning of 'thermal equilibrium'.

Specification References 5.1.1 and 5.1.2

You can observe Brownian motion in the lab. Start by putting some smoke in a glass jar. Use a glass rod to focus light from a lamp into the glass jar to illuminate it, and observe the particles using a microscope — see Figure 3. The smoke particles appear as bright specks moving haphazardly from side to side, and up and down (see Figure 4).

***Figure 2**: British botanist Robert Brown, after whom Brownian motion is named.*

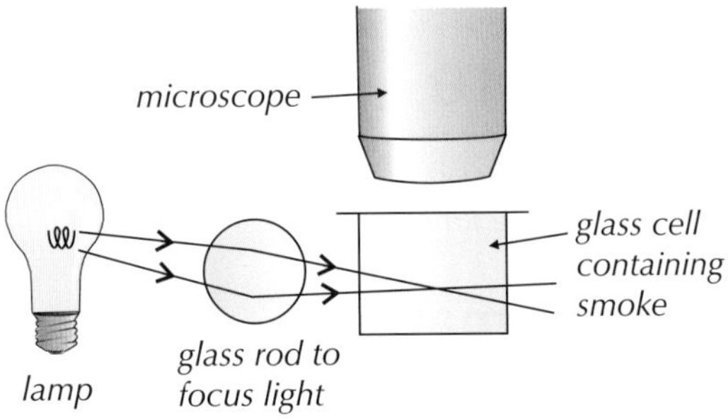

***Figure 3**: Apparatus used to observe the Brownian motion of smoke particles in air.*

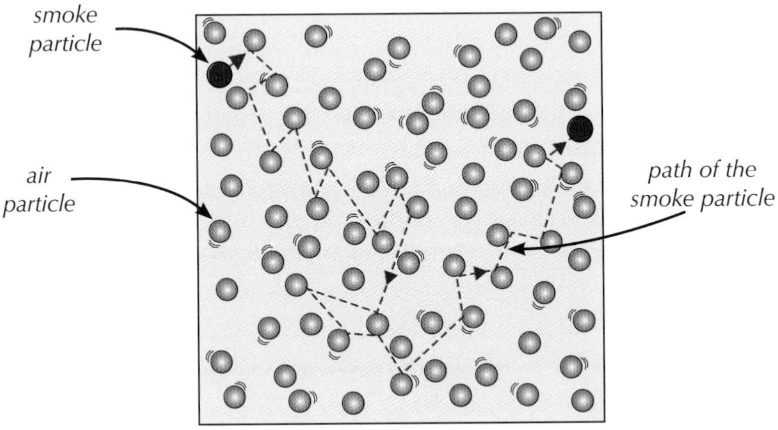

***Figure 4:** Brownian motion of a smoke particle in air.*

Tip: Brownian motion can happen in any fluid — i.e. any liquid or gas.

Internal energy

Particles in a gas have a range of kinetic and potential energies. The kinetic energy of a particle depends on its mass and speed. Potential energy is caused by interactions between particles and is based on their positions relative to each other. These energies are randomly distributed amongst the particles.

The amount of energy contained in a system is called its **internal energy** — it's found by summing the kinetic and potential energies of all the particles within it.

> Internal energy is the sum of the random distribution of kinetic and potential energies associated with the molecules of a system.

Heating a substance means supplying that substance with thermal energy. This energy can be transferred to the kinetic and potential energies associated with the molecules, which increases the temperature of the substance. So as a body's temperature increases, its internal energy increases.

The particles in an ideal gas don't have potential energy, so the internal energy of an ideal gas depends only on the average kinetic energy of its particles — see p.33-34.

Tip: The hotter a gas is, the faster its particles move. Remember — the kinetic energy of an object is equal to $\frac{1}{2}mv^2$, where m is the object's mass and v is the object's speed. So the average kinetic energy of the particles in a gas increases with temperature.

Change of phase

If you heat or cool a substance, a change of phase may take place. A change of phase, sometimes also referred to as a change of state, occurs when a substance changes between a solid, liquid or gas. When a substance changes phase its internal energy changes, but the total kinetic energy of the particles and its temperature stay the same (see Figure 6). This is because the change of phase is altering the bonds between, and therefore potential energies of, the particles, while the kinetic energy of the particles stays constant.

For example, in a pan of boiling water, the sum of the potential energies of the water molecules increases as they break free of the liquid. But the water in both phases is at 100 °C.

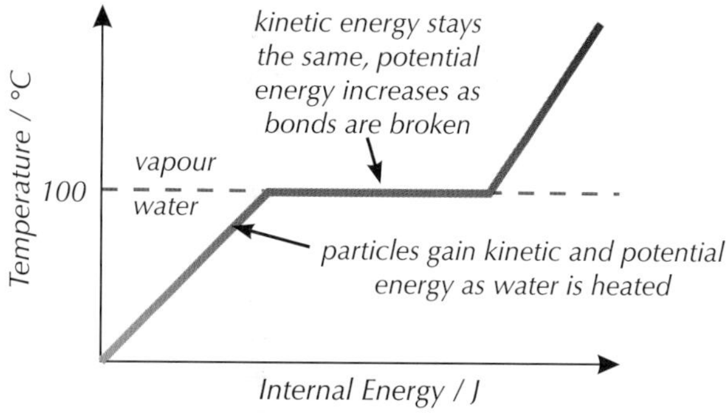

Figure 6: *Graph of temperature against internal energy for a sample of water being heated at a constant rate to its boiling point and beyond.*

Figure 5: *When boiling water, the energy is used to convert the water into steam rather than to increase the temperature of the water.*

The absolute scale of temperature

The Celsius scale (with the freezing and boiling points of water at 0 °C and 100 °C) can easily be used for day-to-day activities. However, scientists use the Kelvin scale (the **absolute scale of temperature**) for all equations in thermal physics. It's also known as the thermodynamic scale, and it does not depend on the properties of any particular substance (like the freezing and boiling points of water). The unit of this scale is the kelvin (K).

Zero kelvins is the lowest possible temperature and is called **absolute zero**. At 0 K (or –273 °C, to 3 s.f.) all particles have the minimum possible internal energy. At higher temperatures, particles have more energy. In fact, with the Kelvin scale, a particle's energy is proportional to its temperature (see page 33).

The Kelvin scale is named after Lord Kelvin who first suggested it. A change of 1 K equals a change of 1 °C. To change from degrees Celsius into kelvins you add 273 (or subtract 273 to go the other way).

$$T \approx \theta + 273$$

T = temperature in kelvins (K)

θ = temperature in degrees Celsius (°C)

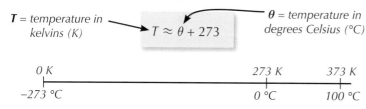

Figure 7: *Equivalent temperatures in the Kelvin and degrees Celsius scales.*

Tip: Absolute zero is rounded to 273 K (3 s.f.) here, so we use an approximately equal sign (≈) rather than an equal sign (=).

Tip: You don't use the degrees sign (°) when writing temperatures in kelvins.

Example — Maths Skills

An object is heated to 75 °C. What is this temperature in kelvin?

$T \approx \theta + 273 = 75 + 273 = 348$ K

Thermal equilibrium

Suppose A, B and C in Figure 8 are three identical metal blocks. A has been in a warm oven, B has come from a refrigerator and C is at room temperature.

Figure 8: *Three identical metal blocks.*

There is a net flow of thermal energy from A to C and C to B until they all reach **thermal equilibrium** and the net flow of energy stops. This happens when the three blocks are at the same temperature.

Tip: Be careful — at thermal equilibrium, energy is still flowing between the bodies, but it's the net flow that stops.

> If body A and body B are both in thermal equilibrium with body C, then body A and body B must be in thermal equilibrium with each other.

Thermal energy is always transferred from regions of higher temperature to regions of lower temperature.

Practice Questions — Application

Q1 A student needs to heat a flask of liquid to 345 K. Calculate the temperature the student needs to heat the flask to in degrees Celsius.

Q2 a) A person making a pie encases some hot filling in cool pastry and leaves it to rest. State the direction of the net flow of thermal energy between the pastry and the filling.

b) The whole pie eventually comes to thermal equilibrium with the room. State how you know that the filling and the pastry are now the same temperature.

Practice Questions — Fact Recall

Q1 Describe the order, spacing and movement of the particles in:
a) a solid b) a liquid c) an ideal gas

Q2 When pollen particles are suspended in water, they appear to move with a zigzag, random motion. State the name of this type of motion and explain why this occurs.

Q3 Explain why the average kinetic energy of particles in a substance remains constant when the substance changes phase whilst being heated.

Q4 Sketch a graph of temperature against internal energy to show how the internal energy of water changes as it is heated at a constant rate to its boiling point and beyond. Label the energy changes of the particles on the graph.

Tip: It's easier to draw the graph in Q4 using degrees Celsius, but you could also use kelvins.

Q5 a) What is the lowest possible temperature an object can theoretically reach called?

b) What is the value of this temperature in kelvins and degrees Celsius?

Q6 What is meant by the internal energy of a body?

Q7 When are two objects said to be in thermal equilibrium?

2. Thermal Properties of Materials

Specific heat capacity and specific latent heat are quantities used to calculate the energy required to heat up or change the state of different substances.

Specific heat capacity

When you heat something, the amount of energy needed to raise its temperature depends on its specific heat capacity.

The **specific heat capacity** (c) of a substance is the amount of energy needed to raise the temperature of 1 kg of the substance by 1 K (or 1 °C).

Or put another way:

energy change = mass × specific heat capacity × change in temperature

$$E = mc\Delta\theta$$

E *= energy change in J*

m *= mass of substance in kg*

c *= specific heat capacity in* $J\,kg^{-1}\,K^{-1}$ *or* $J\,kg^{-1}\,°C^{-1}$

$\boldsymbol{\Delta\theta}$ *= temperature change in K or °C*

Example — Maths Skills

The specific heat capacity of water is 4180 $J\,kg^{-1}\,K^{-1}$. If 172 kJ of energy is supplied to 5.00 kg of water at 300 K (correct to 3 significant figures), what will its final temperature be?

First find the value of $\Delta\theta$:

$$E = mc\Delta\theta \Rightarrow \Delta\theta = \frac{E}{mc} = \frac{172 \times 10^3}{5.00 \times 4180}$$
$$= 8.229...\text{ K}$$

Then add this to the initial temperature:

$$\text{final temperature} = 300 + 8.229... = 308.229...$$
$$= 308\text{ K (to 2 s.f.)}$$

Measuring specific heat capacity

You can use this experiment to measure specific heat capacity yourself. The method is the same for solids and liquids, but the set-up is a little bit different:

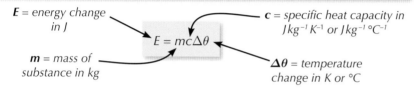

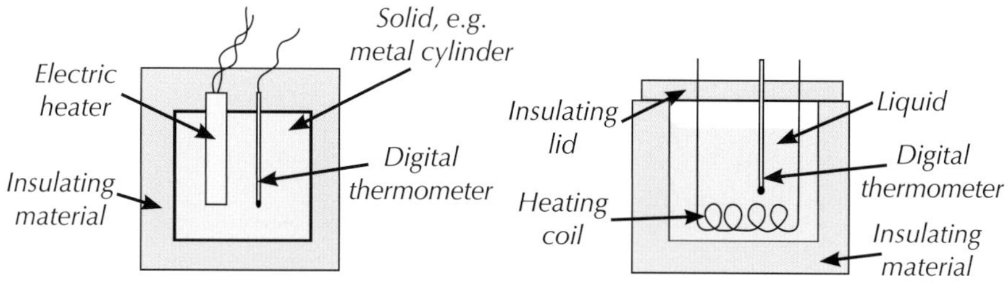

Figure 1: *Set-ups used to determine the specific heat capacity of a solid or liquid.*

Learning Objectives:

- Know what is meant by the specific heat capacity of a substance and be able to use $E = mc\Delta\theta$.
- Be able to describe an electrical experiment to determine the specific heat capacity of a metal block or liquid.
- Know what specific latent heat of fusion and specific latent heat of vaporisation are, and be able to use the formula $E = mL$.
- Be able to describe an electrical experiment to determine the specific latent heat of fusion and vaporisation.

Specification Reference 5.1.3

Tip: Q is sometimes used instead of E for the energy change. You might also see $\Delta\theta$ written as ΔT.

Tip: Energy is supplied to the water, so the temperature goes up.

Tip: As always, do a full risk assessment before beginning this experiment, and take extra care if you are placing an electrical heater in water.

Measure the mass of the substance and its initial temperature. Heat the substance with the heater — you need a temperature rise of about 10 K. Measure the new temperature of the substance and subtract the initial temperature from it to get the change in temperature. Use an ammeter and voltmeter attached to your electric heater to measure the current through, I, and voltage across, V, the heater, and time how long the heater is on for, t, using a stopwatch. You can then calculate the energy (E) supplied by the heater using $E = W = VIt$. Plug your data into $E = mc\Delta\theta$ to calculate c.

Tip: The value you end up with for c will probably be too high by quite a long way. That's because some of the energy from the heater gets transferred to the air and the container. You can reduce the error by starting below and finishing above room temperature to cancel out gains and losses.

Tip: You met the equation $W = VIt$ in Year 1 of A-level Physics.

Tip: You don't need to convert between °C and K here, since the change in temperature will be the same for both.

Example — Maths Skills

0.250 kg of water is heated from 12.1 °C to 22.9 °C with an electric immersion heater. The heater has a voltage of 11.2 V and a current of 5.30 A, and is on for 205 s. Calculate the specific heat capacity of water.

$E = VIt = 11.2 \times 5.30 \times 205 = 12\,168.8$ J

$\Delta\theta = 22.9 - 12.1 = 10.8$ °C

$$E = mc\Delta\theta \text{ so } c = \frac{E}{m\Delta\theta} = \frac{12168.8}{0.250 \times 10.8} = 4506.9... = 4510 \text{ J kg}^{-1}\,°\text{C}^{-1} \text{ (to 3 s.f.)}$$

(or $4510 \text{ J kg}^{-1}\text{K}^{-1}$)

Estimating specific heat capacity

If it's difficult to use the method above for a particular object, you could try using the method of mixtures to estimate the specific heat capacity. It works on the principle that when a hot substance is mixed, or combined, with a cold substance, they will eventually reach thermal equilibrium and end up at the same temperature. It relies on the assumption that no heat is lost from the system to the surroundings. Of course, some heat will be lost, so the specific heat capacity obtained is only an estimate.

Tip: Go to page 16 if you need a refresher on thermal equilibrium.

In order to use this method to find the specific heat capacity of a substance, you must know the specific heat capacity of the other substance it's being mixed with. For example, to find the specific heat capacity of a metal block you can use water, which has a specific heat capacity of 4180 $\text{Jkg}^{-1}\text{K}^{-1}$:

- Heat the metal block of known mass, m_b, up to a temperature T_b.
- Quickly transfer this block into an insulated container containing a mass of water, m_w, at a temperature T_w.
- The hot block will heat the water. Measure the temperature of the water once it has reached a steady value, T_s.

Figure 2: Water has a very high specific heat capacity, so its temperature goes down very slowly as it gives out energy. That's why a mug of tea stays warm long enough for you to drink it.

The heat (energy) gained by the water is equal to the heat lost by the block, so:

$$m_w c_w \Delta\theta_w = m_b c_b \Delta\theta_b$$

which becomes:

$$m_w c_w (T_s - T_w) = m_b c_b (T_b - T_s)$$

Rearrange for c_b, the specific heat capacity of the block:

$$c_b = \frac{m_w c_w (T_s - T_w)}{m_b (T_b - T_s)}$$

So you can plug the specific heat capacity of water into the formula with your measurements to find c_b.

Tip: You could also use this experiment to estimate the specific heat capacity of a liquid if you know the specific heat capacity of a metal block.

Example — Maths Skills

A 0.982 kg steel bar is heated to 75.0 °C then placed in water, which is initially at a temperature of 20.5 °C. The water has a mass of 3.12 kg and a specific heat capacity of 4180 J kg^{-1} K^{-1}. The steel and water come to thermal equilibrium at a temperature of 22.4 °C. Estimate the specific heat capacity of steel.

$$c_b = \frac{m_w c_w (T_s - T_w)}{m_b (T_b - T_s)} = \frac{3.12 \times 4180 \times (22.4 - 20.5)}{0.982 \times (75.0 - 22.4)} = 479.719...$$

$$= 480 \text{ J kg}^{-1} \text{ K}^{-1} \text{ (to 3 s.f.)}$$

Specific latent heat

To melt a solid, you need to break the bonds that hold the particles in place. The energy needed for this is called the latent heat of fusion. Similarly, when you boil or evaporate a liquid, energy is needed to pull the particles apart completely. This is the latent heat of vaporisation.

The larger the mass of the substance, the more energy it takes to change its state. That's why the specific latent heat is defined per kg:

The **specific latent heat** (L) of fusion or vaporisation is the quantity of thermal energy required to change the state of 1 kg of a substance.

Which gives:

energy change = mass × specific latent heat

Or in symbols:

$$E = mL$$

E = *energy change in J*

L = *specific latent heat in J kg^{-1}*

m = *mass of substance in kg*

Example — Maths Skills

Find the energy needed to turn 1.00 kg of water at 90.0 °C to 1.00 kg of steam at 110.0 °C. c_{water} = 4180 J kg^{-1} K^{-1}, c_{steam} = 1890 J kg^{-1} K^{-1}, and the latent heat of vaporisation of water is 2.26 × 10^6 J kg^{-1}.

First find the energy needed to heat the water by 10.0 °C:

$E = mc\Delta\theta = 1.00 \times 4180 \times 10.0$

$= 41\,800$ J (or 41.8 kJ)

Then find the energy needed to turn the water to steam:

$E = mL = 1.00 \times (2.26 \times 10^6)$

$= 2.26 \times 10^6$ J (or 2260 kJ)

Then find the energy needed to heat the steam by 10.0 °C:

$E = mc\Delta\theta = 1.00 \times 1890 \times 10.0$

$= 18\,900$ J (or 18.9 kJ)

Then just add all these numbers together:

Energy needed = 41.8 + 2260 + 18.9 = 2320.7 = 2320 kJ (to 3 s.f.)

Tip: The specific latent heat of fusion is used when a substance is melting or freezing. The specific latent heat of vaporisation is used when a substance is boiling or condensing.

Tip: Energy has to be added for a material to melt or boil. When a material freezes or condenses, energy is lost from the material.

Tip: You'll usually see the latent heat of vaporisation written L_v and the latent heat of fusion written L_f.

Tip: Water turns to steam at 100 °C.

Measuring specific latent heat of fusion

Figure 4 shows a set-up you can use to measure the specific latent heat of fusion of a substance such as ice. First connect a heating coil up to an ammeter and a voltmeter, and put equal masses of ice in two identical funnels above beakers. Put the heating coil in one of the funnels and turn it on for three minutes, using a stopwatch to time it. Measure the current through, I, and voltage across, V, the coil during this time. Calculate the energy transferred by the heater in the three minutes using the equation $E = W = VIt$.

At the end of the three minutes, measure the mass of water collected in the beakers. The unheated funnel of ice is there so you can measure how much ice melts due to the ambient temperature of the room. Subtract the mass of water collected from the unheated funnel from the mass of water collected from the heated funnel. This gives the mass of ice, m, that melted solely due to the presence of the heater. $E = mL$, so to find the specific latent heat of fusion for water just divide the energy supplied by the heater by the mass of ice that the heater melted: $L = E \div m$.

Tip: Make sure that the ammeter is in series with the coil and that the voltmeter is in parallel with the coil.

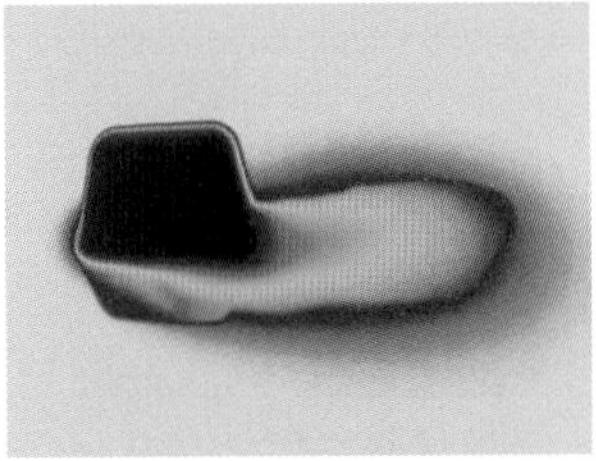

Figure 3: *A thermogram of a melting ice cube shows cool areas in blue and warm areas in red and yellow. Once the specific latent heat of fusion has been transferred to the ice, it melts into water and begins to rise to the same temperature as its surroundings.*

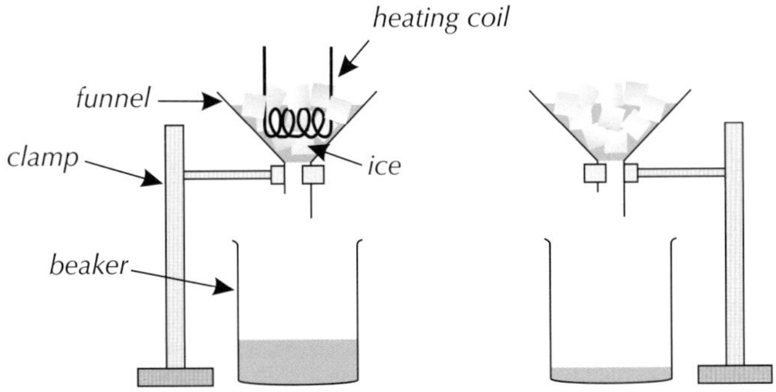

Figure 4: *Experimental set-up for determining the specific latent heat of fusion of water.*

Measuring specific latent heat of vaporisation

Another experiment can be carried out to determine the specific latent heat of vaporisation of a liquid such as water. Figure 5 shows the set-up for the experiment.

Tip: This experiment doesn't require a control apparatus (like the extra funnel of ice in the experiment above) since water won't boil at room temperature.

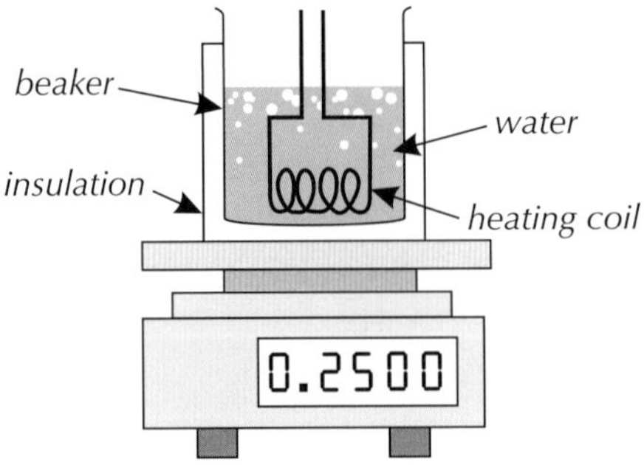

Figure 5: *Experimental set-up for determining the specific latent heat of vaporisation of water.*

Place insulation around the outside of a small beaker, but leave the top of the beaker open to the air. Fill the beaker part-way up with water. Connect a voltmeter and an ammeter to a heating coil and insert the coil into the water. Then place the beaker and its contents on a balance.

Without removing the beaker from the balance, switch the heating coil on. Once the water is boiling, record the mass on the mass balance and start the timer. Measure the voltage across and current through the heating coil. When the mass has decreased by approximately 15 g, stop the timer and turn the heating coil off.

Record the new mass of the beaker and its contents. Subtract this from the original mass to get the decrease in mass of the water. Calculate the energy transferred by the heater to the water by multiplying the recorded voltage and current together with the time taken (since $E = W = VIt$). Rearrange $E = mL$ to $L = E \div m$, then plug in the energy transferred and the change in the mass of water to get the latent heat of vaporisation of water.

Tip: Ensure that the balance remains on for the duration of the experiment.

Tip: Your value for L_v will probably be too high, since some of the heat energy supplied will be lost to the surroundings.

Practice Questions — Application

Q1 How much energy is required to raise the temperature of 0.45 kg of a liquid with specific heat capacity 244 J kg^{-1} K^{-1} by 3.0 kelvins?

Q2 A 270 g metal block with a temperature of 82 °C and a specific heat capacity of 890 J kg^{-1} K^{-1} is placed into a beaker of liquid of mass 910 g and temperature 34 °C. If the specific heat capacity of the liquid is 3600 J kg^{-1} K^{-1}, calculate the temperature at which the solid and liquid reach thermal equilibrium.

Q3 A bowl containing 100.0 g of water at 25.0 °C is placed in a freezer. A few hours later, all of the water has become ice at –5.00 °C. Find how much energy the water lost to its surroundings.
$c_{water} = 4180$ J kg^{-1} K^{-1}, $c_{ice} = 2110$ J kg^{-1} K^{-1}, $L_f = 334\,000$ J kg^{-1}.

Tip: Head back to page 15 for a reminder about the Kelvin temperature scale.

Practice Questions — Fact Recall

Q1 What is the specific heat capacity of a substance?

Q2 Describe an experiment that can be used to measure the specific heat capacity of a solid material.

Q3 What is the specific latent heat of fusion?

Q4 Describe an experiment that can be used to find the specific latent heat of fusion for water.

3. The Gas Laws

An 'ideal gas' follows the three gas laws, which describe how a fixed mass of gas behaves when you change its temperature, pressure or volume. Be careful though — they work on the Kelvin temperature scale, not traditional °C.

Learning Objectives:

- Know techniques and procedures used to investigate PV = constant (Boyle's Law) and $\frac{P}{T}$ = constant (PAG8).
- Know how to make an estimation of absolute zero using the variation of gas temperature with pressure (PAG8).

Specification Reference 5.1.4

Boyle's law

A (theoretical) ideal gas obeys all three gas laws. Each of the three gas laws was worked out independently by careful experiment with fixed masses of gases. The first of them is **Boyle's law**, which says that:

At a constant temperature the pressure p and volume V of an ideal gas are inversely proportional.

Inversely proportional means that as one variable increases, the other decreases by the same proportion, i.e. $p \propto \frac{1}{V}$. For example, if you reduce the volume of a gas, its particles will be closer together and will collide with each other and the container more often, so the pressure increases.

Boyle's law means that at any given temperature the product of p and V will always be the same:

$$pV = \text{constant}$$

Tip: Most gases can usually be assumed to act as ideal gases — see the assumptions on page 29.

Tip: If the volume halves, the pressure doubles, and so on.

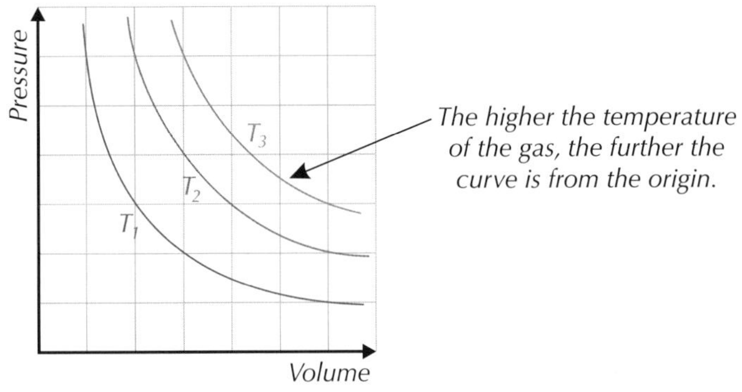

Figure 1: *Graphs of pressure against volume for an ideal gas at different temperatures where $T_1 < T_2 < T_3$.*

Figure 2: *Irish chemist and physicist Robert Boyle.*

Example — Maths Skills

The diagram shows a sealed container with a divider at its centre. An ideal gas fills the left half of the container, and the right half is a vacuum. The gas is contained at a pressure of 350 kPa. The divider is then removed so that the gas fills the whole container. Calculate the new pressure of the gas, assuming that the temperature of the gas remains constant.

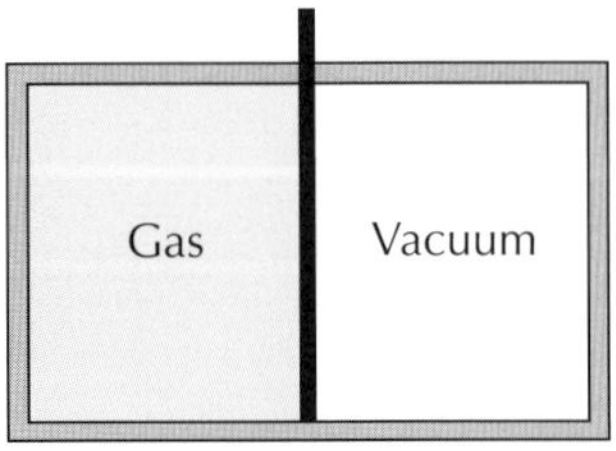

pV = constant so $p_1V_1 = p_2V_2$.

Rearranging: $p_2 = p_1\left(\frac{V_1}{V_2}\right)$

The volume containing the gas is now twice as big, so $V_2 = 2V_1 \Rightarrow \frac{V_1}{V_2} = \frac{1}{2}$.

So $p_2 = 350 \times 10^3 \times \left(\frac{1}{2}\right) = 175 \times 10^3 = 175\,\text{kPa}$

Experiment to investigate Boyle's law

PRACTICAL ACTIVITY GROUP 8

You can investigate the effect of pressure on volume by setting up the experiment shown in Figure 3.

- The oil traps a pocket of air in a sealed tube with fixed dimensions.
- Use a tyre pump to increase the pressure on the oil.
- Use the Bourdon gauge to record the pressure. As the pressure increases, more oil will be pushed into the tube, the oil level will rise, and the air will compress. The volume occupied by air in the tube will reduce.
- Measure the volume of air when the system is at atmospheric pressure by multiplying the length of the part of the tube containing air by $\pi \times$ radius of tube squared.
- Gradually increase the pressure by a set interval. Make sure you wait a few moments each time you change the pressure — this will give the temperature time to stabilise before you take any readings.
- Note down both the pressure and the volume of air as it changes. Multiplying these together at any point should give the same value.
- Repeat the experiment twice more and take a mean for each reading.

If you plot a graph of p against $\frac{1}{V}$ you should get a straight line.

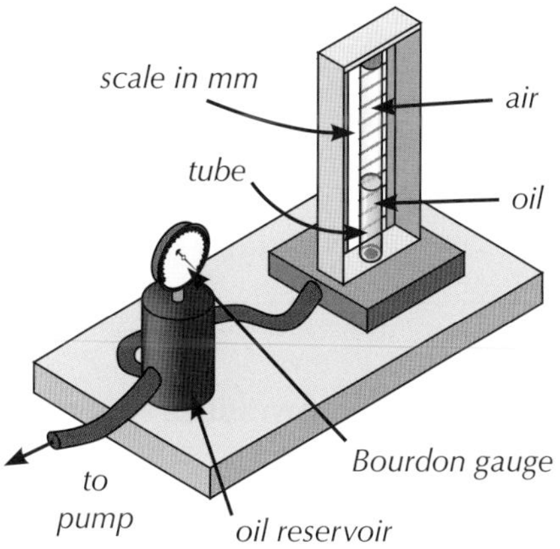

Figure 3: *Experimental set-up for investigating Boyle's Law.*

Tip: Do a full risk assessment before starting any experiments involving gases (including the experiment on the next page). This experiment involves pressurising a glass container — so wear goggles and don't use too high a pressure.

Tip: If your gauge has units of bar, you can multiply by 10^5 to get your readings in Pa.

The pressure law

The **pressure law** states that:

> At constant volume, the pressure p of an ideal gas is directly proportional to its absolute temperature T.

For example, if you heat a gas, the particles gain kinetic energy. This means they move faster. If the volume doesn't change, the particles will collide with each other and their container more often and at higher speeds, increasing the pressure inside the container.

At absolute zero the pressure is also zero. If the pressure law is obeyed, the pressure divided by the temperature for a fixed volume is a constant:

$$\frac{p}{T} = \text{constant}$$

Figure 4: *The low pressure in a plane cabin causes a crisp packet sealed at atmospheric pressure to expand to a larger volume.*

Tip: If you use this relation, make sure the temperature is in K.

So a graph of pressure against temperature will be a straight line (its gradient will be constant). This is shown in Figure 5.

Tip: The lines always reach the temperature axis at absolute zero.

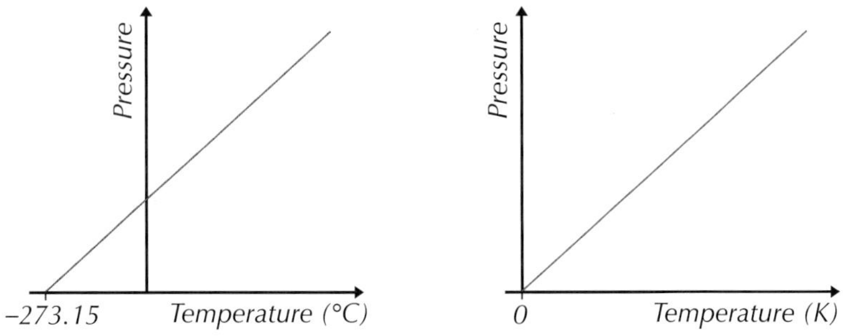

Figure 5: *Graphs of pressure against temperature for an ideal gas at constant volume on the degrees Celsius scale and on the Kelvin scale.*

Tip: To roughly convert from degrees Celsius to kelvins, add 273 (see page 15).

Example — Maths Skills

A gas is contained in a sealed box with a fixed volume at a temperature of 21°C and pressure of 110 000 Pa. Calculate the temperature the gas would need to be heated to in order to increase the pressure by 15 000 Pa.

$\frac{p}{T} = \text{constant}$ so $\frac{p_1}{T_1} = \frac{p_2}{T_2} \Rightarrow T_2 = \frac{p_2}{p_1}T_1$

$T_1 \approx \theta_1 + 273 = 21 + 273 = 294$ K

$p_2 = 110\,000 + 15\,000 = 125\,000$ Pa

So $T_2 = \frac{125\,000 \times 294}{110\,000} = 334.09... =$ **330 K (= 57°C) (to 2 s.f.)**

Tip: Absolute zero is the lowest possible temperature, where all particles have the minimum possible internal energy (see page 15).

Tip: In practice a real (non-ideal) gas would condense before reaching 0 K.

Tip: If the markings on your thermometer are quite far apart, you could interpolate between them (e.g. if the temperature is halfway between the markings for 24 °C and 25 °C you could record it as 24.5 °C). But it's better to use one with a finer scale if you can.

Experiment to investigate the pressure law and estimate absolute zero

PRACTICAL ACTIVITY GROUP 8

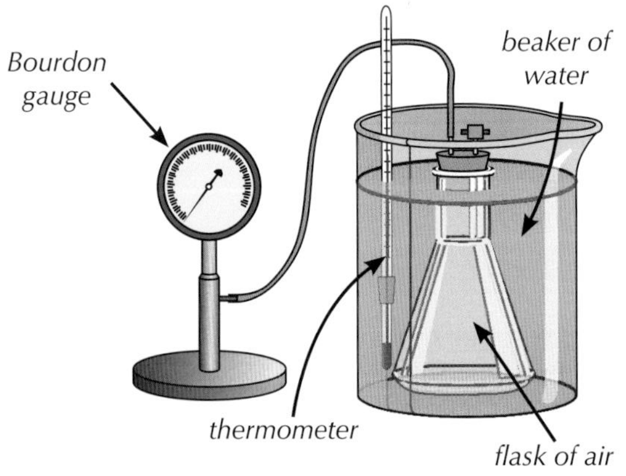

Figure 6: *Experimental set-up for investigating the pressure law.*

- Immerse a stoppered flask of air in a beaker of water so that as much of the flask as possible is submerged.
- Connect the stopper to a Bourdon gauge using a short length of tube — the volume of the tubing must be much smaller than the volume of the flask.
- Record the temperature of the water and the pressure on the gauge.

- Insert an electric heater and switch on for a few minutes to heat the water, then remove it. Stir the water to ensure it is at a uniform temperature and allow some time for the heat to be transferred from the water to the air.
- Record the pressure on the gauge and the temperature, then repeat several times until the water starts to boil.
- Repeat your experiment twice more with fresh cool water.
- Multiplying pressure and temperature together at any point should give you a constant.
- Plot your results on a graph of pressure against temperature. Draw a line of best fit.
- Estimate the value of absolute zero by continuing (extrapolating) your line of best fit until it crosses the *x*-axis.

Example

A student heats a beaker of water that contains a sealed flask of air. He measures the water temperature and air pressure at several points as he heats the water from 0 °C to 100 °C. He plots his results on a graph as shown below. The student then extrapolates back his line of best fit until it intercepts the *x*-axis (shown by the dashed line in the graph below). At this point of interception the pressure is zero and the temperature is therefore at its theoretical minimum value, i.e. absolute zero.

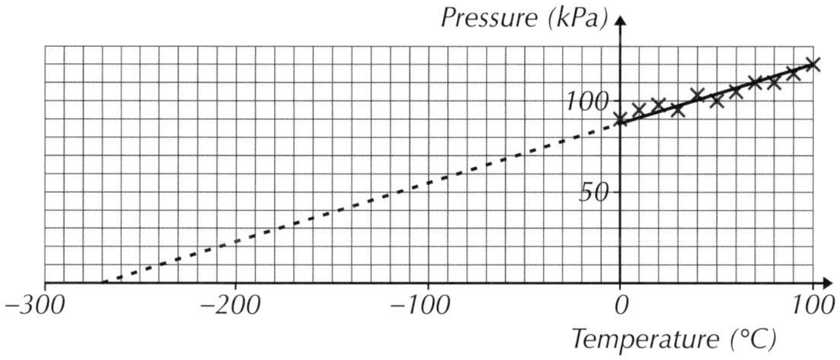

So, using his results, the student finds the value of absolute zero to be –270 °C.

Charles' law

An ideal gas also obeys Charles' law. Charles' law states that:

> At constant pressure, the volume V of an ideal gas is directly proportional to its absolute temperature T.

Tip: If the temperature doubles, the volume doubles, and so on.

At the lowest theoretically possible temperature (0 K, see page 15) the volume is zero. If Charles' law is obeyed, the volume divided by the temperature is a constant:

$$\frac{V}{T} = \text{constant}$$

Practice Questions — Application

Q1 A student conducts an experiment to estimate absolute zero. She immerses a flask of air in water and gradually heats the water. She takes measurements of the temperature of the water and the pressure of the air. The table below shows her results.

Temperature (°C)	Pressure (kPa)
20	102
40	107
60	114
80	122

Plot a graph of her results and use it to estimate absolute zero.

Q2 A gas syringe is filled with 30.0 cm^3 of an ideal gas at 27 °C and the pressure inside the syringe is measured as 1.4×10^5 Pa.

a) The plunger of the syringe is pushed in so that the volume inside the syringe becomes 15 cm^3. If the temperature remains constant, what will the pressure inside the syringe be?

b) The syringe is then cooled to –173 °C. Assuming the volume remains constant, what will the pressure inside the syringe be now?

Q3 A pair of students trap a pocket of air in a sealed tube of radius 3.0 mm by filling the bottom with oil. They use a pump to increase the pressure on the oil which pushes more oil into the tube and decreases the volume of the trapped air. They use a Bourdon gauge to measure the pressure of the oil at certain intervals and measure the length of tube occupied by the air at these pressures.

a) At a pressure of 350 kPa, the length of the trapped air is 92 mm. Calculate the volume occupied by the air at this pressure.

b) Calculate the volume of the air at a pressure of 420 kPa.

c) Explain why the students should wait a few moments each time they increase the pressure before taking any readings.

d) Describe the type of graph the students would expect to see if they plotted their values of p against $\frac{1}{V}$.

e) Which gas law is this experiment investigating?

Practice Questions — Fact Recall

Q1 What must be true of a gas's mass for it to obey the three gas laws?

Q2 What is the pressure law?

4. The Ideal Gas Equation

You can use the gas laws from the last topic to form the equation of state of an ideal gas. It's pretty important in thermal physics, so you'll see it quite a lot.

Learning Objectives:

- Know that the amount of a substance can be measured in moles.
- Know that the Avogadro constant N_A is equal to 6.02×10^{23} mol^{-1} and is the number of particles in one mole.
- Be able to use the equation of state of an ideal gas $pV = nRT$, where n is the number of moles.
- Know that the Boltzmann constant, k, is calculated using $k = \frac{R}{N_A}$.
- Be able to use the equation $pV = NkT$.

Specification Reference 5.1.4

Avogadro's constant

One **mole** of any material contains the same number of particles, no matter what the material is. This number is called **Avogadro's constant** and has the symbol N_A. The value of N_A is 6.02×10^{23} particles per mole. The number of particles, N, in an amount of gas is given by the number of moles, n, multiplied by Avogadro's constant, N_A:

$$N = nN_A$$

N *= number of particles*
***N**$_A$ = Avogadro's constant (= 6.02×10^{23} mol^{-1})*
n *= number of moles*

n is measured in 'mol', so N_A has the unit mol^{-1}.

The equation of state

The three gas laws from the previous topic can be combined to give the equation:

$$\frac{pV}{T} = \text{constant}$$

The constant in the equation depends on the amount of gas used. The amount of gas can be measured in moles. Putting in values for 1 mole of an ideal gas at room temperature and atmospheric pressure gives the constant a value of 8.31 J K^{-1} mol^{-1}. This is the molar gas constant, R.

The value of $\frac{pV}{T}$ increases or decreases if there's more or less gas present — the more gas you have, the more space it takes up. The amount of gas is measured in moles, n, so the constant in the equation above becomes nR, where n is the number of moles of gas present. Plugging this into the equation gives:

$$p\frac{V}{T} = nR$$

Which can be rearranged to give the **equation of state of an ideal gas**:

$$pV = nRT$$

V *= volume in m^3*
n *= number of moles of gas*
T *= temperature in K*
p *= pressure in Pa*
R *= molar gas constant (= 8.31 J mol^{-1} K^{-1})*

Tip: Don't forget to convert the units — remember the Kelvin scale is used for temperature in ideal gas calculations.

This equation works well (i.e. a real gas approximates to an ideal gas) for gases at low pressures and fairly high temperatures.

Example — Maths Skills

What's the volume occupied by exactly one mole of an ideal gas at room temperature (20.0 °C) and at a pressure of 1.00×10^5 Pa?

Rearranging the equation of state: $pV = nRT \Rightarrow V = \frac{nRT}{p}$

So $V = \frac{1 \times 8.31 \times (20.0 + 273)}{1.00 \times 10^5} = 0.0243483 = 2.43 \times 10^{-2}$ m^3 (to 3 s.f.)

Tip: You can think of the Boltzmann constant as the gas constant for one particle of gas, while R is the gas constant for one mole of gas.

Figure 1: *Austrian physicist Ludwig Boltzmann.*

Tip: There are 1000 litres in 1 m^3.

Tip: $R = 8.31\ Jmol^{-1}K^{-1}$ and $k = 1.38 \times 10^{-23}\ JK^{-1}$. You'll be given these in the data and formulae booklet in the exam.

The Boltzmann constant

The **Boltzmann constant**, k, is given by:

$$k = \frac{R}{N_A}$$

k *= the Boltzmann constant (= $1.38 \times 10^{-23}\ JK^{-1}$)*

R *= molar gas constant (= $8.31\ JK^{-1}mol^{-1}$)*

N_A *= Avogadro's constant (= $6.02 \times 10^{23}\ mol^{-1}$)*

If you combine this with $N = nN_A$ from the previous page, you'll see that $Nk = nR$. First rearrange $N = nN_A$ for N_A to get $N_A = \frac{N}{n}$. Then substitute this into the equation for Boltzmann's constant:

$$k = \frac{R}{N_A} \quad \Rightarrow \quad k = \frac{R}{\left(\frac{N}{n}\right)} = \frac{nR}{N} \quad \Rightarrow \quad nR = Nk$$

Substituting this into the equation of state (from the previous page) gives the equation of state for N molecules:

$$pV = NkT$$

V *= volume in m^3*

p *= pressure in Pa*

N *= number of molecules of gas*

T *= temperature in K*

k *= the Boltzmann constant (= $1.38 \times 10^{-23}\ JK^{-1}$)*

Example — Maths Skills

An ideal gas at 303 K and 1.00×10^5 Pa occupies 23.2 litres. Find how many molecules of the gas are present.

Rearrange equation: $N = \frac{pV}{kT} = \frac{(1.00 \times 10^5) \times (23.2 \times 10^{-3})}{(1.38 \times 10^{-23}) \times 303}$

$= 5.5483... \times 10^{23} = 5.55 \times 10^{23}$ (to 3 s.f.)

Practice Questions — Application

Q1 Find the volume occupied by 23 moles of an ideal gas at 25 °C and a pressure of 2.4×10^5 Pa.

Q2 A sealed, airtight container contains 8.21×10^{24} molecules of an ideal gas at a fixed volume of 4.05 m^3. The gas is heated to 500 K (correct to 3 s.f.). What's the pressure inside the container?

Q3 A sealed, airtight container is filled with 1.44×10^{25} molecules of an ideal gas and kept at a constant pressure of 1.29×10^5 Pa. The container is heated so that its volume expands to 0.539 m^3. What temperature is the gas inside the container at this point?

Q4 A sealed, airtight container with a fixed volume of 0.39 m^3 is filled with 20 moles of an ideal gas. The maximum pressure the container can withstand is 2.3×10^5 Pa. What's the highest temperature the container can be heated to before the pressure on it would be too high?

Practice Questions — Fact Recall

Q1 What is Avogadro's constant?

Q2 Write down the equation of state for n moles of an ideal gas, and state what each term in the equation represents.

Q3 Give the equation for calculating the Boltzmann constant, stating what each symbol represents.

Q4 Write down the equation of state for N molecules of an ideal gas, and state what each term in the equation represents.

5. The Pressure of an Ideal Gas

Kinetic theory (p.13) can be used to derive another formula for the pressure of an ideal gas, but first we need some assumptions of ideal gases.

Assumptions made in kinetic theory

In kinetic theory, physicists picture gas particles moving at high speed in random directions. To obtain equations that describe the properties of a gas, some simplifying assumptions are needed:

- The gas contains a large number of particles.
- The particles move rapidly and randomly.
- The volume of the particles is negligible when compared to the volume of the gas.
- Collisions between particles themselves or between particles and the walls of the container are perfectly elastic.
- The duration of each collision is negligible when compared to the time between collisions.
- There are no forces between particles except for the moment when they are in a collision.

An ideal gas obeys all of these assumptions. Real gases behave like ideal gases as long as the pressure isn't too big and the temperature is reasonably high (compared with their boiling points).

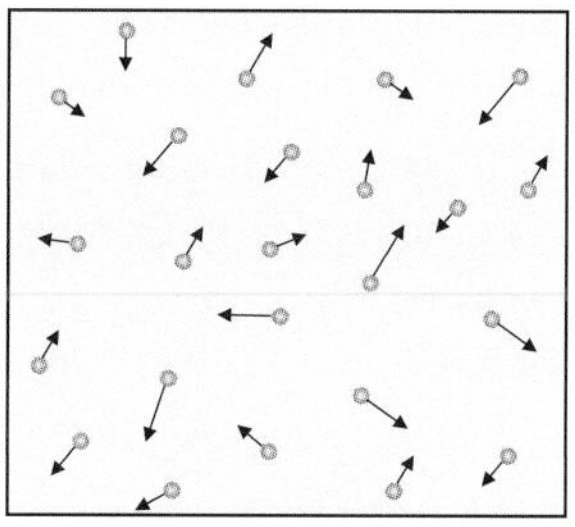

Figure 1: *A visual representation of an ideal gas. All particles are identical, have negligible volume and move randomly.*

Newton's laws and ideal gas pressure

You can use Newton's laws and the assumptions of kinetic theory to explain the pressure of an ideal gas. Imagine a cubic box containing N particles of an ideal gas, each with a mass m.

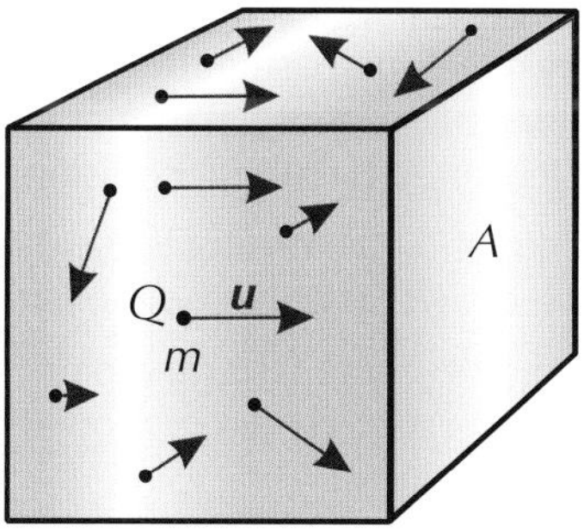

Figure 2: *A cubic box with sides of area A, containing N particles each of mass m.*

Learning Objectives:

- Know the model of kinetic theory of gases.
- Be able to explain pressure in terms of this model.
- Be able to use the equation $pV = \frac{1}{3}Nm\overline{c^2}$, where N is the number of particles (atoms or molecules) and $\overline{c^2}$ is the mean square speed.
- Know what root mean square (r.m.s.) speed and mean square speed are.

Specification Reference 5.1.4

Tip: Perfectly elastic collisions are collisions in which no kinetic energy is lost.

Tip: Remember — particles can mean atoms or molecules in a gas.

Tip: This is an example of using a model to develop a scientific explanation.

HOW SCIENCE WORKS

Tip: You covered Newton's laws in year 1 of A-level.

The particles of the gas are free to move around with constant random motion. There are no forces of attraction between the particles, so according to Newton's 1st law, they continue to move with constant velocities until they collide with another particle or the box itself.

When a particle collides with a wall of the box, it exerts a force on the wall, and the wall exerts an equal and opposite force on the particle. This is Newton's 3rd law in action. The size of the force exerted by the particle on the wall can be calculated using Newton's 2nd law, which says the force is equal to the rate of change of momentum.

For example, if particle Q is travelling directly towards wall A with velocity $\mathbf{u}$, its momentum is $m\mathbf{u}$. When it hits the wall, the force of the impact causes it to rebound in the opposite direction, at the same speed. Its momentum is now $-m\mathbf{u}$, which means the change in momentum is $-2m\mathbf{u}$.

Tip: Imagine putting a single grain of sugar into a lunchbox and shaking it around — the box might as well be empty. Now put a couple of spoonfuls of sugar in, and shake that around — this time you'll hear the sugar thumping against the walls of the box as it collides. You don't notice the effect of the individual grains, only the combined action of loads of them.

So, the force a particle exerts is proportional to its mass and its velocity. The mass of a single gas particle is tiny (for example, an atom of helium gas is only 6.6×10^{-27} kg), so each particle can only exert a minuscule force.

But, there isn't just one particle in the box — there's probably billions of them. The combined force from so many tiny particles is much bigger than the contribution from any individual particle. Because there are so many particles in the box, a significant number will be colliding with each wall of the box at any given moment. And because the particles' motion is random, the collisions will be spread all over the surface of each wall. The result is a steady, even force on all the walls of the box — this is pressure.

So, the pressure in a gas is a result of all the collisions between particles and the walls of the container. The pressure of an ideal gas can be calculated using this equation:

$$pV = \frac{1}{3}Nm\overline{c^2}$$

- $\mathbf{p}$ = *pressure in Pa*
- $\mathbf{V}$ = *volume in m^3*
- $\mathbf{N}$ = *number of particles of gas*
- $\mathbf{m}$ = *mass of a gas particle in kg*
- $\overline{\mathbf{c}^2}$ = *mean square speed of gas particles in m^{2}s^{-2}*

Tip: The mean square speed represents the mean of the squared speeds of all the particles, i.e. if 3 particles had velocities v_1, v_2 and v_3, their mean square speed would be $(v_1^2 + v_2^2 + v_3^2) \div 3$.

The equation shows that the pressure exerted by a gas depends on four things:

- The volume, V, of the container — increasing the volume of the container decreases the frequency of collisions because the particles have further to travel in between collisions. This decreases the pressure.
- The number of particles, N — increasing the number of particles increases the frequency of collisions between the particles and the container, so increases the total force exerted by all the collisions.
- The mass, m, of the particles — according to Newton's 2nd law, force is proportional to mass, so heavier particles will exert a greater force.
- The speed, c, of the particles — the faster the particles are going when they hit the walls, the greater the change in momentum and force exerted.

Figure 3: *Pumping more air into a ball increases the number of particles in the ball, which raises the pressure inside the ball.*

Example — Maths Skills

61.0 moles of a gas are enclosed in a 0.750 m³ container. If the pressure in the container is 101 kPa and each particle has a mass of 2.65×10^{-26} kg, calculate the mean square speed of the gas particles.

Rearrange $pV = \frac{1}{3}Nm\overline{c^2}$ to $\overline{c^2} = \frac{3pV}{Nm}$

Number of particles in the gas $N = nN_A$ (see page 27), so

$$\overline{c^2} = \frac{3pV}{nN_A m} = \frac{3 \times 101 \times 10^3 \times 0.750}{61.0 \times 6.02 \times 10^{23} \times 2.65 \times 10^{-26}} = 2.335... \times 10^5$$
$$= 2.34 \times 10^5 \text{ m}^2\text{s}^{-2} \text{ (to 3 s.f.)}$$

Tip: N_A is 6.02×10^{23} mol^{-1} (see page 27).

Root mean square speed

It often helps to think about the motion of a typical particle in kinetic theory. $\overline{c^2}$ is the **mean square speed** and has units m²s⁻². $\overline{c^2}$ is the average of the squared speeds of all the particles, so the square root of it gives you the typical speed.

This is called the **root mean square speed** or, usually, the r.m.s. speed. It's often written as c_{rms}. The unit is the same as any speed — ms^{-1}.

$$c_{rms} = \sqrt{\text{mean square speed}} = \sqrt{\overline{c^2}}$$

Example — Maths Skills

The particles in a gas have a mean square speed of 2.4×10^5 m²s⁻². Calculate the r.m.s. speed of the particles.

$c_{rms} = \sqrt{\overline{c^2}} = \sqrt{2.4 \times 10^5} = 489.89... = 490$ ms^{-1} (to 2 s.f.)

Practice Questions — Application

Q1 A sealed, rigid container with a volume of 1.44 m³ is filled with exactly 5 moles of an ideal gas. If the molecules of the gas have a mean square speed of 8.11×10^6 m²s⁻² and a mass of 5.31×10^{-26} kg, find the pressure inside the container.

Q2 A container with a volume of 905 cm³ is filled with 0.310 moles of gas. The gas exerts a pressure of 7.40×10^5 Pa on the container walls. Calculate the root mean square speed of the gas molecules if they each have a mass of 5.10×10^{-26} kg.

Practice Questions — Fact Recall

Q1 State four assumptions made about an ideal gas in kinetic theory.

Q2 Explain why a contained gas exerts pressure on the walls of its container.

Q3 What does c_{rms} represent?

6. Internal Energy of an Ideal Gas

Learning Objectives:

- Know what is meant by the internal energy of an ideal gas.
- Be able to derive and use $\frac{1}{2}m\overline{c^2} = \frac{3}{2}kT$.

Specification Reference 5.1.4

Particles in gases all have different amounts of kinetic energy, and it all depends on the absolute temperature.

Maxwell-Boltzmann distribution

The particles in a gas don't all have the same kinetic energy. Some particles have a lot of kinetic energy and move at high speeds, but others don't have much kinetic energy and move much more slowly. Most will travel around the average speed.

If you plot a graph of the numbers of molecules in an ideal gas with different kinetic energies you get a Maxwell-Boltzmann distribution. It looks like this:

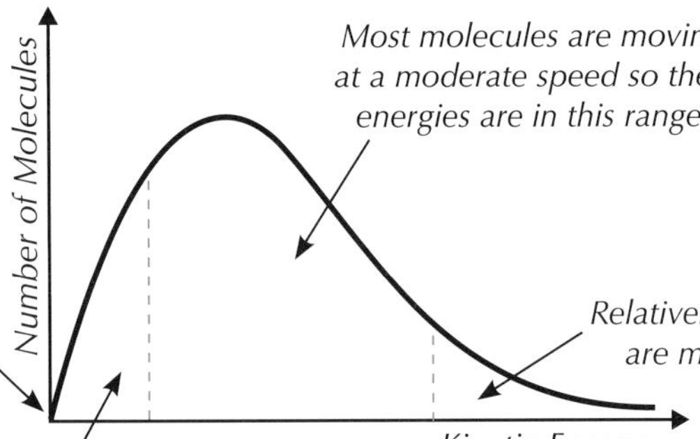

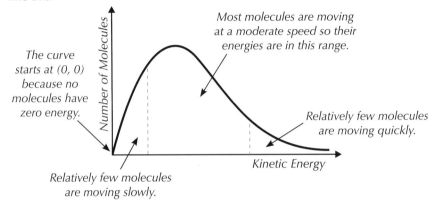

Figure 1: *The Maxwell-Boltzmann distribution of kinetic energies in a gas.*

Tip: The Maxwell-Boltzmann distribution is a theoretical model that has been developed to explain scientific observations. HOW SCIENCE WORKS

Tip: You get pretty much the same distribution in solids and liquids too.

Speed distribution

Kinetic energy is proportional to (speed)^2, so you can also plot the speed distribution of gas particles, which is also a Maxwell-Boltzmann distribution. The shape of the Maxwell-Boltzmann distribution depends on the temperature of the gas. Figure 3 shows two different speed distribution curves. Both curves represent the same number of particles, but the cooler curve has a higher, steeper peak at a lower speed. As the temperature of the gas increases:

- The average particle speed increases.
- The maximum particle speed increases.
- The distribution curve becomes more spread out.

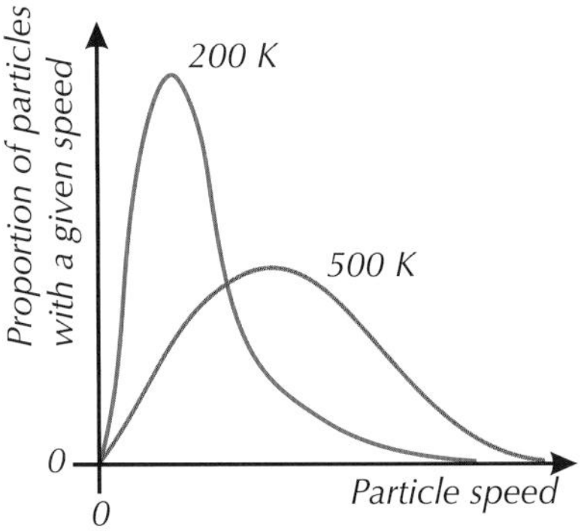

Figure 3: *The Maxwell-Boltzmann distribution of particle speeds for a gas at 200 K and at 500 K.*

Figure 2: *The speed distributions of particles in a gas can be compared to shoppers on a busy street—some will be walking slowly, others will be in a hurry, but most will be walking with a moderate speed.*

Tip: A kinetic energy curve would differ with temperature in the same way.

Energy changes between particles

The particles of a gas collide with each other all the time. Some of these collisions will be 'head-on' (particles moving in opposite directions) while others will be 'shunts from behind' (particles moving in the same direction). As a result of the collisions, energy will be transferred between particles. Some particles will gain speed in a collision and others will slow down. Between collisions, the particles will travel at constant speed.

Although the energy of an individual particle changes at each collision, the collisions don't alter the total energy of the system. So, the average speed of the particles will stay the same provided the temperature of the gas stays the same.

Average kinetic energy and internal energy

You already know that $pV = NkT$ (from page 28) and that $pV = \frac{1}{3}Nm\overline{c^2}$ (page 30). You can combine these to derive the kinetic energy of gas particles in an ideal gas:

$$\frac{1}{3}Nm\overline{c^2} = NkT$$

and you can cancel N to give:

$$\frac{1}{3}m\overline{c^2} = kT$$

which you can rearrange to give:

$$m\overline{c^2} = 3kT$$

$\frac{1}{2}m\overline{c^2}$ is the average kinetic energy of an individual particle. (Remember $\overline{c^2}$ is a measure of speed squared, so this is just like the equation for kinetic energy, $E_k = \frac{1}{2}mv^2$, that you'll know already.)

So multiplying both sides by $\frac{1}{2}$ gives you:

$$\frac{1}{2}m\overline{c^2} = \frac{3}{2}kT$$

m = mass of a gas particle in kg

$\overline{c^2}$ = mean square speed of gas particles in m^2s^{-2}

k = the Boltzmann constant ($= 1.38 \times 10^{-23}$ JK^{-1})

T = temperature in K

So the average kinetic energy, E, of one gas particle is given by:

$$E = \frac{3}{2}kT$$

Exam Tip
You could be tested on this derivation, so make sure you know it.

Tip: The kinetic energy of gas particles is always an average value because the particles are all travelling at different speeds.

Example — Maths Skills

What's the average kinetic energy of the molecules in an ideal gas at 100 °C (correct to 3 significant figures)?

Just put the numbers into the equation above, making sure to convert the temperature to kelvin.

$$\frac{1}{2}m\overline{c^2} = \frac{3}{2}kT$$

$$= \frac{3}{2} \times (1.38 \times 10^{-23}) \times (100 + 273)$$

$$= 7.7211 \times 10^{-21} = 7.72 \times 10^{-21}\ \text{J (to 3 s.f.)}$$

In an ideal gas the potential energy is 0 J because there are no forces between the particles. This means that the internal energy is equal to the total random kinetic energy only (see p.14).

So you can multiply E by the number of gas particles, N, to get an equation for the internal energy, U, of an ideal gas:

$$U = \frac{3}{2}NkT$$

N = *number of particles of gas*

k = *the Boltzmann constant in* JK^{-1}

U = *internal energy of ideal gas in J*

T = *temperature in K*

Example — Maths Skills

A container is filled with 2.5 moles of an ideal gas. The temperature of the gas is 310 K. Calculate the internal energy of the gas.

Number of particles = $N = nN_A = 2.5 \times 6.02 \times 10^{23} = 1.505 \times 10^{24}$

$U = \frac{3}{2}NkT = \frac{3}{2} \times 1.505 \times 10^{24} \times 1.38 \times 10^{-23} \times 310 = 9657.585$ J

$= 9.7$ kJ (to 2 s.f.)

Tip: $N = nN_A$ was covered on page 27.

The equation above and the equation on the previous page show that average kinetic energy and internal energy are both directly proportional to the absolute temperature — a rise in absolute temperature will cause an increase in the kinetic energy of the particles, meaning a rise in internal energy.

Tip: Remember — absolute temperature is just the temperature in kelvin — see page 15.

Practice Questions — Application

Q1 The graph below shows the Maxwell-Boltzmann distributions for a gas at two different temperatures. State whether the higher temperature is shown by curve A or curve B.

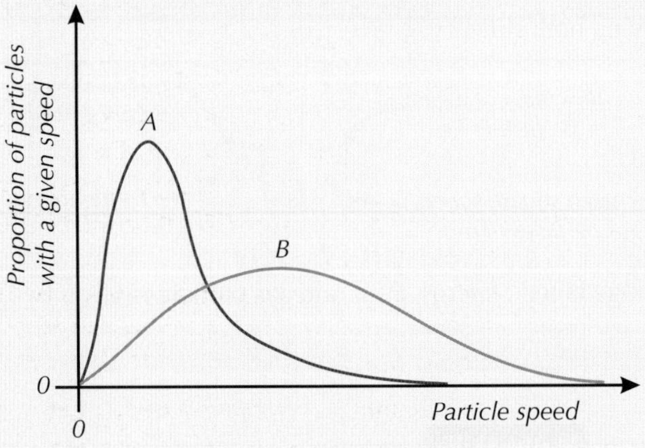

Q2 What's the average kinetic energy of the molecules of an ideal gas at 112 K?

Practice Questions — Fact Recall

Q1 State the two equations from which $\frac{1}{2}m\overline{c^2} = \frac{3}{2}kT$ can be derived.

Q2 For an ideal gas, what can you assume about its potential energy?

Section Summary

Make sure you know...

- That solids, liquids and gases can be described by a simple kinetic model in terms of the spacing, ordering and motion of particles.
- That Brownian motion describes the zigzag, random motion of small particles in air, and be able to describe how a microscope can be used to observe Brownian motion in smoke particles.
- That internal energy is the sum of the random distribution of kinetic and potential energies associated with the molecules of a system, and that it increases with temperature.
- That internal energy changes during a change of phase while temperature remains constant.
- That the absolute scale of temperature does not depend on the property of any particular substance.
- That the lowest limit for temperature is absolute zero (equal to 0 K or –273 °C), and that at this temperature a substance has minimum internal energy.
- That temperature measurements can be in Celsius (°C) or kelvin (K), and how to convert between them using $T(\text{K}) \approx \theta\,(°\text{C}) + 273$.
- That two objects in thermal equilibrium are at the same temperature with no net flow of energy between them.
- That if body A and body B are both in thermal equilibrium with body C, then body A and body B must be in thermal equilibrium with each other.
- That the specific heat capacity of a substance is the amount of energy needed to raise the temperature of 1 kg of the substance by 1 K, and how to calculate the energy change using $E = mc\Delta\theta$.
- How the specific heat capacity of a metal or liquid can be determined using a stopwatch and an electrical heater connected to a voltmeter and an ammeter.
- How the method of mixtures can be used to estimate the specific heat capacity of a metal block and a liquid.
- That the specific latent heat of fusion or vaporisation is the quantity of thermal energy required to change the state of 1 kg of a substance, and how to use $E = mL$ to calculate the energy needed to change the state of a substance.
- How the specific latent heat of fusion or the specific latent heat of vaporisation of a substance can be determined using a stopwatch and a heating coil connected to a voltmeter and an ammeter.
- That Boyle's law is PV = constant and experimental techniques that can be used to investigate it.
- That the pressure law is $\frac{P}{T}$ = constant and experimental techniques that can be used to investigate it.
- How the variation of gas temperature with pressure can be used to estimate the value for absolute zero.
- That Avogadro's constant, N_A, is the number of particles contained in one mole of a substance and is equal to $6.02 \times 10^{23}\ \text{mol}^{-1}$.
- How to use the equation of state of an ideal gas $pV = nRT$, where n is the number of moles and R is the molar gas constant, equal to $8.31\ \text{J mol}^{-1}\,\text{K}^{-1}$.
- That the Boltzmann constant, k, can be calculated using $k = \frac{R}{N_A}$.
- How to use the equation of state for N particles of gas, $pV = NkT$.
- The assumptions for the model of kinetic theory of gases, and be able to explain pressure in terms of this model.
- How to use the equation $pV = \frac{1}{3}Nm\overline{c^2}$.
- That the mean square speed, $\overline{c^2}$, is the average of the squared speeds of all the particles in a gas, and that the square root of it gives you the root mean square (r.m.s.) speed, c_{rms}.
- That the distribution of the kinetic energies of particles in an ideal gas can be described by a Maxwell-Boltzmann curve.
- How to use $\frac{1}{2}m\overline{c^2} = \frac{3}{2}kT$ and be able to derive it from $pV = NkT$ and $pV = \frac{1}{3}Nm\overline{c^2}$.

Exam-style Questions

1 A syringe contains a fixed amount of an ideal gas at a pressure of 120 kPa. The plunger is pulled out, extending the tube length containing the gas from 30 mm to 50 mm, whilst the temperature is kept constant. What is the new gas pressure?

A 43 kPa

B 72 kPa

C 200 kPa

D 330 kPa

(1 mark)

2 A cold spoon is put in some hot water. Which of the following is true?

A There is no net energy flow from the water to the spoon.

B The spoon and water can never reach thermal equilibrium.

C Energy won't flow from the water to the spoon when they reach the same temperature.

D The spoon and water are in thermal equilibrium when they reach the same temperature.

(1 mark)

3 A sealed container with a fixed volume of 0.51 m^3 is filled with an ideal gas at low temperature. The gas is heated so that the pressure inside the container increases. At 0.0 °C the pressure inside the container is 8.1×10^5 Pa.

(a) By considering the molecules of the gas, explain why increasing the temperature leads to an increase in pressure inside the container.

(3 marks)

(b) Using the axes below, show how the pressure of the gas varies with temperature between –273 °C and 100 °C.

(2 marks)

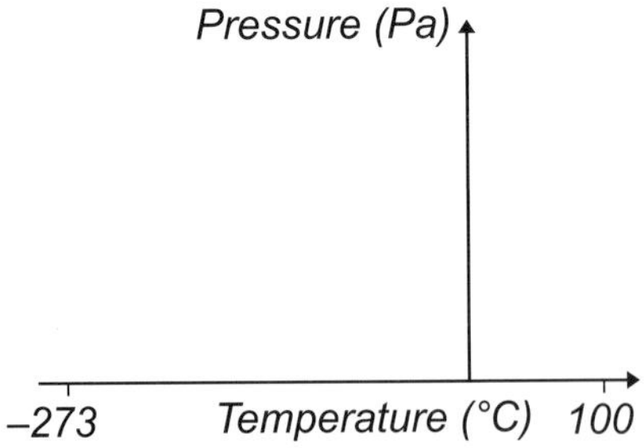

(c) Calculate the number of molecules of gas present in the container.

(2 marks)

(d) The specific heat capacity of the gas is 2.2×10^3 $J\,kg^{-1}K^{-1}$ and the mass of a molecule of the gas is 2.7×10^{-26} kg. If the gas was heated from –150 °C to 0.0 °C, calculate the energy transferred to the gas.

(2 marks)

4 A glass beaker contains a cylindrical block of ice at –25 °C. A heating element is placed inside the block of ice and turned on, supplying energy at a rate of 50.0 $\mathrm{J\,s^{-1}}$. The ice has a mass of 92 g, a specific heat capacity of 2110 $\mathrm{J\,kg^{-1}\,K^{-1}}$ and a specific latent heat of fusion of 3.3×10^5 $\mathrm{J\,kg^{-1}}$.

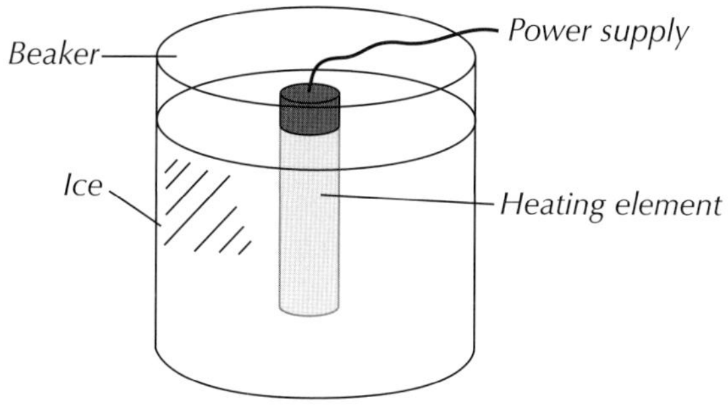

(a) Calculate the energy needed to heat the ice from –25 °C to its melting point, 0.0 °C.

(1 mark)

(b) Calculate the energy needed to melt all of the ice once it reaches 0.0 °C.

(1 mark)

(c) Assuming the ice is heated uniformly and there is no heat transfer between the ice and the surroundings, calculate how long it will take for all of the ice to melt.

(2 marks)

(d) In practice, heat can be transferred between the ice and the surroundings. State whether insulating the beaker would speed up or slow down the melting of the ice. Explain your answer.

(2 marks)

5 The kinetic model of matter is used to describe substances in different states.

(a)* Describe solids, liquids and gases in terms of the kinetic model of matter. Include an explanation of Brownian motion and describe a simple demonstration that can be used to observe the Brownian motion of smoke particles in air.

(6 marks)

(b) A sealed container with a fixed volume of 4.18 m^3 is filled with 54.0 moles of an ideal gas. Each particle has a mass of 3.40×10^{-26} kg.

(i) Calculate the total number of molecules of gas in the container.

(1 mark)

(ii) Calculate the temperature of the gas if the pressure is 1.00×10^5 Pa.

(2 marks)

(iii) The container is uniformly heated to 855 °C. Calculate the r.m.s. speed of the gas particles.

(3 marks)

(iv) Sketch a graph showing the distribution of the particle speeds in the gas at the temperature found in **(ii)** and at 855 °C, and identify each curve.

(2 marks)

* The quality of your response will be assessed in this question.

Learning Objectives:
- Know how to use radians as a measure of angle.
- Know what is meant by angular velocity, ω.
- Be able to use the formula for constant speed in a circle, $v = \omega r$.
- Understand what is meant by the period and frequency of an object in circular motion.
- Know how to calculate angular velocity using $\omega = \frac{2\pi}{T}$ and $\omega = 2\pi f$.

Specification References 5.2.1 and 5.2.2

Exam Tip
You're given formulas for the circumference (and other properties) of a circle in the data and formulae booklet.

Exam Tip
You should remember the equation to convert between radians and degrees, but it's a good idea to memorise these common angles too.

Tip: Angular velocity is a vector quantity (i.e. it has a size and a direction). However, you're not usually interested in its direction — just the size.

1. Circular Motion

In circular motion, you'll see that all angle measurements use radians, so make sure you're comfortable with them before tackling this section.

Radians

Objects in circular motion travel through angles — these angles are usually measured in radians.

The angle in **radians**, θ, is equal to the arc-length divided by the radius of the circle (see Figure 1).

In other words:

$$\text{arc length} = r\theta$$

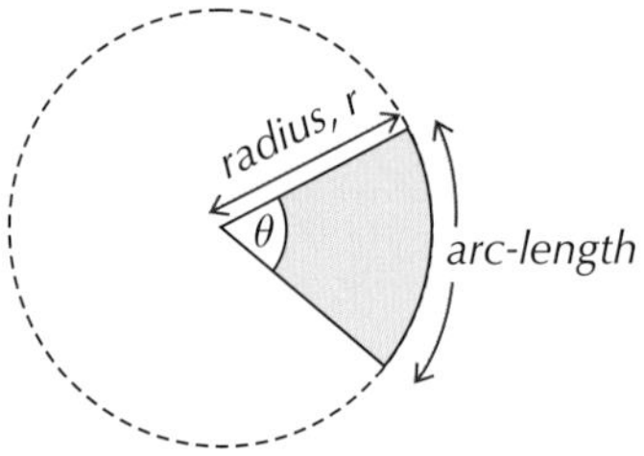

Figure 1: *The radius of a circle, and the angle and arc-length of a sector.*

For a complete circle (360°), the arc-length is just the circumference of the circle ($2\pi r$). Dividing this by the radius (r) gives 2π. So there are 2π radians in a complete circle (see Figure 2). 1 radian is equal to about 57°.

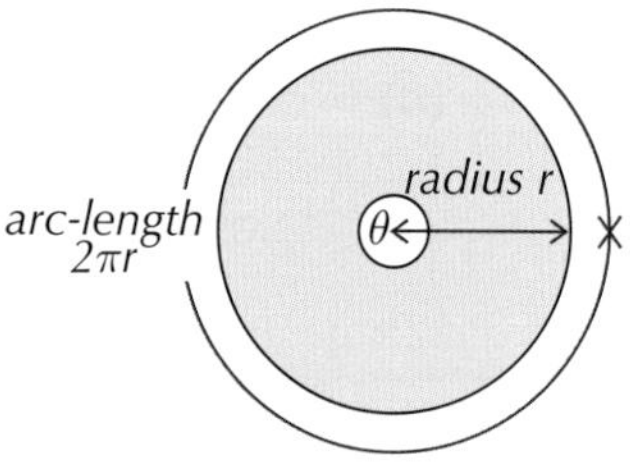

Figure 2: *For a complete circle, the angle $\theta = 2\pi r$.*

To convert from degrees to radians, you multiply the angle by $\frac{\pi}{180°}$:

$$\text{angle in radians} = \frac{\pi}{180°} \times \text{angle in degrees}$$

Figure 3 shows some common angles, given in both degrees and radians.

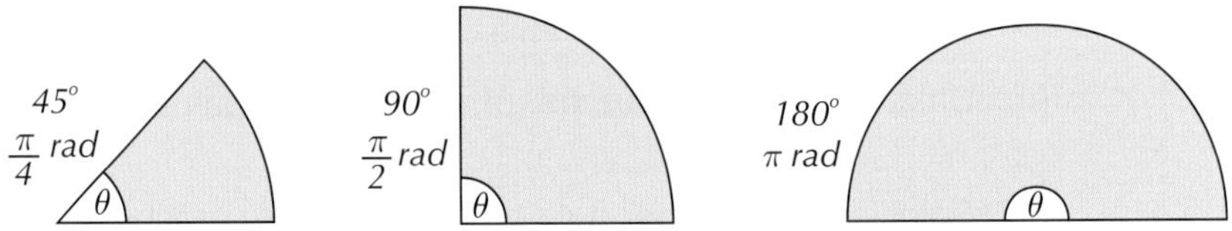

Figure 3: *Angles of sectors in degrees and radians.*

Angular velocity

Angular velocity is the angle an object rotates through per second. Just as linear velocity, ***v***, is defined as displacement ÷ time, the angular velocity, ω, is defined as angle ÷ time. The unit is rad s^{-1} (radians per second).

$$\omega = \frac{\theta}{t}$$

ω = angular velocity in rad s^{-1}
θ = angle that the object turns through in rad
t = time in s

The magnitude of the angular velocity (angular speed) can be written in terms of the magnitude of the linear velocity (linear speed), v. Consider an object moving in a circular path of radius r that moves an angle of θ radians in t seconds. The linear speed is equal to distance ÷ time, where the distance travelled is the arc length that the object moves through in its circular motion. So the equation for speed can be written as:

$$v = \frac{\text{arc length}}{t}$$

The arc length of a sector of angle θ was given on the previous page as:

$$\text{arc length} = r\theta$$

Substituting this into the equation for v gives:

$$v = \frac{r\theta}{t}$$

From the equation on the previous page, you can see that angular speed (ω) can be written as $\omega = \frac{\theta}{t}$, so the equation for linear speed can be written as:

$$v = \omega r$$

v = linear speed in ms^{-1}
r = radius of circle of rotation in m
ω = angular speed in $rad\,s^{-1}$

Tip: Angular velocity is the same at any point on a solid rotating object.

Tip: The linear speed v is how fast an object would be travelling in a straight line if it broke off from circular motion.

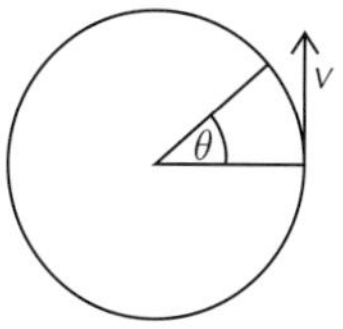

Tip: You don't need to know the derivation of $v = \omega r$ for your exam. And you'll only need to use it in situations where the speed v is constant.

Example

A cyclotron is a type of particle accelerator. Particles start off in the centre of the accelerator, and electric and magnetic fields cause them to move in circles of increasing size.

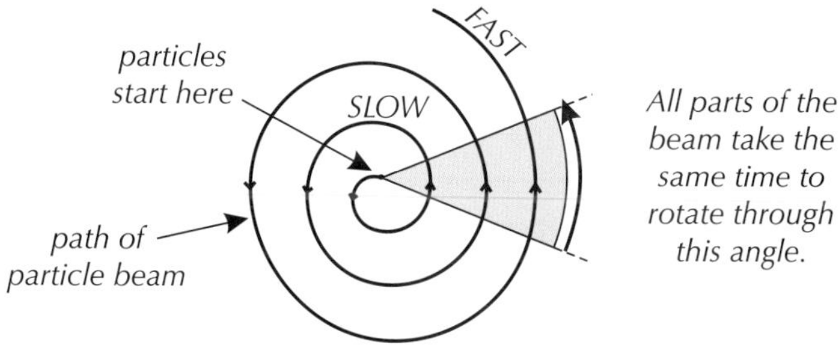

***Figure 4:** The path of a particle in a cyclotron.*

Different parts of the particle beam are rotating at different linear speeds, v. But all the parts rotate through the same angle in the same time — so they have the same angular velocity, ω.

***Figure 5:** One of the first cyclotrons, built by Ernest Lawrence at the University of California.*

Example — Maths Skills

A cyclist is travelling at a speed of 28.8 km h⁻¹ along a road. The diameter of his front wheel is 68 cm. What is the angular speed of the wheel? How long does it take for the wheel to turn one complete revolution?

Linear speed in ms^{-1} = $28.8 \times 1000 \div 3600 = 8.00\ ms^{-1}$

Angular speed = $\omega = \frac{v}{r}$,

where $r = \frac{1}{2} \times \text{diameter} = \frac{1}{2} \times 0.68 = 0.34$ m

So angular speed: $\omega = \frac{8.00}{0.34} = 23.5294... = 24\ rad\,s^{-1}$ (to 2 s.f.)

Time to complete one revolution can be found by rearranging $\omega = \frac{\theta}{t}$:

$$t = \frac{\theta}{\omega} = \frac{2\pi}{23.5294...} = 0.2670... = 0.27 \text{ s (to 2 s.f.)}$$

Tip: The cyclist's speed is the same as the linear speed of the tyre, as long as there's enough friction between the tyre and ground so that the wheel doesn't slip.

Frequency and period

Circular motion has a frequency and period. The **frequency**, f, is the number of complete revolutions per second (measured in hertz, Hz, or s^{-1}). The **period**, T, is the time taken for a complete revolution (in seconds). Frequency and period are linked by the equation:

$$f = \frac{1}{T}$$

f = *frequency in Hz*; **T** = *period in s*

For a complete circle, an object turns through 2π radians in a time T. So the equation for angular velocity becomes:

$$\omega = \frac{2\pi}{T}$$

ω = *angular velocity in rad* s^{-1}; **T** = *period in s*

By replacing $\frac{1}{T}$ with frequency, f, you get an equation that relates ω and f:

$$\omega = 2\pi f$$

ω = *angular velocity in rad* s^{-1}; **f** = *frequency in Hz*

Exam Tip
The equations $\omega = \frac{2\pi}{T}$ and $\omega = 2\pi f$ will be given in your data and formulae booklet, but make sure you know how to use them.

Exam Tip
Exam questions may ask you to leave an answer in terms of π rather than converting it to a horrible fraction or decimal. Always read the question carefully.

Example — Maths Skills

A wheel is turning at a frequency of 20 revolutions per second. Calculate the period and angular velocity of its rotation. Leave your answer in terms of π.

$f = \frac{1}{T}$, so $T = \frac{1}{f} = \frac{1}{20} = 0.05$ s

Angular velocity = $2\pi f = 2 \times \pi \times 20 = 40\pi$ rad s^{-1}

Practice Questions — Application

Q1 The Moon orbits the Earth at a distance of roughly 384 000 km. If it takes the Moon 28 days to complete a full orbit, find its angular velocity and its linear speed.

Q2 An observation wheel of diameter 125 m turns exactly four times every hour. What angle does it rotate through each minute? What is the linear speed of one of the passenger cars?

Q3 A CD is spinning at a frequency of 460 rpm. What are the angular and linear speeds at points 2.0 and 4.0 cm from the centre of the CD?

Q4 A ball with mass m is spun at a constant speed in a circle on the end of a string of length l with time period T. Find the kinetic energy of the ball in terms of m, l and T.

Tip: The radii of the Earth and Moon are much smaller than the distance between them, so you can just treat them as two points.

Tip: 'rpm' is 'revolutions per minute'.

Tip: The equation for kinetic energy is $\frac{1}{2}mv^2$.

Practice Questions — Fact Recall

Q1 How would you convert an angle from degrees to radians?

Q2 What is the definition of angular velocity?

Q3 Write down the formula that links angular velocity and linear speed, defining any symbols you use.

Q4 What is meant by the period and frequency of rotation?

Q5 How would you calculate the angular velocity of an object in circular motion using its frequency? Define any symbols you use.

2. Centripetal Force and Acceleration

Objects moving in a circle are accelerating even if their speed isn't changing. This might sound strange, but it's because velocity is a vector quantity and the direction of the object's velocity is constantly changing.

Centripetal acceleration

Objects travelling in circles are accelerating since their velocity is changing. Even if the car shown in Figure 1 below is going at a constant speed, its velocity is changing since its direction is changing. Since acceleration is defined as the rate of change of velocity, the car is accelerating even though it isn't going any faster. This acceleration is called the **centripetal acceleration** and is always directed towards the centre of the circle.

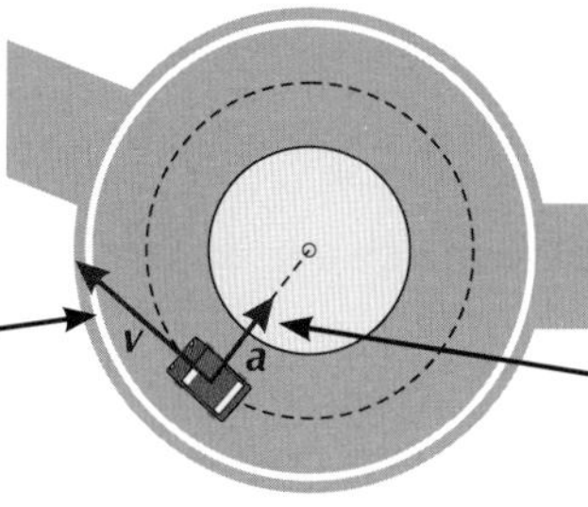

v is at a tangent to the circle.
v
a
The acceleration of the car is directed towards the centre of the circle.

Figure 1: *A car moving in a circle.*

Derivation of formula for centripetal acceleration

A ball is moving at a constant speed in a circle (see Figure 2). Because the ball is always changing direction, the linear velocity is always changing. However, the magnitude of the linear velocity, v, is always the same (sometimes this is called linear speed). During time Δt, the ball moves from a point on the circle, A, to another point, B. The ball turns through the angle θ (the angle between the lines OA and OB).

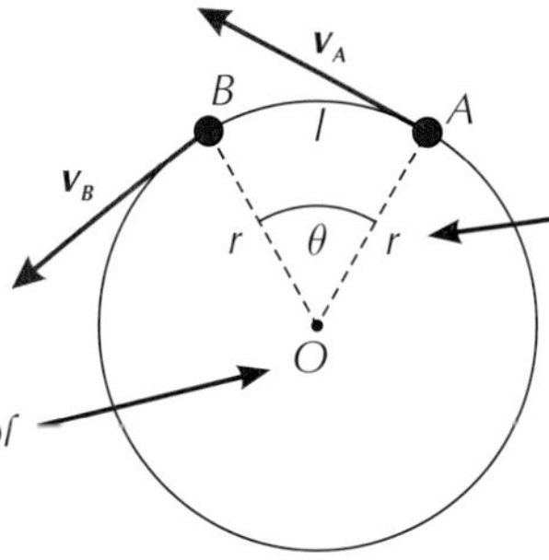

v_A
B
A
l
v_B
r
θ
r
r is the radius of the circle.
O
O is the centre of the circle.

Figure 2: *A ball travelling in a circular path travels through angle θ when it moves from point A to point B.*

The distance, l, that the ball travels along the circle from A to B equals the ball's linear speed multiplied by the time it takes to move that distance:

$$l = v\Delta t$$

At point A the ball has linear velocity $\mathbf{v}_A$ and at point B it has linear velocity $\mathbf{v}_B$. The change in linear velocity, $\Delta\mathbf{v}$, is:

$$\Delta\mathbf{v} = \mathbf{v}_B - \mathbf{v}_A$$

You can draw a triangle made up of the velocity vectors $\mathbf{v}_A$, $\mathbf{v}_B$ and $\Delta\mathbf{v}$ (see Figure 3). The linear velocity is always at a tangent to the radius, so the angle between $\mathbf{v}_A$ and $\mathbf{v}_B$ is also θ. This triangle is the same shape as the triangle ABO, since both are isosceles triangles with the same angle θ between the two sides of identical length.

Learning Objectives:

- Understand that a constant net force perpendicular to the velocity of an object causes it to travel in a circular path.
- Know how to use the equations for centripetal acceleration, $\mathbf{a} = \frac{v^2}{r}$ and $\mathbf{a} = \omega^2 r$, and centripetal force, $\mathbf{F} = \frac{mv^2}{r}$ and $\mathbf{F} = m\omega^2 r$.
- Know and understand techniques and procedures used to investigate circular motion using a whirling bung.

Specification Reference 5.2.2

Tip: You don't need to know this derivation for the exam, it's just here so you can see where the equations for centripetal acceleration come from. These equations are on the next page.

Tip: Linear velocity has the same magnitude as linear speed, but is a vector that is a tangent to the circle. $\mathbf{v}_A$ and $\mathbf{v}_B$ have the same magnitude but different directions.

Tip: Isosceles triangles have two sides the same length.

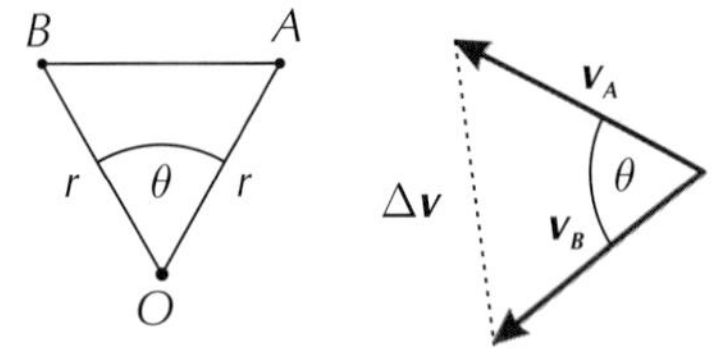

Figure 3: *The velocity vectors* $\mathbf{v}_A$, $\mathbf{v}_B$ *and* $\Delta\mathbf{v}$ *form an isosceles triangle.*

Tip: These two triangles are similar — they're the same shape, just different sizes and orientations.

Tip: If θ is small, $l \approx AB$.

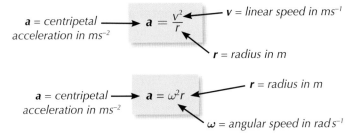

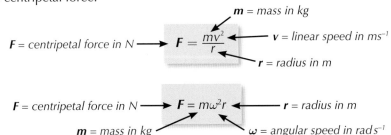

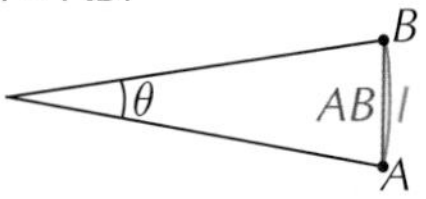

If θ is small, then the length of the straight line AB is approximately equal to the length of the arc l. Because the two triangles in Figure 3 have the same shape, the ratio of l and r is the same as the ratio of $\Delta\mathbf{v}$ and $\mathbf{v}_A$.

$$\frac{l}{r} = \frac{\Delta\mathbf{v}}{\mathbf{v}_A} = \frac{\Delta\mathbf{v}}{v}$$

Tip: The length of each of the vectors $\mathbf{v}_A$ and $\mathbf{v}_B$ is just v.

The previous equation, $l = v\Delta t$, can be substituted in:

$$\frac{v\Delta t}{r} = \frac{\Delta\mathbf{v}}{v} \Rightarrow \frac{\Delta\mathbf{v}}{\Delta t} = \frac{v \times v}{r} = \frac{v^2}{r}$$

Since acceleration, $\mathbf{a}$, is equal to the change in velocity over time, this gives:

$$\mathbf{a} = \frac{\Delta\mathbf{v}}{\Delta t} = \frac{v^2}{r}$$

Tip: Remember — linear speed and the magnitude of the linear velocity are the same thing.

The formula for centripetal acceleration can then be written in terms of either the linear speed or angular speed:

$$\mathbf{a} = \frac{v^2}{r}$$

$\mathbf{a}$ = centripetal acceleration in ms^{-2}
$\mathbf{v}$ = linear speed in ms^{-1}
$\mathbf{r}$ = radius in m

$$\mathbf{a} = \omega^2 r$$

$\mathbf{a}$ = centripetal acceleration in ms^{-2}
$\mathbf{r}$ = radius in m
$\boldsymbol{\omega}$ = angular speed in $rad\,s^{-1}$

Exam Tip
You'll be given both of these formulas in the exam data and formulae booklet.

Tip: $v = r\omega$, so $v^2 = \omega^2 r^2$. This gives $\frac{v^2}{r} = \frac{\omega^2 r^2}{r} = \omega^2 r$.

Centripetal force

Newton's first law of motion says that an object's velocity will stay the same unless there's a force acting on it. Since an object travelling in a circle has a centripetal acceleration, there must be a force causing this acceleration. Circular motion is caused by a constant net force perpendicular to the velocity (i.e. acting towards the centre of the circle), called a **centripetal force**.

Tip: You met Newton's laws of motion in year 1 of A-level.

Although the force changes the direction of the motion, the object never moves towards or away from the centre of the circle, so there is no motion in the direction of the force. Hence no work is done on the object, and the object's kinetic energy (and therefore speed) remains constant.

Tip: The centripetal force is what keeps the object moving in a circle — remove the force and the object would fly off at a tangent with velocity $\mathbf{v}$.

Newton's second law says $\mathbf{F} = m\mathbf{a}$, so substituting this into the equations above for the centripetal acceleration gives you equations for the centripetal force:

$$\mathbf{F} = \frac{mv^2}{r}$$

$\mathbf{F}$ = centripetal force in N
$\mathbf{m}$ = mass in kg
$\mathbf{v}$ = linear speed in ms^{-1}
$\mathbf{r}$ = radius in m

$$\mathbf{F} = m\omega^2 r$$

$\mathbf{F}$ = centripetal force in N
$\mathbf{m}$ = mass in kg
$\mathbf{r}$ = radius in m
$\boldsymbol{\omega}$ = angular speed in $rad\,s^{-1}$

Tip: Don't confuse the centripetal force with the centrifugal force. The centrifugal force is the outwards reaction force you experience when you're spinning.

Example — Maths Skills

A car of mass 890 kg is moving at 15 ms^{-1}. It travels around a circular bend of radius 18 m. Calculate the centripetal acceleration and centripetal force experienced by the car.

Acceleration = $\boldsymbol{a} = \frac{v^2}{r} = \frac{15^2}{18} = 12.5 = 13$ ms^{-2} (to 2 s.f.)

Force = $\boldsymbol{F} = \frac{mv^2}{r} = m\boldsymbol{a} = 890 \times 12.5 = 11\ 125 = 11\ 000$ N (to 2 s.f.)

Example — Maths Skills

A satellite orbits the Earth twice every 24 hours. The acceleration due to the Earth's gravity is 0.57 ms^{-2} at the satellite's orbiting altitude. How far above the Earth's surface is it orbiting? The radius of the Earth is 6400 km.

1 orbit takes 12 hours, so period in seconds = $T = 12 \times 3600 = 43\ 200$ s

Angular speed = $\omega = \frac{2\pi}{T}$

$= \frac{2\pi}{43\ 200} = 1.4544... \times 10^{-4}$ rad s^{-1}

Acceleration $\boldsymbol{a} = \omega^2 r$ so radius $r = \frac{\boldsymbol{a}}{\omega^2}$

$= \frac{0.57}{(1.4544... \times 10^{-4})^2} = 26\ 945.2...$ km

Height above Earth = radius of orbit – radius of Earth

= 26 945.2... – 6400 = 20 545.2... = 21 000 km (to 2 s.f.)

Tip: The acceleration due to the Earth's gravity decreases as you move further away from the centre of the planet — only at the surface is it 9.81 ms^{-2}. See page 65 for more.

Investigating circular motion

You can investigate circular motion using a rubber bung, some washers, some string and a clear plastic tube, as shown in Figure 4.

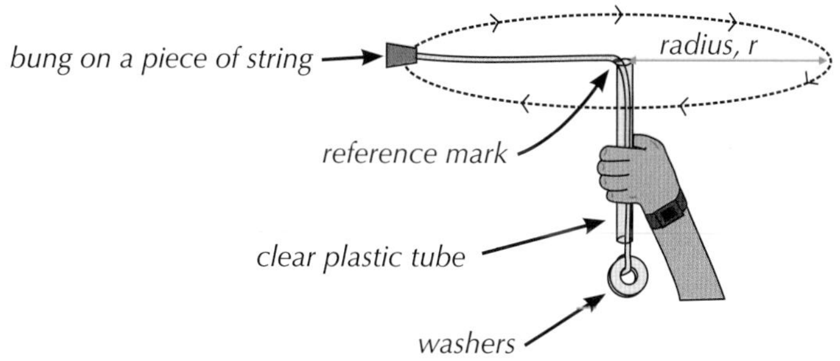

Figure 4: *The experimental set-up for investigating circular motion.*

Tip: Make sure that the bung and the washers are securely fastened to the string, that you're not standing too close to anyone and that there's nothing breakable nearby. And remember to wear safety goggles.

Measure the mass of the bung (m_b) and the mass of the washers (m_w), then attach the bung to the string. Thread the string through the plastic tube, and weigh down the free end using the washers.

Make a reference mark on the string, then measure the distance from the mark to the centre of the bung. Pull the string taut to make sure this measurement is as accurate as possible.

Line the reference mark up with the top of the tube, then begin to spin the bung in a horizontal circle, as shown in Figure 4. You'll need to spin it at the right speed to keep the reference mark level with the top of the tube (if you spin too quickly it'll move outwards, but if you spin too slowly it'll move down). Try to keep your hand as still as possible whist you spin.

Tip: Keeping the reference mark level with the top of the tube is a bit tricky, so you may need to repeat the experiment quite a few times to get an accurate measurement of T.

Figure 5: *When a hammer thrower is spinning, the centripetal force acting along the chain keeps the hammer moving in a circle. When the thrower lets go, this force vanishes and the hammer flies off with an initial speed equal to its linear speed.*

Tip: For Q2, start by finding expressions for ω in terms of $\mathbf{a}$ and r and in terms of T. You can then rearrange and combine them to find the answer.

Measure the time taken for the bung to make one complete circle. This is the time period, T. In practice, this may be too small to time accurately, so you might need to measure the time taken to complete ten circles and divide to get an average.

Once you've got an accurate value for T, use the formula $\omega = \frac{2\pi}{T}$ to find the angular speed of the bung. You can then find the centripetal force exerted on the bung using $\mathbf{F} = m_b\omega^2 r$. In this equation, r is the radius of the bung's circular path, which will equal the distance from the reference mark on the string to the centre of the bung.

The centripetal force should be equal to the weight of the washers ($\mathbf{W} = m_w\mathbf{g}$). This weight is what causes the tension in the string, which acts as the centripetal force.

Repeat this experiment for different values of r (i.e. different distances between the bung and the reference mark) — you should find that as r gets bigger, the time period gets longer but the centripetal force stays the same.

Practice Questions — Application

Q1 On a ride at a theme park, riders are strapped into seats attached to the edge of a horizontal wheel of diameter 8.5 m. The wheel rotates 15 times a minute. Calculate the force felt by a rider of mass 60.0 kg.

Q2 A planet follows a circular orbit of radius r around a star. The planet experiences a constant centripetal acceleration $\mathbf{a}$. How long does it take for the planet to orbit the star 3 times? Choose the correct option.

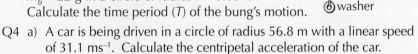

A $3\pi\sqrt{\frac{\mathbf{a}}{r}}$ B $6\pi\sqrt{\frac{\mathbf{a}}{r}}$ C $6\pi\sqrt{\frac{r}{\mathbf{a}}}$ D $\frac{2}{3}\pi\sqrt{\mathbf{a}r}$

Q3 A student is investigating circular motion using the experimental set-up shown. She attaches a washer of mass $m_w = 310$ g to the end of the string, and spins a bung of mass $m_b = 22$ g in a circle of radius $r = 0.60$ m. Calculate the time period (T) of the bung's motion.

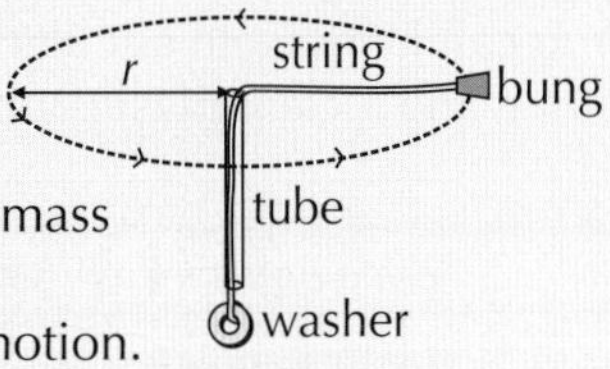

Q4 a) A car is being driven in a circle of radius 56.8 m with a linear speed of 31.1 ms^{-1}. Calculate the centripetal acceleration of the car.

b) The linear speed of the car decreases to half its original value. If the original centripetal force experienced by the car was F, what is the new centripetal force experienced by the car in terms of F?

Q5 A biker rides in a vertical circle around the inside of a cylinder of radius 5.0 m (so he's upside down at the top of the cylinder). For the biker to not fall off at the top of the loop, he must be going fast enough that his centripetal acceleration does not drop below the acceleration due to gravity, 9.81 ms^{-1}. What's the minimum speed he must be travelling at the top of the loop?

Practice Questions — Fact Recall

Q1 What is meant by centripetal acceleration and centripetal force?

Q2 Give the equation for centripetal acceleration in terms of angular velocity. Define all the symbols you use.

Q3 Give the equation for centripetal force in terms of linear speed. Define all the symbols you use.

3. Simple Harmonic Motion

A swinging pendulum moves with simple harmonic motion — this topic is all about what simple harmonic motion is and where you might see it occurring.

What is simple harmonic motion?

An object moving with **simple harmonic motion** (SHM) oscillates to and fro, either side of an equilibrium position (see Figure 1). This equilibrium position is the midpoint of the object's motion. The distance of the object from the equilibrium is called its displacement.

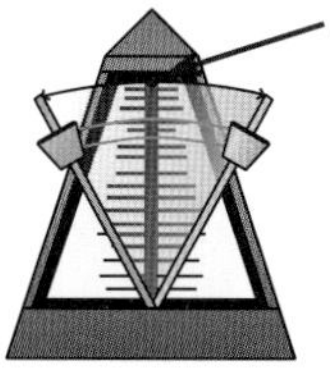

Figure 1: *A metronome moves with simple harmonic motion about an equilibrium position.*

There is always a restoring force pulling or pushing the object back towards the equilibrium position. The size of the restoring force depends on the displacement (see Figure 2). The restoring force makes the object accelerate towards the equilibrium.

small displacement, therefore small force
large displacement, therefore large force

Figure 2: *The size of the restoring force for an object moving with simple harmonic motion depends on its displacement from its equilibrium position.*

SHM can be defined as:

> An oscillation in which the acceleration of an object is directly proportional to its displacement from the midpoint, and is directed towards the midpoint.

This definition can be written as the relation:

$$\boldsymbol{a} = -\omega^2 \boldsymbol{x}$$

Here, $\boldsymbol{a}$ is the acceleration, $\boldsymbol{x}$ is the displacement, and ω^2 is a constant. The minus sign shows that the acceleration is always opposing the displacement (that's the "directed towards the midpoint" bit in the definition).

Learning Objectives:

- Know what is meant by simple harmonic motion.
- Know that the defining equation of SHM is $\boldsymbol{a} = -\omega^2\boldsymbol{x}$.
- Understand the terms displacement, amplitude, period, frequency, angular frequency and phase difference in the context of simple harmonic motion.
- Know what angular frequency (ω) is, and be able to calculate it using $\omega = 2\pi f$ and $\omega = \frac{2\pi}{T}$.
- Be able to use graphical methods to relate the changes in displacement, velocity and acceleration during simple harmonic motion.
- Know that the period of a simple harmonic oscillator is independent of its amplitude (isochronous oscillator).
- Be able to describe the interchange between kinetic and potential energy during simple harmonic motion.
- Understand energy-displacement graphs for a simple harmonic oscillator.

Specification References 5.3.1 and 5.3.2

Tip: As acceleration is proportional to displacement, the force is also proportional to displacement.

Graphs of simple harmonic motion

You can draw graphs to show how the displacement, velocity and acceleration of an object oscillating with SHM change with time (see Figure 3). There's more information about where these graphs come from on pages 49-50.

Displacement

Displacement, $\mathbf{x}$, varies as a cosine or sine wave with a maximum value, A (the amplitude).

Velocity

Velocity, $\mathbf{v}$, is the gradient of the displacement-time graph. It has a maximum value of ωA, where ω is the angular frequency of the oscillation. The angular frequency can be calculated using $\omega = 2\pi f$ or $\omega = \frac{2\pi}{T}$ (see page 40).

Acceleration

Acceleration, $\mathbf{a}$, is the gradient of the velocity-time graph. It has a maximum value of $\omega^2 A$.

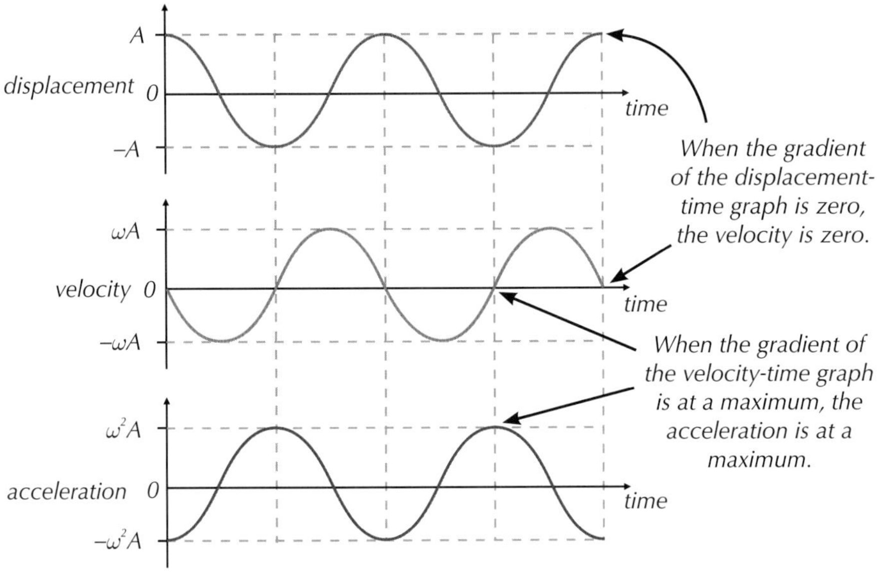

Figure 3: *Graphs showing how displacement, velocity and acceleration of an object experiencing SHM change with time.*

It might seem odd that the acceleration should be a maximum at zero velocity, but it might be helpful to remember that acceleration is the rate of change of velocity — acceleration is found from the gradient of the velocity-time graph. The steepest part of the velocity-time graph (and therefore the largest acceleration) is when the curve is passing through zero.

Also, remember that $\mathbf{a} = -\omega^2\mathbf{x}$. So the maximum acceleration occurs at the maximum negative displacement (the turning points of the displacement-time graph), which coincides with where velocity is zero.

Phase difference

Phase difference is a measure of how much one wave lags behind another wave, and can be measured in radians, degrees, or fractions of a cycle. Two waves that are in phase with each other have a phase difference of zero (or 2π radians) — i.e. the maxima and minima in each wave occur at the same time. If two waves are exactly out of phase ('in antiphase'), they have a phase difference of π radians or 180° — one wave's maximum occurs at the same time as the other's minimum.

Tip: Sine and cosine waves are graphs plotted of the functions sin and cos of some changing value (like time).

Tip: ω is called angular frequency in SHM, and angular velocity in circular motion (see page 38).

Tip: The equations that determine the displacement-time graph are on p.49. The velocity-time graph is derived from the gradient of the displacement-time graph because $\mathbf{v} = \frac{\Delta \mathbf{x}}{\Delta t}$. Similarly, $\mathbf{a} = \frac{\Delta \mathbf{v}}{\Delta t}$.

Tip: See page 50 to see where these maximum values of velocity and acceleration come from.

Tip: Kinetic energy (E_K) will vary in a similar way to $\mathbf{v}$, since $E_K = \frac{1}{2}mv^2$ — except the graph for E_K will always be positive (as v^2 is always positive).

Tip: These graphs can be plotted from experiments by using position sensors and data loggers to measure the motion of an object undergoing SHM — see page 52.

The velocity is a quarter of a cycle in front of the displacement in the velocity-time and displacement-time graphs for an object in SHM (see Figure 3) — it's $\frac{\pi}{2}$ radians out of phase. The acceleration is another $\frac{\pi}{2}$ radians ahead of the velocity, and so is in antiphase with the displacement.

Tip: Remember there are 2π radians in a full cycle, so $\frac{\pi}{2}$ radians is a quarter cycle.

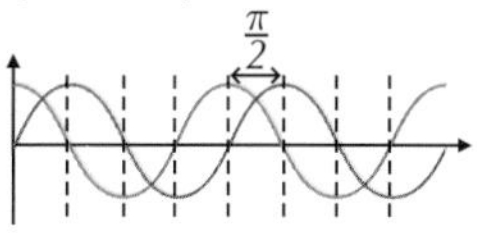

Frequency and period

From maximum positive displacement (e.g. maximum displacement to the right) to maximum negative displacement (e.g. maximum displacement to the left) and back again is called a cycle of oscillation. The frequency, f, of the SHM is the number of cycles per second (measured in hertz, Hz). The period, T, is the time taken for a complete cycle (in seconds).

Tip: Remember, $f = \frac{1}{T}$ (see p.40).

The amplitude of an oscillation is the maximum magnitude of the displacement. In SHM, the frequency and period are independent of the amplitude (i.e. they're constant for a given oscillation). So a pendulum clock will keep ticking in regular time intervals even if its swing becomes very small. This kind of oscillator is called an isochronous oscillator.

Potential and kinetic energy

An object in SHM exchanges **potential energy** and **kinetic energy** as it oscillates. The type of potential energy (E_P) depends on what it is that's providing the restoring force. This will be gravitational E_P for pendulums and elastic E_P (elastic potential energy) and possibly gravitational E_P for masses on springs.

- As the object moves towards the equilibrium position, the restoring force does work on the object and so transfers some E_P to E_K.
- When the object is moving away from the equilibrium, all that E_K is transferred back to E_P again.
- At the equilibrium, the object's E_P is said to be zero and its E_K is maximum — therefore its velocity is maximum.
- At the maximum displacement (the amplitude) on both sides of the equilibrium, the object's E_P is maximum, and its E_K is zero — so its velocity is zero.

Tip: Gravitational potential energy is gained by moving away from a mass (page 71), and elastic potential energy is stored by elastic objects when they're taut.

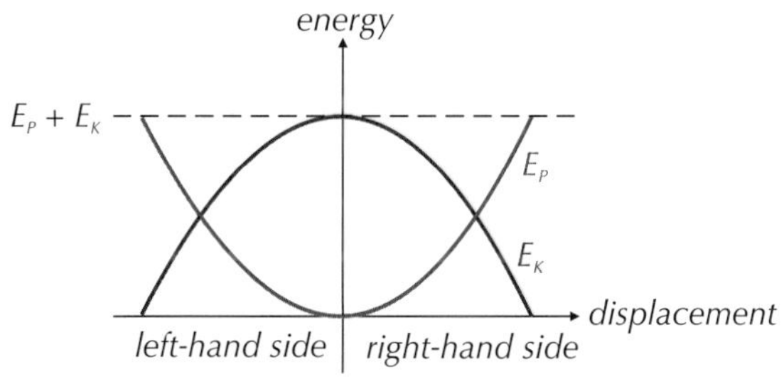

Figure 5: Graph to show how the kinetic and potential energy of an object in SHM change with displacement.

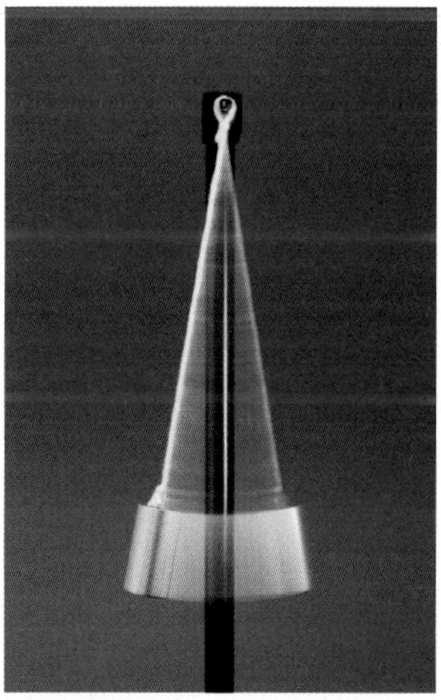

Figure 4: An oscillating pendulum moves fastest when it passes through its equilibrium position — when all its energy is kinetic.

The sum of the potential and kinetic energy is called the mechanical energy and stays constant (as long as the motion isn't damped — see page 56). The energy transfer for one complete cycle of oscillation (see Figure 6) is: E_P to E_K to E_P to E_K to E_P ... and then the process repeats...

Exam Tip
Make sure you know the graphs in Figures 5 and 6 well — you could be asked to recognise, sketch and apply them to other things.

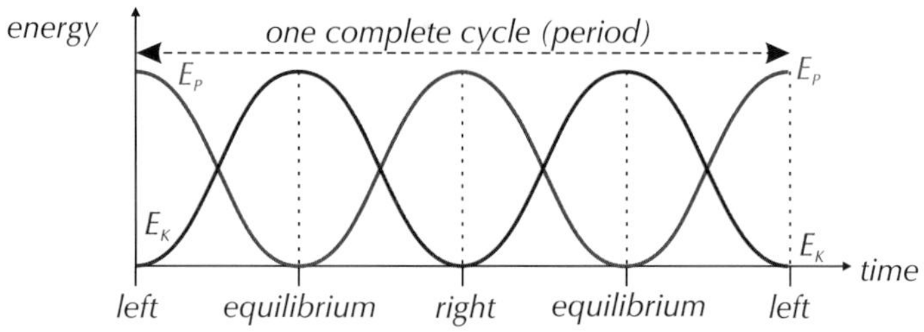

Figure 6: *Graph to show how the kinetic and potential energy of an object in SHM change with time.*

Practice Questions — Application

Q1 This graph shows displacement against time for a mass on the end of a spring. Which of the point(s) A-E shows:

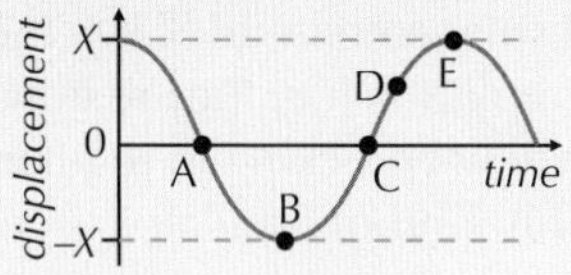

a) when the magnitude of the velocity of the mass is at a maximum?

b) when the magnitude of the acceleration of the mass is at a maximum?

Q2 A girl is sitting on a swing undergoing simple harmonic motion. To start the motion, she raised the swing so it was higher than its lowest position, then lifted her feet off the ground. Sketch a graph of the girl's kinetic energy against time for two complete oscillations of the swing, where $t = 0$ at the point where she lifted her off the ground. You can assume no energy is lost to the surroundings.

Q3 A pendulum is undergoing simple harmonic motion. The maximum displacement of the pendulum is 0.60 m and the maximum speed of the pendulum is 0.90 ms^{-1}.

a) Calculate the frequency of the oscillation.

b) Calculate the time taken for the pendulum to complete one oscillation.

c) Calculate the maximum acceleration of the pendulum.

Tip: Remember that speed has a maximum value of ωA or $2\pi fA$.

Practice Questions — Fact Recall

Q1 Write down the defining equation of simple harmonic motion.

Q2 What is meant by the frequency and period of an oscillation?

Q3 What is the phase difference between the displacement and the velocity of an object moving with SHM?

Q4 For an object undergoing simple harmonic motion, state whether the following occur at equilibrium or maximum displacement:

a) maximum velocity

b) minimum velocity

c) maximum acceleration

d) minimum acceleration

Q5 Sketch the graphs of displacement, velocity and acceleration against time for an object moving with SHM. Mark on your graphs the maximum and minimum values that each can take.

Q6 Describe how the kinetic and potential energy of an object moving with SHM change with time.

4. Calculations with SHM

You saw the defining equation of simple harmonic motion ($\mathbf{a} = -\omega^2\mathbf{x}$) on p.45. The next few pages are about solutions to that equation, which will be equations for $\mathbf{x}$ in terms of time (t). There's some maths here, but it's hopefully not too bad.

Learning Objectives:

- Know solutions to the equation $\mathbf{a} = -\omega^2\mathbf{x}$, e.g. $\mathbf{x} = A\cos(\omega t)$ or $\mathbf{x} = A\sin(\omega t)$.
- Be able to use the equation $\mathbf{v} = \pm\omega\sqrt{A^2 - \mathbf{x}^2}$.
- Know that $v_{max} = \omega A$.

Specification Reference 5.3.1

Equations for simple harmonic motion

To understand the maths behind SHM, it's useful to think about what circular motion in a horizontal plane would look like when viewed from the side. Imagine a ball spinning in a horizontal circle (see Figure 1). From above, you'd see the ball's circular path, but from the side (in the plane of rotation) it will look like the ball's oscillating from side to side, and moving with SHM. Its speed, v, will appear fastest when $\mathbf{x} = 0$ and appear slowest when $\mathbf{x} = \pm A$.

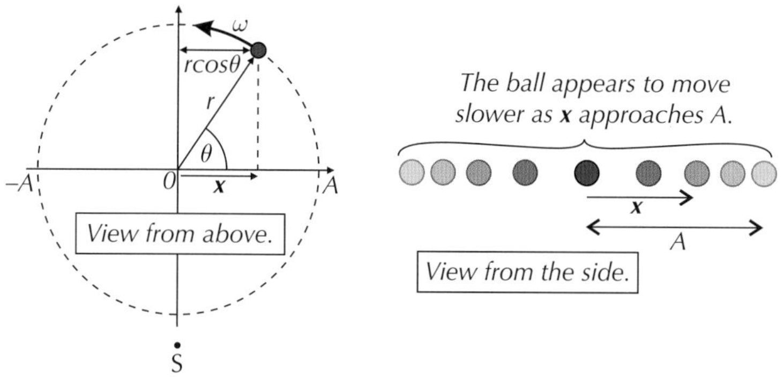

Figure 1: *A ball following a circular path, pictured from above (left) and from the side at point S (right).*

Tip: See pages 38-44 for more about circular motion.

Displacement

- Figure 1 shows that $\mathbf{x}$, the displacement of the ball when viewed from point S, is just equal to the 'left-to-right' component of the ball's position, i.e. $\mathbf{x} = r\cos\theta$.
- When viewed from the side, the amplitude of the ball's simple harmonic motion, A, is the same as the radius of the circle, r.
- If the ball's displacement (as seen from the side) is at its maximum when $t = 0$ (i.e. when $t = 0$, $\theta = 0$ and $\mathbf{x} = A$) and its angular speed is ω, then $\theta = \omega t$. Note that the equivalent 'component' to ω in SHM is called **angular frequency**. It still has the same value, the same units and the same symbol, it's just got a different name.
- Combining these equations gives you an expression for the displacement, $\mathbf{x}$, of an object undergoing SHM:

$$\mathbf{x} = r\cos\theta = A\cos(\omega t)$$

- If the displacement (as seen from the side) is <u>not</u> at its maximum when $t = 0$, then the displacement graph (see p.46) would just be 'shifted' horizontally. In particular, if the ball is passing through its equilibrium position ($\mathbf{x} = 0$) when $t = 0$, the equation for displacement becomes:

$$\mathbf{x} = A\sin(\omega t)$$

- So for an object undergoing SHM, the displacement can be described by:

$$\mathbf{x} = A\cos(\omega t) \quad \text{or} \quad \mathbf{x} = A\sin(\omega t)$$

$\mathbf{x}$ = displacement in m, ***A** = amplitude in m*, ***t** = time in s*, *ω = angular frequency in rads^{-1}*

Tip: A is the maximum displacement — it's not acceleration.

Tip: Angular frequency (ω) and frequency (f) are similar but not the same. ω is measured in rads^{-1} and f is measured in Hz (or s^{-1}). The relation between the two is $\omega = 2\pi f$.

Tip: Remember to set your calculator to radians when using the equations here and on the next page.

Tip: Remember, the only difference between the graphs for sin and cos is a 'horizontal shift'.

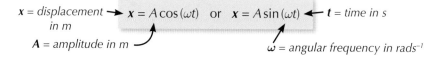

Tip: You don't need to know the derivations of all these equations — they're just here to help you understand SHM a little more.

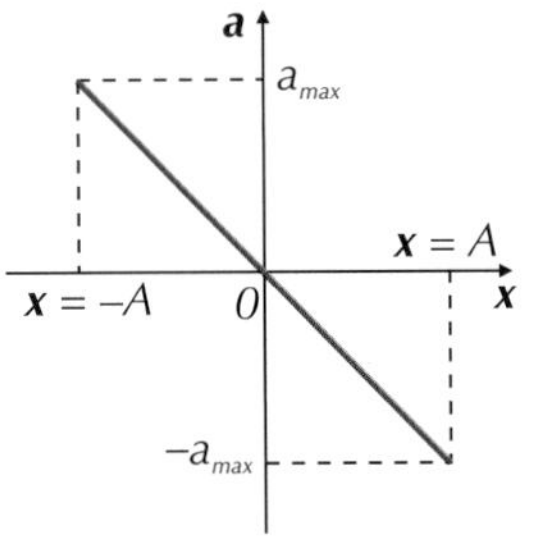

Figure 2: *Graph showing how acceleration varies with displacement for an object moving with SHM.*

Tip: You'll be given the formulas for ***x***, ***v*** and ***a*** in the data and formulae booklet, but you need to understand what they mean and how to use them.

You won't be given the formulas for v_{max} and a_{max} though, so you either need to learn them, or be able to work them out.

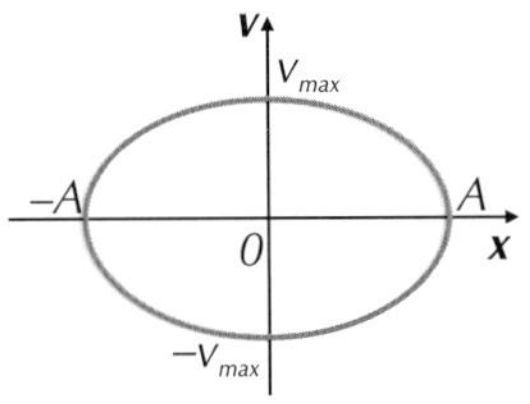

Figure 3: *Graph showing how velocity varies with displacement for an object moving with SHM.*

Tip: Remember — ω is called angular velocity in circular motion and angular frequency in SHM. It's just the name that's different, so the equation for ω is always: $\omega = 2\pi f$ or $\omega = \frac{2\pi}{T}$.

Acceleration

The acceleration of an object moving with SHM can also be related to the acceleration of an object in circular motion. Look again at Figure 1.

- The ball is moving in a circle, so its acceleration is $\omega^2 r$.
- Seen from point S (and assuming $\theta = 0$ at $t = 0$), the acceleration of the ball, ***a***, would appear to be just the 'left-to-right' component of $\omega^2 r$:

  $$\mathbf{a} = -\omega^2 r\cos\theta = -\omega^2 A\cos(\omega t)$$

  There's a minus sign because the acceleration is always acting towards the centre of the circle (e.g. if ***x*** is positive, then ***a*** is in the negative direction).
- But $A\cos(\omega t)$ is equal to the ball's displacement, ***x***. So the acceleration of the object is:

$$\mathbf{a} = -\omega^2 \mathbf{x}$$

a = acceleration in ms^{-2}; x = displacement in m; ω = angular frequency in $rads^{-1}$

- This is the defining equation of SHM (see page 45), which means the formula $\mathbf{x} = A\cos(\omega t)$ is a solution to this equation.
- Similarly, if at $t = 0$ the ball had been passing through its equilibrium position, its acceleration would still be given by the 'left-to-right' component of $\omega^2 r$. But this would now be $\mathbf{a} = -\omega^2 A\sin(\omega t) = -\omega^2\mathbf{x}$. So the formula $\mathbf{x} = A\sin(\omega t)$ is also a solution to the SHM equation.
- Since $\mathbf{a} = -\omega^2 A\cos(\omega t)$ or $\mathbf{a} = -\omega^2 A\sin(\omega t)$, the object's acceleration has a maximum magnitude of $a_{max} = \omega^2 A$. This occurs when the object's displacement is at its maximum magnitude — i.e. when $\mathbf{x} = \pm A$:

$$a_{max} = \omega^2 A$$

a_{max} = magnitude of the max acceleration in ms^{-2}; A = amplitude in m; ω = angular frequency in $rads^{-1}$

Velocity

You also need to know the equation for the velocity of an object moving with SHM. Don't worry about the derivation of ***v***, it's pretty complicated. You just need to know that it is given by:

$$\mathbf{v} = \pm\omega\sqrt{A^2 - \mathbf{x}^2}$$

v = velocity in ms^{-1}; x = displacement in m; ω = angular frequency in $rads^{-1}$; A = amplitude in m

The velocity is positive if the object is moving in the positive direction (e.g. to the right), and negative if it's moving in the negative direction (e.g. to the left) — that's why there is a ± sign in there. The maximum speed (magnitude of velocity) is when the object is passing through the equilibrium, where $\mathbf{x} = 0$:

$$\text{Max speed} = v_{max} = \omega A$$

v_{max} = max speed in ms^{-1}; A = amplitude in m; ω = angular frequency in $rads^{-1}$

Example — Maths Skills

A mass is attached to a horizontal spring. It is pulled 7.5 cm from its equilibrium position and released. It begins oscillating with SHM, and takes 1.2 s to complete a full cycle.

a) What is the frequency of oscillation of the mass?

b) What is the magnitude of its maximum acceleration?

c) What is its speed when it is 2.0 cm from its equilibrium position?

a) Frequency $= \frac{1}{T} = \frac{1}{1.2} = 0.8333... = 0.83$ Hz (to 2 s.f.)

b) Amplitude in m $= 7.5 \div 100 = 0.075$ m

$\omega = 2\pi f = 2 \times \pi \times 0.8333... = 5.2359...$ rads^{-1}

$a_{max} = \omega^2 A = (5.2359...)^2 \times 0.075 = 2.056... = 2.1$ ms^{-2} (to 2 s.f.)

c) Speed at $\boldsymbol{x} = 0.02$ m

$v = \omega\sqrt{A^2 - \boldsymbol{x}^2} = 5.2359... \times \sqrt{0.075^2 - 0.02^2} = 0.3784...$

$= 0.38$ ms^{-1} (to 2 s.f.)

Tip: It's always helpful to draw a quick sketch for questions like this — it can help you visualise what's happening.

Tip: Part c) of this example asks you to find the speed. You can use $v = \pm\omega\sqrt{A^2 - \boldsymbol{x}^2}$ and just ignore the ± sign, as this shows the direction of the motion — which is only important for velocity, not speed.

Practice Questions — Application

Q1 A pendulum oscillating with SHM has an angular frequency of 1.5 rads^{-1} and an amplitude of 1.6 m.

a) Calculate its acceleration when it has a displacement of 1.6 m.

b) How long does it take to complete 15 oscillations?

Q2 A mass attached to a spring oscillates with SHM. It has a period of 0.75 s, and moves with speed 0.85 ms^{-1} when passing through its equilibrium position.

a) What is the amplitude of its oscillation?

b) What will its velocity be when it is 8.0 cm to the right of its equilibrium position?

Q3 A pendulum is pulled a distance 0.45 m from its equilibrium position and is released at time $t = 0$. If it takes 15.5 s to complete exactly 5 oscillations, how far will it be from its equilibrium position after 10.0 s?

Q4 A mechanical metronome produces a ticking sound every time a pendulum arm moving with SHM passes through its equilibrium position. Its maximum displacement from its equilibrium position is 6.2 cm. If it is set to produce 120 ticks per minute, what is the magnitude of the arm's maximum acceleration?

Tip: Remember the time period is just the inverse of the frequency.

Tip: The speed is at its maximum value when the mass passes through its equilibrium position.

Tip: See pages 53-54 for more on pendulums.

Practice Questions — Fact Recall

Q1 Write down the solution to the SHM equation $\boldsymbol{a} = -\omega^2\boldsymbol{x}$, assuming that at $t = 0$, the object is at its maximum displacement. Give your answer as a function of time, and define any symbols you use.

Q2 Give the equations for the acceleration and velocity of an object moving with SHM as a function of displacement. Define any symbols you use.

Q3 What is the maximum speed of an object moving with SHM? Define any symbols you use.

Q4 What is the magnitude of the maximum acceleration of an object moving with SHM? Define any symbols you use.

5. Investigating SHM

PRACTICAL ACTIVITY GROUP 10

Simple harmonic motion (SHM) is very common. This means there are various different experimental set-ups you can use to investigate it — for example, with a mass on a spring, or with a pendulum.

Learning Objective:

- Understand techniques and procedures used to determine the period and frequency of simple harmonic oscillations (PAG10).

Specification Reference 5.3.1

Tip: Before carrying out any experiment, you should always carry out a risk assessment.

Using sensors and a data logger

Data loggers and sensors are a useful way of investigating simple harmonic motion, as they allow you to make very precise measurements. To investigate the motion of a mass oscillating on a spring, set up the equipment as shown in Figure 1. If you don't have a long spring you can connect a few shorter ones together.

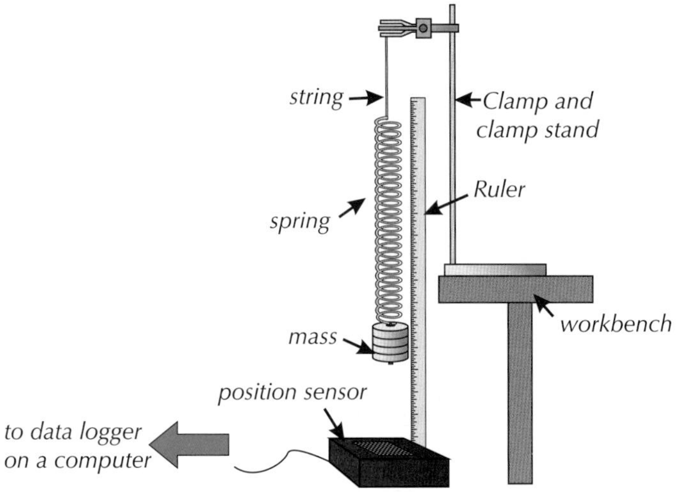

Figure 1: *Experimental set-up for investigating the motion of a mass oscillating on a spring.*

- Lift the mass slightly and release it — this will cause the mass-spring system to start oscillating with simple harmonic motion.
- To make sure your experiment is repeatable, place a ruler behind the spring to measure how far you raise the mass. Make sure your eye is level with the mass when you take the measurement.
- As the mass oscillates, the position sensor will measure the displacement of the mass over time. The computer can be set to record this data automatically.
- Let the experiment run until you've got a good amount of data (at least ten complete oscillations).

Once you've collected your data, you can use the computer to generate a displacement-time graph, like the one in Figure 2. From the graph, you can measure T, the time period of the oscillation, and A, the amplitude of the oscillation. The frequency of the oscillations (f) is given by $f = \frac{1}{T}$.

You should find that the amplitude of the oscillations gets smaller over time, but the time period and the frequency remain constant. This is because energy is lost to overcoming air resistance as the mass moves up and down.

Tip: You should try to lift the mass straight up, to stop the mass from swinging from side to side. The string also helps to stop the mass from swinging from side to side.

Tip: You could also use a set square to help get an accurate reading for the initial displacement of the mass. Using a set square helps you read the mark on the ruler that's exactly level with the mass.

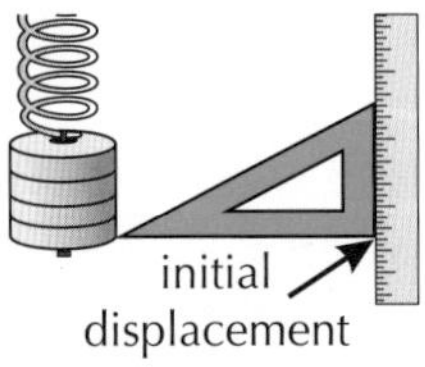

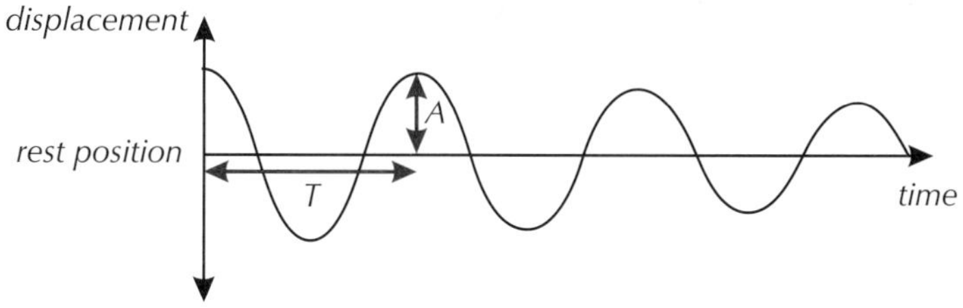

Figure 2: *A displacement-time graph for a mass oscillating on a spring. You can use the graph to measure the time period of the oscillations (T) and their amplitude (A).*

Tip: Depending on how your position sensor is set up, the rest position may not be shown as zero.

Tip: If you leave the system oscillating for long enough, the amplitude will decrease until the mass eventually comes to rest.

You can use this system to investigate how different variables affect the time period of the oscillation. You just have to change one variable at a time (the independent variable), while keeping all the others fixed. You should repeat the experiment for each different value of the independent variable you're investigating. For example, you could:

- Change the weight of the mass.
- Change the stiffness of the spring you're using.
- Change the size of the initial displacement.

You should find that a heavier mass leads to a longer time period and a stiffer spring leads to a shorter time period, while the initial displacement has no effect on the time period.

Tip: If the oscillations happen so quickly that it's difficult to measure the time period accurately, then you could use a stroboscope. A stroboscope is a light that flashes at a frequency which the user can set. When the frequency of the stroboscope matches the frequency of the oscillation, then the mass will appear to be stationary. You can find the time period using $T = \frac{1}{f}$, where f is the frequency of the stroboscope (and therefore the frequency of the oscillation).

Without using sensors and a data logger

If you don't have a data logger, you won't be able to generate a displacement-time graph, but you can still investigate the time period of an oscillation. For example, you can investigate the simple harmonic motion of a pendulum using the experimental set-up in Figure 3 and a stopwatch.

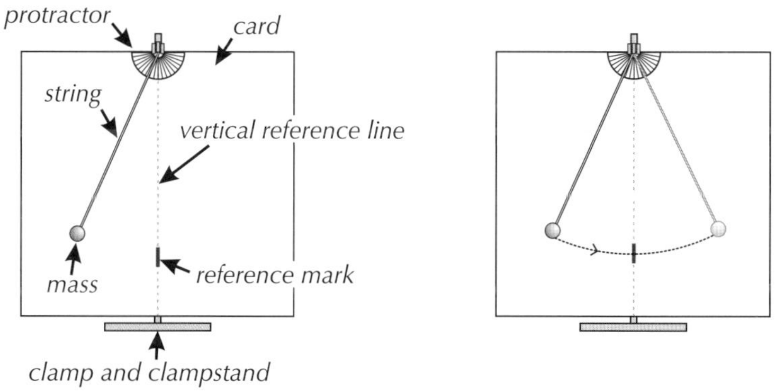

Figure 3: *Experimental set-up for investigating the motion of a pendulum.*

Tip: A reference mark like the one shown in Figure 3 is called a fiducial marker.

Figure 4: *To get a perfectly vertical reference line, you could use a plumb bob — a mass on the end of a string.*

- Measure the weight of the mass, and use a ruler to find the string's length.
- Move the mass to the side, keeping the string taut. Measure the angle between the string and the vertical using the protractor. Make sure it's less than 10°, or the mass won't swing with simple harmonic motion when you release it.
- Release the mass. Position your eye level with the reference mark on the card, and start the stopwatch when the mass passes in front of it.

Tip: To get the time period of a complete oscillation, only record times when the mass passes in front of the reference mark from the same direction (e.g. from left to right).

Tip: Measuring the total time for a number of complete oscillations and taking an average means any human error introduced by starting or stopping the stopwatch too late (or too early) will be 'shared' over several oscillations. This means the time period you calculate for a single oscillation will be more accurate.

Tip: Your data will contain more random errors if you don't use a data logger, so you'll need to do more repeats.

- Record the time when the mass passes the mark again, moving from the same direction — this is the time period of the oscillator. Depending on your pendulum, T might be too short to measure accurately from one swing. If so, measure the total time for a number of complete oscillations combined (say 5 or 10) and take an average (or you could use a stroboscope — see previous page).
- Keep recording T at regular intervals as the motion dies away. You can use these values of T to calculate the frequency of the oscillations using the formula $f = \frac{1}{T}$. You should find that T and f remain constant as the amplitude of the swing decreases.

You can investigate how different factors affect the motion of the pendulum. Measure all the variables as accurately as possible, and only change one variable at a time.

- Change the length of the piece of string you use for your pendulum.
- Change the weight of the mass on the end of the pendulum.
- Change the amount by which you initially move the mass to the side before releasing it (making sure you keep the angle below 10°).

You should find that as the length of the string increases, the time period also increases, but that the weight of the mass and the angle of the initial displacement have no effect on the time period of the pendulum.

Practice Question — Application

Q1 The graph below was plotted by a student investigating simple harmonic motion. The graph shows the displacement of a mass on a spring against time.

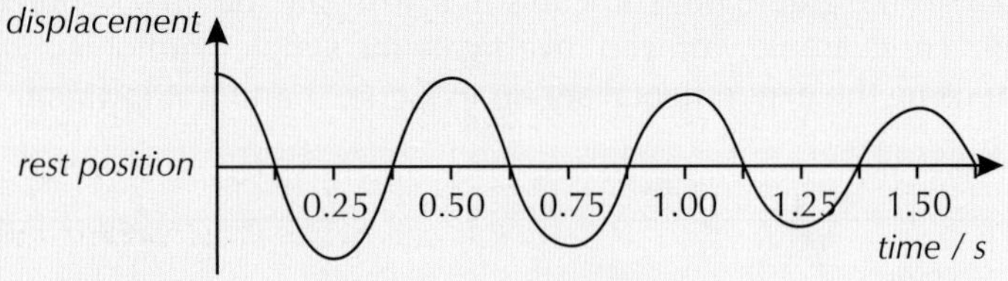

a) Find the time period (T) of the oscillations.

b) Find the frequency (f) of the oscillations.

c) Calculate the angular frequency (ω) of the oscillations.

Practice Questions — Fact Recall

Q1 Describe an experimental set-up you could use to investigate simple harmonic motion.

Q2 If you're investigating simple harmonic motion without using data loggers and sensors, why is it a good idea to:

a) find the time period of a single oscillation by recording the time for multiple oscillations, then dividing by the total number of oscillations?

b) use fiducial markers when recording the time to complete a number of oscillations?

6. Free and Forced Oscillations

An object can be forced to oscillate by providing a driving frequency. If this is near the object's natural frequency, the object will start to resonate — which can be good or bad news, depending on what the object is needed for.

Learning Objectives:

- Understand what is meant by free and forced oscillations.
- Know what resonance is in terms of the natural frequency.
- Understand and be able to sketch amplitude-driving frequency graphs for forced oscillators.
- Be able to give practical examples of forced oscillations and resonance.
- Understand the effects of damping on an oscillatory system.
- Understand observations of forced and damped oscillations for a range of systems.

Specification Reference 5.3.3

Free vibrations

Free vibrations involve no transfer of energy to or from the surroundings. If you stretch and release a mass on a spring, it oscillates at its **resonant frequency** (or **natural frequency**). The same happens if you strike a metal object — the sound you hear is caused by vibrations at the object's natural frequency. If no energy's transferred to or from the surroundings, it will keep oscillating with the same amplitude forever. In practice this never happens, but a spring vibrating in air is called a free vibration (or oscillation) anyway.

Forced vibrations

Forced vibrations happen when there's an external driving force. A system can be forced to vibrate by a periodic external force. The frequency of this force is called the **driving frequency**.

Resonance

When the driving frequency approaches the natural frequency, the system gains more and more energy from the driving force and so vibrates with a rapidly increasing amplitude. When this happens the system is **resonating**. At resonance the phase difference between the driver and oscillator is 90°. Figures 1 and 2 show how the relationship between amplitude and driving frequency can be investigated by experiment.

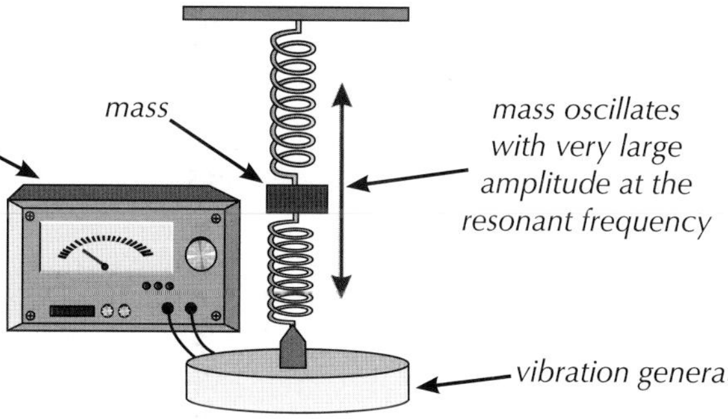

Figure 1: *Using a vibration generator to oscillate a mass-spring system. The system resonates when the driving frequency equals the natural frequency.*

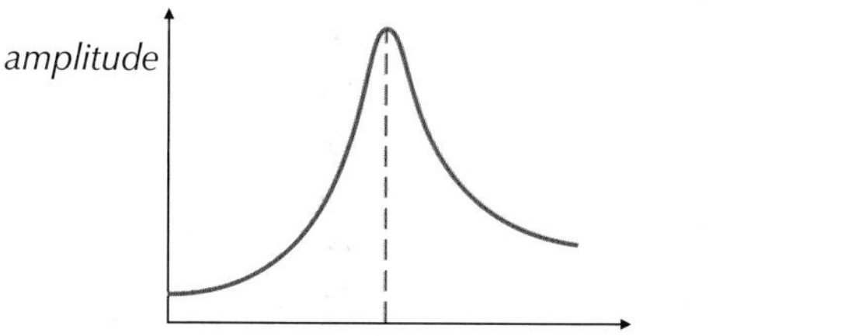

Figure 2: *Graph showing how the amplitude of oscillation of a system changes with driving frequency.*

Tip: You should carry out a risk assessment before starting this experiment.

Tip: At resonance, the driver displacement is 90° out of phase to the displacement of the oscillator — i.e. when the displacement of the driver is at its maximum, the oscillator is passing through its equilibrium point.

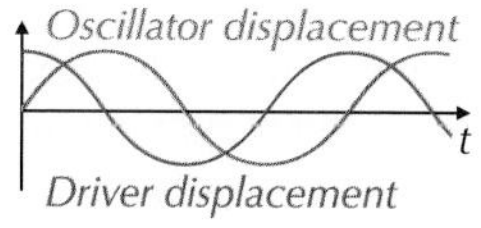

Tip: Make sure you know some examples of resonance.

Tip: Remember, a stationary wave is created by the superposition of two waves (with the same frequency and amplitude) moving in opposite directions.

Examples

Here are some examples of resonance:

A radio is tuned so the electric circuit resonates at the same frequency as the radio station you want to listen to.

A glass resonates when driven by a sound wave at the right frequency.

The column of air resonates in an organ pipe, driven by the motion of air at the base. This creates a stationary wave in the pipe.

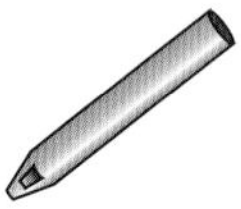

A swing resonates if it's driven by someone pushing it at its natural frequency.

Damping

HOW SCIENCE WORKS

In practice, any oscillating system loses energy to its surroundings. This is usually down to frictional forces like air resistance. These are called **damping** forces. Systems are often deliberately damped to stop them oscillating or to minimise the effect of resonance.

Tip: Another name for damping forces is 'dissipative forces', since they dissipate the energy of the oscillator to its surroundings.

Example

Shock absorbers in a car's suspension provide a damping force by squashing oil through a hole when compressed.

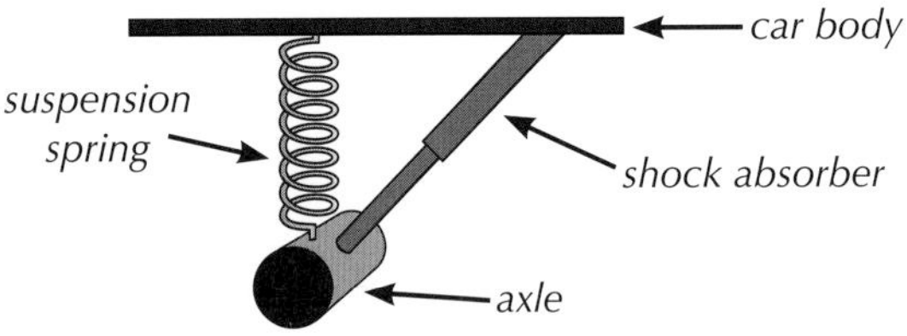

Figure 3: *Damping in a car suspension system.*

Figure 4: *In this car suspension system, the shock absorber is inside the spring. It quickly reduces the amplitude of the spring's oscillations.*

The degree of damping can vary from light damping (where the damping force is small) to overdamping. Damping reduces the amplitude of the oscillation over time. Generally, the heavier the damping, the quicker the amplitude is reduced to zero (although overdamping is an exception). Figure 5 shows how different degrees of damping reduce the amplitude to zero at different speeds.

Light and heavy damping

Lightly damped systems take a long time to stop oscillating, and their amplitude only reduces a small amount each period. Heavily damped systems take less time to stop oscillating, and their amplitude gets much smaller each period.

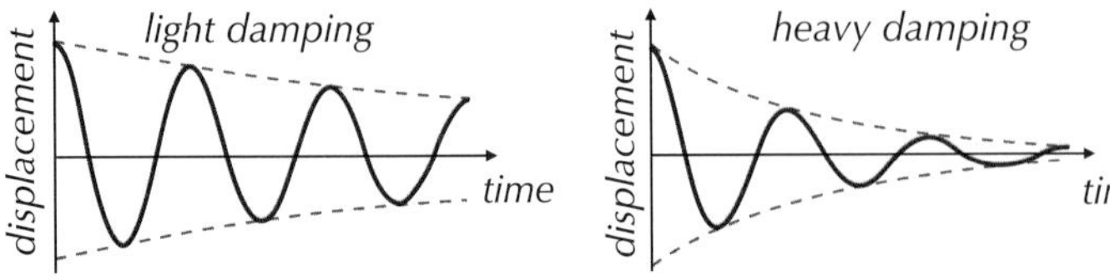

Figure 5: *Graphs showing the effect of light and heavy damping.*

Exam Tip
Make sure you're able to describe the four different types of damping (the next two are on the next page), and sketch these graphs showing how the amplitude of oscillation changes with time.

Example

A pendulum formed of a small bob on a rod is an example of a lightly damped system — air resistance will cause the pendulum to slow down only very slightly each period. If the pendulum bob was removed and replaced with this book, the larger surface area of the book would increase air resistance, and the damping forces would be larger, slowing the oscillation more quickly.

Tip: The period of a pendulum doesn't depend on its mass, so the mass of the book wouldn't affect the SHM in any way.

Critical damping

Critical damping reduces the amplitude (i.e. stops the system oscillating) in the shortest possible time (see Figure 6).

Example

Car suspension systems are critically damped so that they don't oscillate but return to equilibrium as quickly as possible.

Overdamping

Systems with even heavier damping are **overdamped**. They take longer to return to equilibrium than a critically damped system (see Figure 6).

Example

Some heavy doors are overdamped, so that they don't slam shut too quickly, but instead close slowly, giving people time to walk through them.

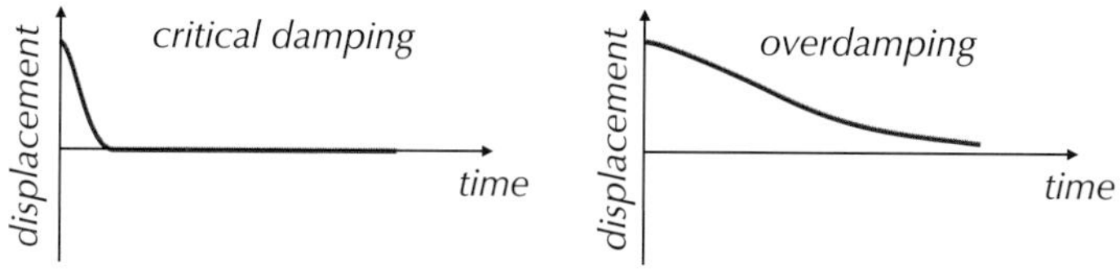

Figure 6: *Graphs showing the effects of different amounts of damping.*

Oscillations can also be damped using objects made from ductile materials. As the shape of the object changes and the object plastically deforms, it absorbs energy and reduces the amplitude of the oscillations (i.e. it damps the oscillations).

Tip: You met plastic deformation in year 1 of A-level — it's when a material is permanently stretched, beyond its elastic limit.

Damping of forced systems

Lightly damped systems have a very sharp resonance peak. Their amplitude only increases dramatically when the driving frequency is very close to the natural frequency. Heavily damped systems have a flatter response. Their amplitude doesn't increase very much near the natural frequency and they aren't as sensitive to the driving frequency. Figure 7 shows the effect of increasing levels of damping on oscillations near the natural frequency.

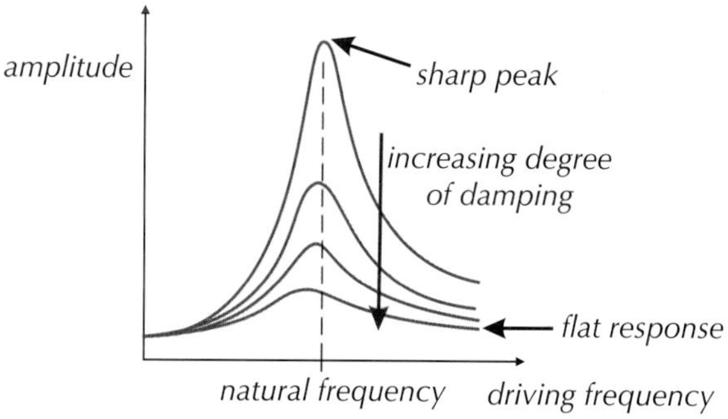

Figure 7: *Graph showing how damping affects resonance.*

Tip: Structures are damped to avoid being damaged by resonance. Loudspeakers are also made to have as flat a response as possible so that they don't 'colour' the sound.

Tip: In general, the more damped a system is, the flatter the graph of amplitude of oscillation against driving frequency.

Tip: For heavier damping, resonance occurs at a slightly lower driving frequency than the natural frequency.

Figure 8: The Taipei 101 building in Taiwan, standing at 508 m tall. A 660-tonne pendulum suspended down the centre of the building acts as a tuned mass damper, reducing the amplitude of the building's oscillations in earthquakes.

Tip: Before carrying out this experiment, you should be aware of any safety risks that are associated with the equipment.

Notice from Figure 7 that when the driving frequency approaches zero, the amplitude of the oscillations are the same, however much the system is damped. If the driving frequency is close to zero, the amplitude of the oscillation must be the same as the amplitude of whatever's driving the oscillation.

Example

Some structures are damped to avoid being damaged by resonance. Some buildings in regions prone to earthquakes have a large mass called a tuned mass damper. When an earthquake causes the building to shake, the mass moves in the opposite direction to the building, damping its oscillation (see Figure 8). This is an example of critical damping.

Figure 9 shows apparatus that can be used to demonstrate the effect of damping on the resonance of a spring-mass system. A flat disc is attached to the set-up you saw in Figure 1. As the mass oscillates, air resistance on the disc acts as a damping force, reducing the amplitude of the oscillation. The larger the disc, the larger the damping force and the smaller the amplitude of oscillation of the system at resonance.

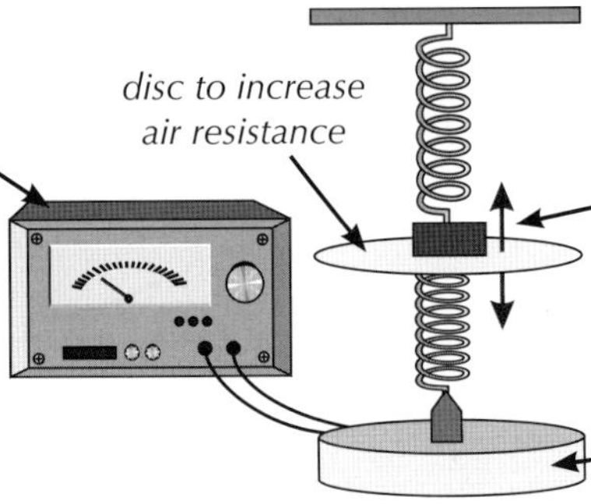
signal generator sets driving frequency
disc to increase air resistance
mass oscillates at a smaller amplitude at the resonant frequency than a free oscillator
vibration generator

Figure 9: Experiment to show how damping affects resonance.

Practice Questions — Fact Recall

Q1 What is the difference between a free vibration and a forced vibration?

Q2 What is resonance, and when does it occur?

Q3 Give three examples of situations where resonance can occur.

Q4 What is meant by a damping force?

Q5 Name and briefly describe the four types of damping.

Q6 Sketch a graph of amplitude against driving frequency for a system to show how the level of damping affects the sharpness of the resonance peak at the natural frequency.

Section Summary

Make sure you know...

- How to use radians as a measure of angles, and how to convert between radians and degrees.
- That angular velocity is the angle an object moving with circular motion rotates through per second.
- How to use $v = \omega r$ to calculate the angular speed of an object moving with circular motion.
- That the frequency of rotation is the number of complete revolutions per second.
- That the period of rotation is the time taken for one complete revolution in seconds.
- The relationship between frequency and period.
- How to use the formulas $\omega = \frac{2\pi}{T}$ and $\omega = 2\pi f$ to calculate angular velocity.
- Objects undergoing circular motion experience a centripetal acceleration towards the circle's centre.
- That a constant net force perpendicular to an object's velocity causes it to travel in a circular path.
- How to use the formulas $\boldsymbol{a} = \frac{v^2}{r}$, $\boldsymbol{a} = \omega^2 r$, $\boldsymbol{F} = \frac{mv^2}{r}$ and $\boldsymbol{F} = m\omega^2 r$ to calculate centripetal acceleration and force.
- Techniques and procedures used to investigate circular motion using a whirling bung.
- That simple harmonic motion is the oscillation of an object with an acceleration that is proportional to its displacement from the midpoint and that is always directed towards the midpoint ($\boldsymbol{a} = -\omega^2\boldsymbol{x}$).
- What is meant by the displacement and amplitude of an object undergoing SHM.
- How to use $\omega = \frac{2\pi}{T}$ and $\omega = 2\pi f$ to calculate the angular frequency of an object undergoing SHM.
- How to sketch the graphs of displacement, velocity and acceleration against time for SHM.
- That the velocity of an object undergoing SHM is given by the gradient of a displacement-time graph and acceleration is given by the gradient of a velocity-time graph.
- The phase differences between displacement, velocity and acceleration for SHM.
- That a cycle of oscillation in SHM is from maximum positive displacement to maximum negative displacement and back again, and how this relates to the frequency and period of SHM.
- That frequency and period are independent of amplitude for an object moving with SHM (i.e. it is an isochronous oscillator).
- How the kinetic and potential energy of an object moving with SHM vary with time and displacement.
- That mechanical energy is the sum of the kinetic and potential energy of an object and that it stays constant for undamped oscillations.
- That solutions to the equation $\boldsymbol{a} = -\omega^2\boldsymbol{x}$ are of the form $\boldsymbol{x} = A\cos(\omega t)$ or $\boldsymbol{x} = A\sin(\omega t)$.
- How to use the formulas $\boldsymbol{x} = A\cos(\omega t)$ and $\boldsymbol{x} = A\sin(\omega t)$ for the displacement of an object in SHM.
- How to use $\boldsymbol{a} = -\omega^2\boldsymbol{x}$ and $\boldsymbol{v} = \pm\omega\sqrt{A^2 - \boldsymbol{x}^2}$ for the acceleration and velocity of an object in SHM.
- How to use the formulas for the magnitude of maximum acceleration and maximum speed.
- How to determine the frequency and period of an object undergoing simple harmonic motion using, e.g. position sensors and data loggers, or by using a stopwatch.
- That free vibrations involve no transfer of energy between an object and its surroundings, and that an object oscillating freely does so at its natural frequency.
- That a forced vibration is driven by a periodic, external driving force at a driving frequency.
- That resonance is a rapid increase in the amplitude of oscillation of an object, and that it usually occurs at the object's natural frequency.
- Some practical examples of forced oscillations and resonance.
- That a damping force causes an oscillator to lose energy to its surroundings and reduce the amplitude of its oscillations.
- How the shape of an amplitude-frequency graph changes as you change the degree of damping.

Exam-style Questions

1 A roundabout with radius 3.0 m is spinning at a rate of 6 revolutions per minute. What is the linear speed of someone standing on the edge of the roundabout?

A 0.3 ms^{-1}

B 18 ms^{-1}

C 0.63 ms^{-1}

D 1.9 ms^{-1}

(1 mark)

2 A mass is attached to a spring and set oscillating with simple harmonic motion. The system is damped. Which of the following is false?

A If there was no damping at all, the mass would oscillate forever.

B The maximum displacement of the mass decreases with each oscillation.

C The mass would take longer to come to rest if it was critically damped compared to if it was overdamped.

D When the amplitude decreases, the frequency remains the same.

(1 mark)

3 A pendulum is oscillating with simple harmonic motion. At time $t = 0$, the pendulum passes through its equilibrium position, and passes through it again 1.5 s later. Which of the following graphs shows how the pendulum's potential energy, E_P, varies with time?

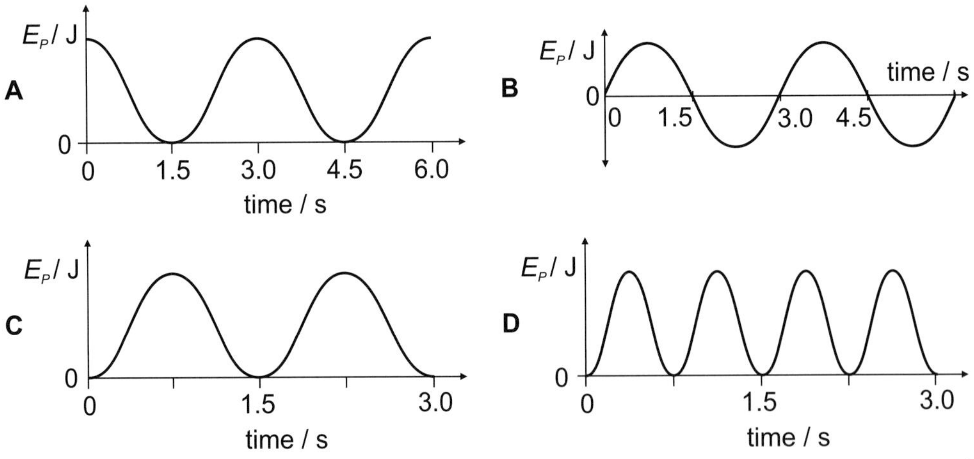

(1 mark)

4 A block of wood on a smooth surface is attached to one end of a horizontal spring. The other end of the spring is attached to a wall. The mass of the block is 0.60 kg and the restoring force (F, in newtons) produced by the spring is given by $F = -18x$, where x is the displacement of the block (in m) from its equilibrium position. When the block is displaced a distance of 20 cm from its equilibrium position and released, it begins to oscillate with a time period of 1.147 s.

(a) Find the velocity of the block when it is 15 cm away from the equilibrium position.

(2 marks)

(b) Calculate the maximum kinetic energy of the block.

(3 marks)

(c) As the block oscillates, a stopwatch is started as the block passes through the equilibrium position. Find the distance of the block from its equilibrium position 3.0 s after the stopwatch is started.

(2 marks)

5 A motorcyclist is riding around a vertical cylindrical track with radius 5.0 m, as shown in **Fig. 5.1**. To avoid sliding down the track and falling off, the centripetal force acting on her must be at least 1500 N.

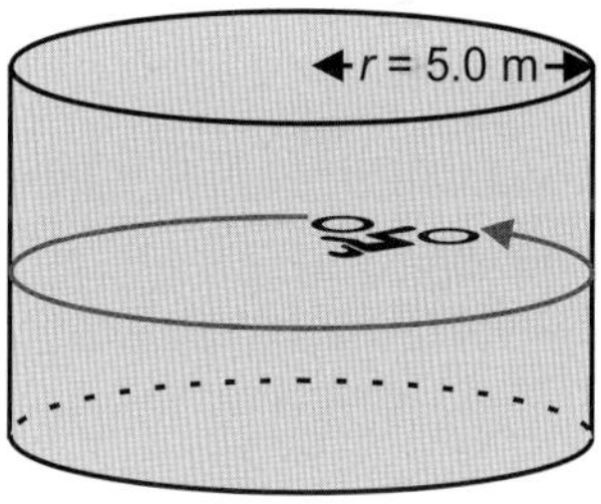

Fig. 5.1

(a) The combined mass of the motorcyclist and her motorcycle is 210 kg. Calculate the minimum speed (v_{min}) that the motorcyclist needs to ride at to avoid sliding down the track.

(2 marks)

(b) Find the motorcyclist's angular speed if she were to ride at speed v_{min}.

(2 marks)

(c) Find the time period of the motorcyclist's motion at speed v_{min}.

(2 marks)

6 A simple pendulum is made to oscillate in a vacuum by a periodic driving force, at a frequency below its natural frequency.

(a) Describe what happens as the driving frequency is increased to the pendulum's natural frequency. Give the name of this phenomenon.

(2 marks)

(b) The same pendulum is placed in a tank of water, and made to oscillate again. Describe and explain what happens this time as the driving frequency is increased up to the pendulum's natural frequency.

(3 marks)

1. Gravitational Fields

Learning Objectives:
- Understand the concept of gravitational fields as being one of a number of forms of field giving rise to a force.
- Know that gravitational fields are due to objects having mass.
- Be able to use gravitational field lines to map gravitational fields.
- Be able to model the mass of a spherical object as a point mass at its centre.
- Be able to apply Newton's law of gravitation for the force between two point masses, $F = -\frac{GMm}{r^2}$.

Specification References 5.4.1 and 5.4.2

So far you've probably only considered forces acting at a specific point, with a specific cause (e.g. the pushing of a swing). Gravitational fields, on the other hand, are regions in which a mass will experience a force.

What is a gravitational field?

A **gravitational field** is a **force field** — a region where an object will experience a non-contact force. Force fields cause interactions between objects or particles — e.g. static or moving charges interact through electric fields (p.133) and objects with mass interact through gravitational fields.

A gravitational field is a force field generated by any object with mass which causes any other object with mass to experience an attractive force.

Any object with mass has a gravitational field (even you have a gravitational field). Any object with mass will experience an attractive force if you put it in the gravitational field of another object. Only objects with a large mass, such as stars and planets, have gravitational fields that produce a significant effect. For example, the gravitational fields of the Moon and the Sun are noticeable here on Earth — they're the main cause of our tides. Smaller objects do still have gravitational fields that attract other masses, but the effect is too weak to detect without specialised equipment.

Force fields can be represented as vectors, showing the direction of the force they would exert on an object placed in that field. **Gravitational field lines**, or "lines of force", are arrows showing the direction of the force that masses would feel in a gravitational field (see Figure 2 below).

Figure 1: *The gravitational fields of the Moon and the Sun are the main causes of our tides.*

Radial and uniform fields

Point masses have radial gravitational fields, where all the field lines meet at the point where the mass is concentrated. Spherical objects of uniform density can be treated as point masses with all of their mass concentrated at their centre.

The Earth's gravitational field is radial — if you extended the lines of force, they would meet at the centre of the Earth (see Figure 2). If you put a small mass, m, anywhere in the Earth's gravitational field, it will always be attracted towards the Earth.

The lines can be used to show the strength of the field at each point, where the lines being closer together shows a stronger gravitational field. The stronger the gravitational field, the larger the force on a mass, m, due to the field (see page 65). If you move mass m further away from the Earth — where the lines of force are further apart — the force it experiences decreases.

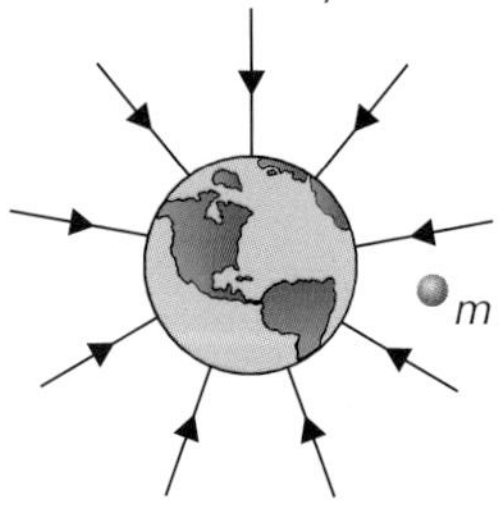

Figure 2: *A small mass, m, in Earth's gravitational field.*

Tip: The smaller mass, m, has a gravitational field of its own. This doesn't have a noticeable effect on Earth though, because the Earth is so much more massive.

Close to Earth's surface, the field is (almost) uniform — the field lines are (almost) parallel and equally spaced.

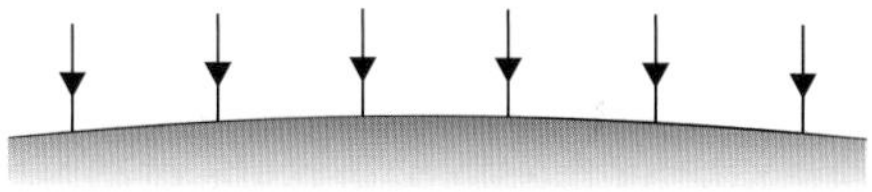

Figure 3: *The gravitational field at Earth's surface is roughly uniform.*

Newton's law of gravitation

Newton's law of gravitation says that the force acting between two point masses (or spherical masses) is proportional to the product of their masses and inversely proportional to the square of the distance between their centres of mass:

$$F = -\frac{GMm}{r^2}$$

G = gravitational constant = 6.67×10^{-11} Nm^2kg^{-2}

M = mass causing the force in kg

F = force acting on mass m due to mass M in N

m = mass the force is acting on in kg

r = distance between the centres of the two masses in m

Remember, force is a vector quantity, so direction is important. For this equation, the positive direction is always defined as from M (the mass causing the force) to m (the mass feeling the force). So, the force on m is negative (and has a minus sign) because its direction is always towards the centre of the M. The force on m due to M is equal and opposite to the force on M due to m.

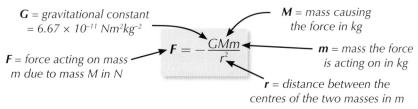

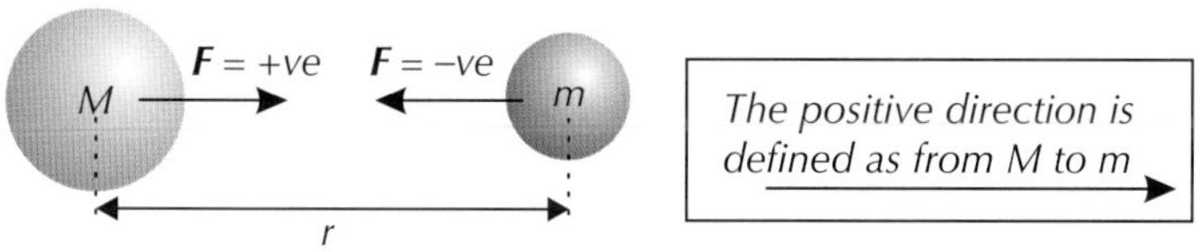

Figure 4: *The forces acting on the two masses are equal but opposite.*

We sometimes only consider the force acting on the smaller object because that's the one that experiences a greater acceleration — $\mathbf{a} = \frac{\mathbf{F}}{m}$, so as m becomes bigger, $\mathbf{a}$ becomes smaller.

Example — Maths Skills

Two planets have masses of 7.55×10^{24} kg and 9.04×10^{24} kg respectively. If the force due to gravity between the two planets is 6.69×10^{17} N, how far apart are the planets?

You want to find r, so you'll need Newton's law of gravitation.

We're not interested in the direction of a particular force in this example, so you can ignore the minus sign in the formula:

$$F = \frac{GMm}{r^2} \Rightarrow r^2 = \frac{GMm}{F} \Rightarrow r = \sqrt{\frac{GMm}{F}}$$

Then just put the numbers in:

$$r = \sqrt{\frac{GMm}{F}} = \sqrt{\frac{(6.67 \times 10^{-11}) \times (7.55 \times 10^{24}) \times (9.04 \times 10^{24})}{6.69 \times 10^{17}}}$$

$= 8.2491... \times 10^{10}$ m

$= 82.5$ million km (to 3 s.f.)

Tip: Remember, spherical objects can be treated as point masses (see previous page).

Tip: G is the gravitational constant — don't get this confused with $\mathbf{g}$, the gravitational field strength (see p.65).

Tip: For these calculations, you have to arbitrarily pick which mass is M and which is m. The bigger mass is usually M. The equation gives the force on m due to M, where the positive direction is from M to m.

Tip: Note that r is the distance between the centres of mass of the objects, not the edges.

Exam Tip
An exam question might ask you to work out the mass of one of the objects first. E.g. you might need to use density = $\frac{\text{mass}}{\text{volume}}$.

Tip: If you'd have used the minus sign in the equation, you'd have had to use a minus value for the force too, so the minus signs would cancel out and you'd get the same answer.

> **Tip:** The $\propto$ symbol means "is proportional to".

Inverse square laws

The law of gravitation is an **inverse square law**: $F \propto \frac{1}{r^2}$

This means if the distance r between the masses increases, then the force F will decrease. Because it's r^2 and not just r, if the distance doubles then the force will be one quarter the strength of the original force.

Example — Maths Skills

The gravitational force between two objects 10 m apart (to 2 s.f.) is 0.291 N. What will the gravitational force between them be if they move to 25 m apart?

25 m is $\frac{25}{10} = 2.5$ times larger than 10 m. So to find the new gravitational force, divide 0.291 N by 2.5^2 (because of the inverse square law):

$$\frac{0.291}{2.5^2} = 4.656 \times 10^{-2} \text{ N} = 4.7 \times 10^{-2} \text{ N (to 2 s.f.)}$$

Practice Questions — Application

> **Tip:** Don't forget to convert any distance measurements into metres before using Newton's law of gravitation.

> **Tip:** Remember, you have to assume that the Earth is a point mass, so r is the distance of the aircraft from the Earth's centre.

Q1 Two stars are orbiting each other with a constant separation of 100 million km. If their masses are 2.15×10^{30} kg and 2.91×10^{30} kg, show that the force they are exerting on each other is approximately 4.2×10^{28} N.

Q2 Two asteroids 2.5 km apart exert a gravitational force on each other of 25 N. Calculate the magnitude of the force they will exert on each other when they're 0.5 km apart.

Q3 An aircraft with a mass of 2500 kg is hovering 10 000 m above ground.
- a) If the radius of Earth is 6370 km, how far is the aircraft from the centre of the Earth?
- b) How much upwards force must be acting on the aircraft to keep it hovering at a constant altitude?
  The mass of Earth is 5.97×10^{24} kg.

Practice Questions — Fact Recall

Q1 What's a gravitational field?

Q2 Draw a diagram showing the gravitational field lines of Earth:
- a) Looking at Earth from a distance.
- b) At Earth's surface.

Q3 Give the formula for Newton's law of gravitation.
Define any symbols you use.

2. Gravitational Field Strength

It's no use being able to draw gravitational field lines if you can't use them to work anything out. The gravitational field strength, g, tells you how strong the force due to gravity is at any point in a gravitational field.

Learning Objectives:
- Know what gravitational field strength is.
- Be able to use $g = \frac{F}{m}$ to calculate gravitational field strength.
- Know that gravitational field strength is uniform close to the surface of the Earth and numerically equal to the acceleration of free fall.
- Be able to use $g = -\frac{GM}{r^2}$ to calculate gravitational field strength for a point mass.

Specification References 5.4.1 and 5.4.2

The gravitational field strength, *g*

Gravitational field strength, ***g***, is the force per unit mass. Its value depends on where you are in the field. There's a really simple equation for working it out:

$$g = \frac{F}{m}$$

g *= gravitational field strength in Nkg*$^{-1}$

F *= force experienced by a mass, m, in the gravitational field in N*

m *= mass in kg*

F is the force experienced by a mass *m* when it's placed in the gravitational field. Divide ***F*** by *m* and you get the force per unit mass.

g is a vector quantity — it has a magnitude and a direction. It's always pointing towards the centre of mass of the object whose field you're describing. Depending on the direction defined to be positive, it could be negative. Since the gravitational field is almost uniform at the Earth's surface, you can assume ***g*** is uniform near the Earth's surface.

The value of ***g*** at the Earth's surface is approximately 9.81 Nkg^{-1}

g can also be seen as the acceleration of a mass in a gravitational field. It's often called the acceleration due to gravity. On Earth, this is approximately 9.81 ms^{-2}.

Tip: From ***F*** = *m****a*** you can get ***a*** = ***F****/m*, which has the units Nkg^{-1} — just another way of measuring acceleration.

Example — **Maths Skills**

An 80.0 kg astronaut feels a force of 130.0 N due to gravity on the Moon. What's the value of *g* on the moon?

Just put the numbers into the formula:

$$g = \frac{F}{m} = \frac{130}{80} = 1.625 = 1.63\,\text{Nkg}^{-1} \text{ (to 3 s.f.)}$$

Radial fields

Point masses (or spherical masses) have radial gravitational fields (see page 62). The value of ***g*** depends on the distance *r* from the centre of the mass *M*.

$$g = -\frac{GM}{r^2}$$

G *= gravitational constant in Nm*^{2}*kg*$^{-2}$

g *= gravitational field strength in Nkg*$^{-1}$

M *= mass of object causing the gravitational field in kg*

r *= distance from the centre of mass M in m*

Tip: Remember, a spherical mass of uniform density can be treated as a point mass with all of its mass concentrated at its centre (p.62).

You can derive this formula by looking at Newton's law of gravitation (p.63): Start with $F = -\frac{GMm}{r^2}$ and substitute this into $g = \frac{F}{m}$, cancelling down where possible:

$$g = \frac{F}{m} = \frac{\left(-\frac{GM\cancel{m}}{r^2}\right)}{\cancel{m}} = -\frac{GM}{r^2}$$

Tip: Remember, for a spherical mass, *M*, *r* is the distance from its centre of mass, not from its surface.

g is negative because the positive direction is defined, again, to be away from *M* (see p.63). ***g*** is a vector towards the centre of *M*, so ***g*** is negative.

Example — Maths Skills

The mass of the Earth is 5.97×10^{24} kg and its radius is 6.37×10^6 m. Find the value of g at the Earth's surface.

Just put the numbers into the equation:

$$\mathbf{g} = -\frac{GM}{r^2} = -\frac{(6.67 \times 10^{-11}) \times (5.97 \times 10^{24})}{(6.37 \times 10^6)^2}$$
$$= -9.8134...$$
$$= -9.81 \text{ Nkg}^{-1} \text{ (to 3 s.f.)}$$

This is another case of the inverse square law (page 64) — as r doubles, **g** decreases to a quarter of its original value.

If you plot a graph of **g** against r for the Earth, you get a curve like this:

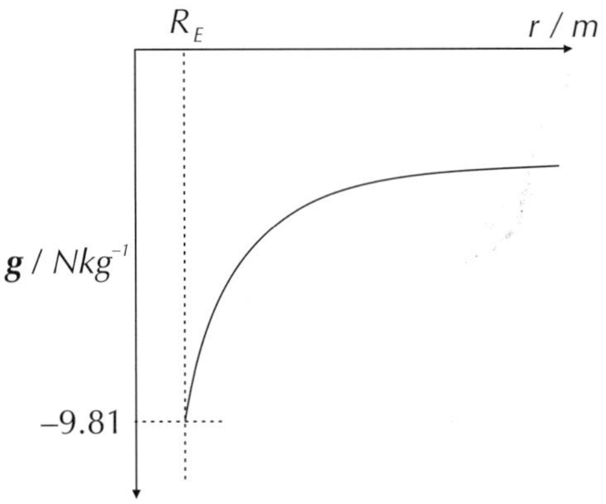

Figure 1: Graph showing the relationship between ***g*** *and r for Earth.*

Figure 2: The value of ***g*** *on top of a mountain is slightly lower than at sea level.*

It shows that **g** is greatest at the surface of the Earth (R_E), but decreases rapidly as r increases and you move further from the centre of the Earth.

Example — Maths Skills

The graph shows how the gravitational field strength, g, varies with distance, r, from the centre of the planet Mars. The radius of Mars is approximately 3.4×10^3 km. Estimate the mass of Mars.

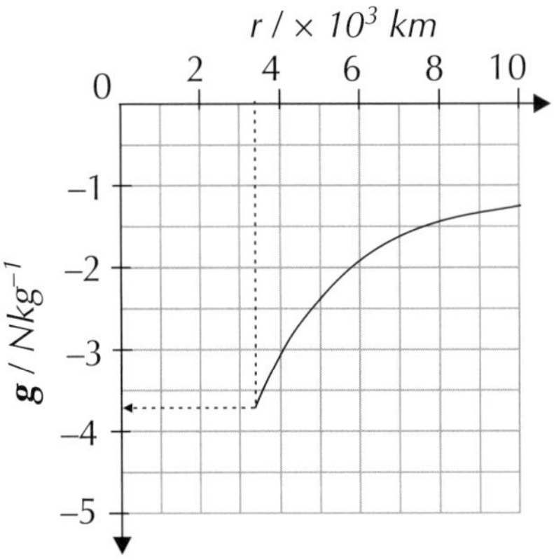

You can see from the graph that the value of **g** at the surface of Mars is about -3.7 Nkg^{-1}.

Rearrange the formula: $\mathbf{g} = -\frac{GM}{r^2}$ to find M, then stick the values in — don't forget to convert to standard units first.

So, $M = -\frac{\mathbf{g}r^2}{G}$

$$= -\frac{-3.7 \times (3.4 \times 10^6)^2}{6.67 \times 10^{-11}}$$
$$= 6.4 \times 10^{23} \text{ kg (to 2 s.f.)}$$

Practice Questions — Application

Q1 A 105 kg object experiences an attractive force due to gravity of 581 N. What's the gravitational field strength?

Q2 Why would an astronaut find it easier to pick up a rock with a mass of 20 kg on the Moon than a rock with a mass of 20 kg on Earth?

Q3 Venus has a radius of 6050 km and a mass of 4.87×10^{24} kg. Calculate the gravitational field strength at the surface of Venus.

Q4 a) A person standing on the surface of the Moon experiences a gravitational force of magnitude 105 N. If the mass of the Moon is 7.34×10^{22} kg and the mass of the person is 65 kg, what is the radius of the Moon?

b) What is the gravitational field strength 640 km above the surface of the Moon?

Tip: The mass of an object is always the same wherever it is.

Practice Questions — Fact Recall

Q1 Other than the acceleration due to gravity, how is $\boldsymbol{g}$ defined and what are its units?

Q2 What does M represent in the formula for gravitational field strength in a radial field?

Q3 Which of the following shows how gravitational field strength changes with distance r?

a)

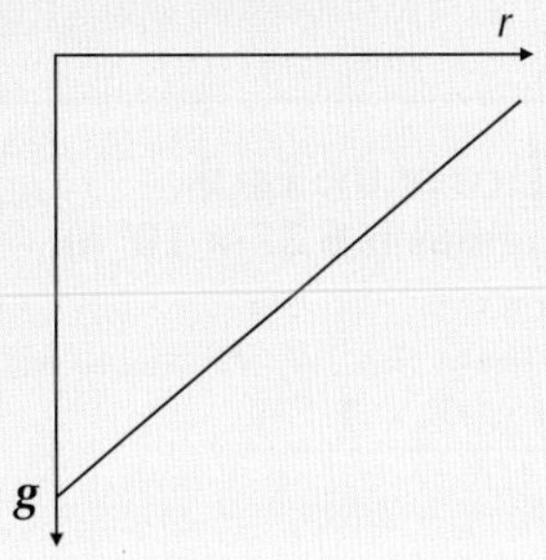

b)

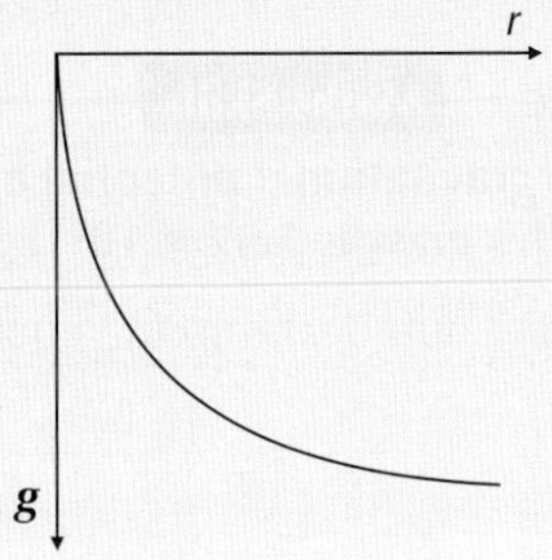

c)

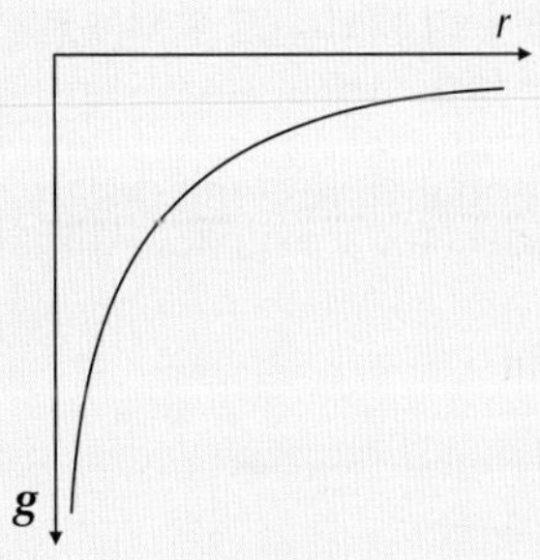

d)

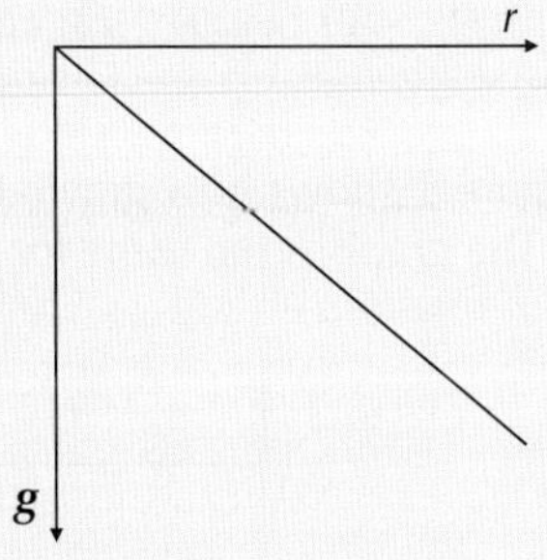

3. Gravitational Potential and Energy

All objects in a gravitational field have a gravitational potential that depends on how far they are from the centre of the field. It can be hard to get your head round at first, but try not to get it confused with gravitational potential energy.

Learning Objectives:

- Know that gravitational potential at a point is the work done in bringing a unit mass from infinity to the point.
- Know that gravitational potential is zero at infinity.
- Be able to calculate gravitational potential using $V_g = -\frac{GM}{r}$ at a distance r from a point mass M.
- Be able to calculate changes in gravitational potential.
- Be able to sketch a force-distance graph for a point or spherical mass.
- Know that the area under the force-distance graph for a point or spherical mass is the work done.
- Be able to calculate gravitational potential energy using $E = mV_g = -\frac{GMm}{r}$ at a distance r from a point mass M.
- Know what escape velocity is.

Specification Reference 5.4.4

What is gravitational potential?

The **gravitational potential**, V_g, at a point is the work done in moving a unit mass from infinity to that point. In a radial field (like the Earth's), the equation for gravitational potential is:

$$V_g = -\frac{GM}{r}$$

G *= gravitational constant in Nm^2kg^{-2}*

M *= mass of the object causing the gravitational field in kg*

V_g *= gravitational potential in Jkg^{-1}*

r *= distance from the centre of the object in m*

Gravitational potential is negative — you have to do work against the gravitational field to move an object out of it. The further you are from the centre of a radial field, the smaller the magnitude of V_g. At an infinite distance from the mass, the gravitational potential will be zero.

Example — Maths Skills

Find the gravitational potential at the surface of the Earth. The Earth's mass is 5.97×10^{24} kg and its radius is 6.37×10^6 m.

Just put the numbers into the equation:

$$V_g = -\frac{GM}{r} = -\frac{(6.67 \times 10^{-11}) \times (5.97 \times 10^{24})}{6.37 \times 10^6}$$

$$= -6.2511... \times 10^7$$

$$= -6.25 \times 10^7 \text{ Jkg}^{-1} \text{ (to 3 s.f.) } (= -62.5 \text{ MJkg}^{-1})$$

Tip: The answer to the example means a 1 kg mass needs 62.6 MJ of energy to be able to fully escape the Earth's gravitational pull.

Figure 1 shows how gravitational potential varies with distance from the Earth. The graph is the same for any sphere, where R_E is the radius of the sphere.

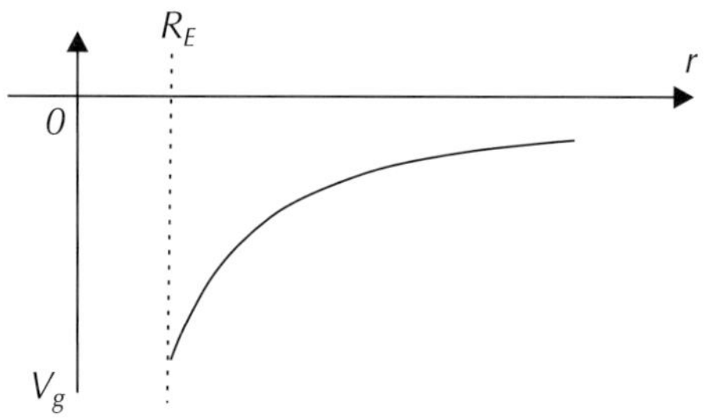

Figure 1: *A graph of gravitational potential against distance for the Earth.*

Tip: Note that V is proportional to $1/r$, not $1/r^2$ like the gravitational field strength.

If you find the gradient of a V_g-r graph at a particular point using a tangent you get the value of $-\mathbf{g}$ at that point. In other words $\mathbf{g} = -\frac{\Delta V_g}{\Delta r}$.

Example — Maths Skills

The graph below shows the gravitational potential V_g against the distance r from the centre of a planet. Find the gravitational field strength g at $r = 15 \times 10^6$ m.

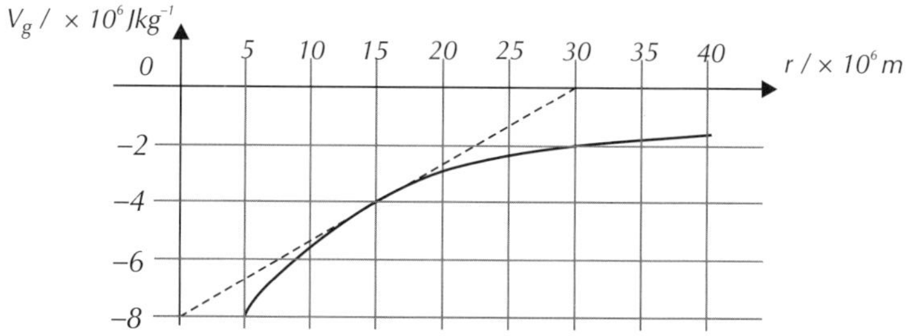

The gravitational field strength is given by the negative of the gradient:

$$g = -\frac{\Delta V_g}{\Delta r} = -\frac{0-(-8\times 10^6)}{(30\times 10^6)-0} = -0.2666 = -0.3\,\text{Nkg}^{-1}\ \text{(to 1 s.f.)}$$

Changes in gravitational potential

Two points at different distances from a mass will have different gravitational potentials (because the magnitude of the gravitational potential decreases with distance) — this means that there is a gravitational potential difference between these two points.

When you move an object against gravity to change the gravitational potential, you do work — the amount of energy you need depends on the mass of the object and the gravitational potential difference you move it through:

$$\Delta W = m\Delta V_g$$

ΔW = *work done in J*

ΔV_g = *gravitational potential difference in Jkg^{-1}*

m = *mass of the object in kg*

Tip: In physics, doing work means using a force to transfer energy from one type to another. For example, if you drop a ball from a height, gravitational potential energy is converted into kinetic energy.

Example — Maths Skills

Show that the work done to move a mass *m* through a gravitational potential difference of ΔV_g can be derived from the equation for gravitational field strength. The mass *m* is being moved away from the mass creating the gravitational field.

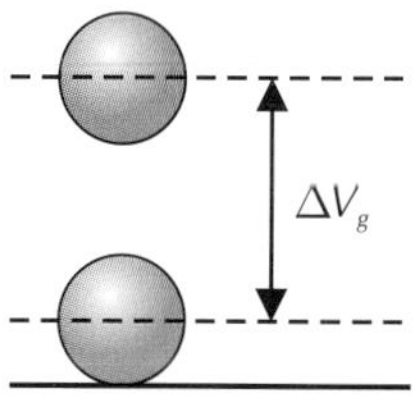

The gravitational field strength is given by:

$$g = -\frac{\Delta V_g}{\Delta r} = \frac{F}{m}$$

g *in a uniform field* (left side); ***g*** *defined as force per unit mass (page 65).* (right side)

This rearranges to give:

$$m\Delta V_g = -F\Delta r$$

By definition, the work done is equal to force × distance moved, where the force is equal and opposite to ***F***, since the work's being done against gravity. So the work done is given by $-F\Delta r$ and so you can write:

$$\text{work done} = -F\Delta r = m\Delta V_g$$

So the energy needed to move a mass *m* against a gravitational potential difference is the same as the work done, which is given by $m\Delta V_g$.

Figure 2: *A boy does work to kick a ball high into the air, where the ball has a higher gravitational potential.*

Example — Maths Skills

A forklift truck lifts an 80 kg pig 2.5 m from the surface of the Earth. The mass of the Earth is 5.97 × 10²⁴ kg and its radius is 6370 km. Calculate the difference in potential, and therefore how much work is done.

$\Delta W = m\Delta V_g$, and m = 80 kg, so you need to find ΔV_g.

$$\Delta V_g = -\frac{GM}{r_2} - -\frac{GM}{r_1} = -\frac{GM}{r_2} + \frac{GM}{r_1}$$

$$= -\frac{(6.67\times10^{-11})\times(5.97\times10^{24})}{(6370\times10^3)+2.5} + \frac{(6.67\times10^{-11})\times(5.97\times10^{24})}{6370\times10^3}$$

$$= 24.533... = 25\ \text{Jkg}^{-1}\ \text{(to 2 s.f.)}$$

So $\Delta W = 80 \times 24.533... = 1962.68... = 2000$ J (to 2 s.f.)

Tip: Check you've got the sign right. The pig is being lifted further from the Earth, so it's gravitational potential should increase. So ΔV_g should be positive. If it isn't you've probably done the subtraction the wrong way round.

The graphs in Figure 3 show how the magnitude of the force on an object, due to the gravitational field of a point mass and a spherical mass, varies with the object's distance, *r*, from the mass. The area under the curve between two values of *r* gives the work done to move the object between those two points.

Exam Tip
Since ***F*** is usually defined to be negative (see page 63), you might see a graph of ***F*** against *r*. This would have the same shape as the graphs in Figure 3, but reflected in the *r* axis instead. Figure 3 just shows the magnitude of ***F***.

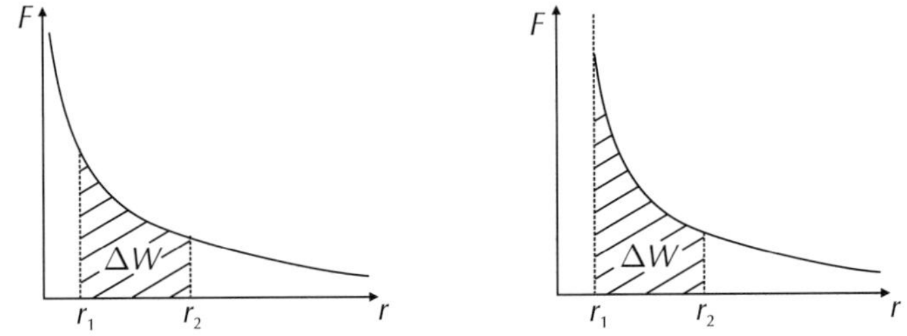

***Figure 3:** A graph of the size of the force on an object due to the gravitational field of a point mass (left) or a spherical mass (right) against distance.*

Example — Maths Skills

The graph below shows the force on an object due to a planet's gravitational field as a function of the distance *r* away from the centre of the planet. The radius of the planet is approximately 2 × 10⁶ m. Use the graph to estimate the work done to move the object from the surface of the planet to a point 3 × 10⁶ m above the planet's surface.

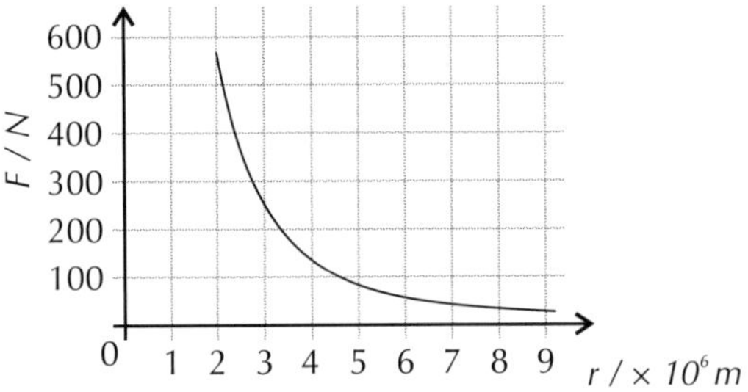

The work done is equal to the area under the graph. The question is asking for an estimate, so the best way to do this is to count the number of squares. 3 × 10⁶ m above the surface is equal to 5 × 10⁶ m from the planet's centre.

The total number of squares under the graph between 2 × 10⁶ m and 5 × 10⁶ m is approximately 6.5.
Each square has an area of 100 × (1 × 10⁶) = 1 × 10⁸ Nm, so the work done ≈ 6.5 × (1 × 10⁸) = 6.5 × 10⁸ J.

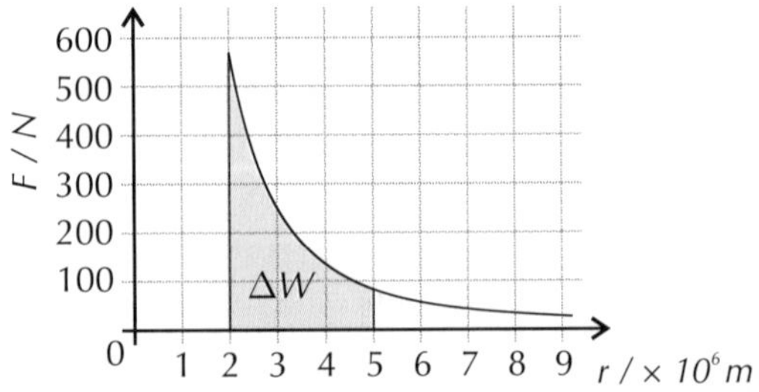

Tip: Remember, joules (J) are equal to Nm. Make sure the area you've calculated is in Nm before writing it in J.

Gravitational potential energy

Gravitational potential is work done per unit mass, so the **gravitational potential energy** (E) of an object at a point in a gravitational field is its gravitational potential multiplied by its mass. You know the formula for V_g so you can substitute it in to get the following formula:

$$E = mV_g = -\frac{GMm}{r}$$

E = gravitational potential energy of object of mass m in J

m = mass of the object in kg

G = gravitational constant $= 6.67 \times 10^{-11}$ Nm^{2}kg^{-2}

M = mass of the object causing the gravitational field in kg

r = the distance between centres of the two masses in m

Example — Maths Skills

Calculate the gravitational potential energy of a person of mass 70.0 kg standing on the surface of the Earth. The mass of the Earth is 5.97 × 10^{24} kg and its radius is 6370 km.

$$E = -\frac{GMm}{r} = -\frac{6.67 \times 10^{-11} \times 5.97 \times 10^{24} \times 70.0}{6370 \times 10^3}$$

$$= -4.375... \times 10^9 = -4.38 \times 10^9 \text{ J (to 3 s.f.)}$$

Tip: E is negative because V_g is negative. It makes sense if you think about it. An object at infinity has zero gravitational potential, and so zero gravitational potential energy. As it is attracted towards the mass, it loses gravitational potential energy. Think of a ball falling to Earth, it's losing g.p.e.

Escape velocity

The **escape velocity** is defined as the velocity needed so an object has just enough kinetic energy to escape a gravitational field. In other words, it's the minimum speed an unpowered object needs in order to leave the gravitational field of a planet and not fall back towards the planet due to gravitational attraction.

To derive the equation for escape velocity, you need to think about the energies involved. At an infinite distance (i.e. the point at which an object has escaped the gravitational field of the planet), the gravitational potential energy of the object is zero. At the surface of the planet, the gravitational potential energy is negative.

The increase in the gravitational potential energy of the object (from the surface to infinity) comes from the initial kinetic energy of the object. If the object is given just enough kinetic energy to escape the gravitational field of the planet, you know from conservation of energy that all of its initial kinetic energy will be converted into gravitational potential energy.

So kinetic energy lost = gravitational potential energy gained, and so:

$$\frac{1}{2}mv^2 = \frac{GMm}{r}$$

Cancelling out m and rearranging for v gives the escape velocity:

$$v = \sqrt{\frac{2GM}{r}}$$

v = escape velocity in ms^{-1}

G = gravitational constant $= 6.67 \times 10^{-11}$ Nm^{2}kg^{-2}

r = distance of the object from the centre of mass of the planet in m

M = mass of the planet in kg

Be careful with the value of r in this equation — it's the radial distance from the centre of the planet to the object. So if the object is initially at the planet's surface, then r is just the planet's radius. But if the object is initially in orbit, then r will be the planet's radius plus the orbital distance.

Figure 4: *Rockets burn a lot of fuel to provide the kinetic energy needed to escape Earth.*

Tip: The escape velocity equation is not dependent on the mass of the object. So, whether you were throwing a tennis ball or projecting a double decker bus into the air, they'd need the same velocity to escape the Earth's gravitational field (assuming there was no air resistance).

Tip: The direction of the escape velocity is always directly away from the mass creating the gravitational field.

Tip: This means you'd have to throw a ball upwards at 11.1 km s^{-1} for it to fully escape Earth's pull.

Example — Maths Skills

Calculate the escape velocity for an object on the surface of the Earth.

Earth's mass = 5.97×10^{24} kg and Earth's radius = 6.37×10^6 m.

Substitute these values into the equation for escape velocity:

$$v = \sqrt{\frac{2GM}{r}}$$

$$= \sqrt{\frac{2 \times (6.67 \times 10^{-11}) \times (5.97 \times 10^{24})}{(6.37 \times 10^6)}}$$

$$= 1.1181... \times 10^4$$

$$= 1.11 \times 10^4 \text{ ms}^{-1} \text{ (to 3 s.f.) (or 11.1 kms}^{-1}\text{)}$$

Practice Questions — Application

Q1 A satellite is orbiting Earth. What's the effect (if any) on the following values of halving the satellite's orbital radius? Explain your answers.

a) G

b) V_g at the satellite's orbital radius

c) m

d) $\mathbf{g}$ at the satellite's orbital radius

Q2 A 1.72 kg brick is dropped off the side of a cliff. If its gravitational potential changes by 531 Jkg^{-1}, how much work is done by gravity on the brick?

Q3 The graph below shows the force on an object due to the gravitational field of a planet against the radial distance from the centre of the planet. The object is moved upwards from 2×10^6 m to 5×10^6 m. Use the graph to estimate the work done moving the object.

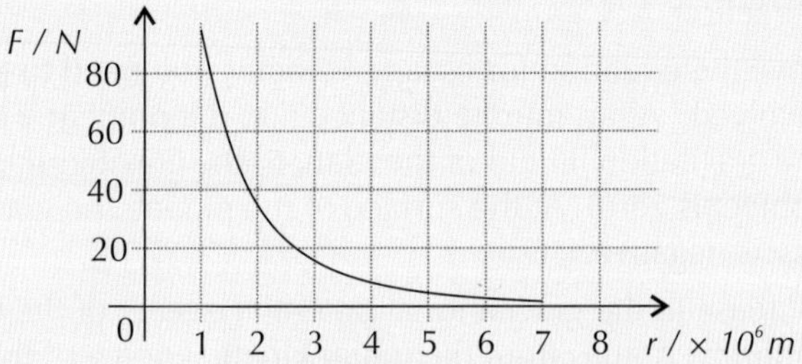

Q4 a) If the mass of the Moon is 7.34×10^{22} kg and the radius of the Moon is 1.74×10^6 m, what is the gravitational potential energy of a 95.0 kg astronaut standing on its surface?

b) What is the minimum velocity with which the astronaut must project a piece of Moon rock so that it is able to escape the Moon's gravitational field?

Practice Questions — Fact Recall

Q1 What is gravitational potential, V_g? What are its units?

Q2 What is the significance of the negative sign in the equation $V_g = -\frac{GM}{r}$?

Q3 Sketch a graph of the gravitational potential (V_g) against distance (r) for an object in Earth's gravitational field.

Q4 Define escape velocity.

Tip: Make sure your graph for Q3 starts at the radius of Earth and not at its centre.

4. Motion of Masses in Gravitational Fields

You saw circular motion earlier on pages 38-44, and here it's covered again for objects in gravitational fields. Gravity provides the centripetal force that keeps objects in orbit around a much larger body.

Learning Objectives:

- Know that the centripetal force on a planet is provided by the gravitational force between it and the Sun.
- Know and be able to derive the equation $T^2 = \left(\frac{4\pi^2}{GM}\right)r^3$.
- Know what is meant by a geostationary orbit, and uses of geostationary satellites.
- Know Kepler's three laws of planetary motion.
- Know the relationship for Kepler's third law $T^2 \propto r^3$ and that it can be applied to systems other than our solar system.

Specification Reference 5.4.3

Satellites

Satellites are kept in orbit by gravitational forces. A satellite is just any smaller mass which orbits a much larger mass — the Moon is a satellite of the Earth, planets are satellites of the Sun, etc.

Satellites are kept in orbit by the gravitational 'pull' of the mass they're orbiting. In our Solar System, the planets have nearly circular orbits, so you can use the equations of circular motion (pages 38-44) to investigate their **orbital speed** and **orbital period**.

Orbital speed

Any object undergoing circular motion is kept in its path by a centripetal force. What causes this force depends on the object — in the case of satellites it's the gravitational attraction of the mass they're orbiting. This means that in this case the centripetal force is the gravitational force.

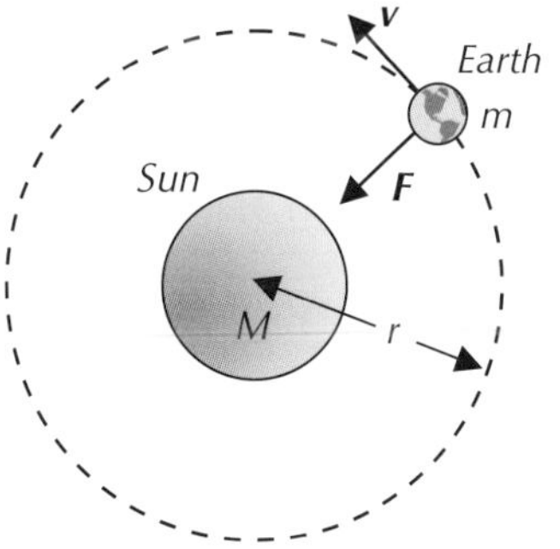

***Figure 1:** Diagram showing the Earth's orbit around the Sun.*

The Earth feels a force due to the gravitational 'pull' of the Sun — see Figure 1. The magnitude of this force is given by Newton's law of gravitation (p.63):

$$F = \frac{GMm}{r^2}$$

Tip: The minus sign is missing in Newton's law of gravitation because it's just the magnitude of the force.

The Earth has velocity ***v***. Its linear speed is constant but its direction is not — so it's accelerating. The magnitude of the centripetal force (p.42) causing this acceleration is:

$$F = \frac{mv^2}{r}$$

The centripetal force on the Earth must be a result of the gravitational force due to the Sun, and so these forces must be equal:

$$\frac{mv^2}{r} = \frac{GMm}{r^2} \Rightarrow v^2 = \frac{GMmr}{r^2m}$$

So the orbital speed is:

$$v = \sqrt{\frac{GM}{r}}$$

G = gravitational constant = 6.67×10^{-11} Nm^{2}kg^{-2}

M = mass of the object being orbited in kg

r – distance from the centre of the object being orbited to the centre of the orbiting satellite in m

v = the orbital speed in ms^{-1}

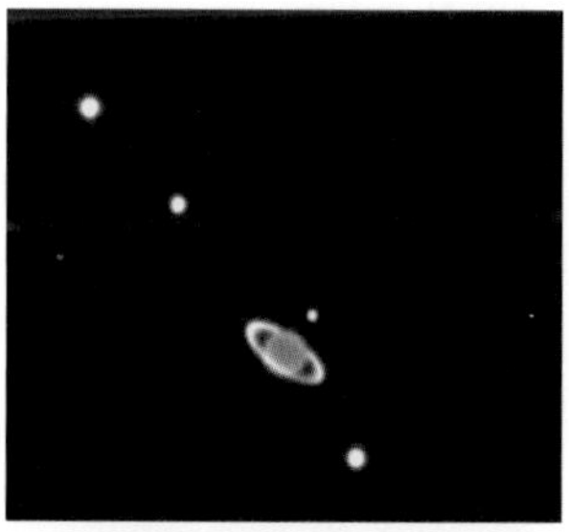

***Figure 2:** Near-infrared image of Uranus and some of its moons. The moons are natural satellites of Uranus, kept in orbit by the planet's gravity.*

Orbital period

The time taken for a satellite to make one orbit is called the orbital period, T. Remember, speed = $\frac{\text{distance}}{\text{time}}$, and the distance for a circular orbit is $2\pi r$, so $v = \frac{2\pi r}{T}$:

$$v = \frac{2\pi r}{T} \Rightarrow T = \frac{2\pi r}{v}$$

Then substitute the expression for v on the previous page and rearrange:

$$T = \frac{2\pi r}{v} = \frac{2\pi r}{\left(\sqrt{\frac{GM}{r}}\right)} = \frac{2\pi r\sqrt{r}}{\sqrt{GM}}$$

Taking the square of the expression gives:

$$T^2 = \left(\frac{4\pi^2}{GM}\right)r^3$$

T *= period in s*

r *= distance from the centre of the object being orbited to the centre of the orbiting satellite in m*

G *= gravitational constant = 6.67 × 10^{-11} Nm^2kg^{-2}*

M *= mass of the object being orbited in kg*

Tip: This is where Kepler's third law comes from (see next page).

Example — Maths Skills

The Moon takes 27.3 days to orbit the Earth. Calculate its distance from the Earth. Take the mass of the Earth to be 5.97 × 10^{24} kg.

You're trying to find the radius of the orbit, r. Use the formula for period, T, and rearrange for r^3:

$$T^2 = \frac{4\pi^2 r^3}{GM} \Rightarrow r^3 = \frac{T^2 GM}{4\pi^2}$$

Then just put the numbers in:

$$r^3 = \frac{(2.35... \times 10^6)^2 \times (6.67 \times 10^{-11}) \times (5.97 \times 10^{24})}{4\pi^2}$$

$$= 5.6116... \times 10^{25}\ \text{m}^3$$

$$r = \sqrt[3]{5.6116... \times 10^{25}}$$

$$= 3.8285... \times 10^8\ \text{m} = 3.83 \times 10^5\ \text{km (to 3 s.f.)}$$

Tip: This is the distance from the centre of the Earth to the centre of the Moon.

Tip: T is given in days, so you need to convert it to seconds first. 1 day = 8.64 × 10^4 s (this will be in your data and formulae booklet in the exam), so 27.3 days = 2.35... × 10^6 s.

Tip: Just take the cube root of r^3 here — you'll probably have a button for it on your calculator ($\sqrt[3]{\square}$).

Geostationary satellites

HOW SCIENCE WORKS

Geostationary satellites orbit directly over the equator and are always above the same point on Earth. A geostationary satellite travels at the same angular speed as the Earth turns below it and in the same direction (west to east). A geostationary orbit takes exactly one day.

These satellites are really useful for sending TV and telephone signals and have improved communication around the world. Communications satellites are stationary relative to a certain point on the Earth, so you don't have to alter the angle of your receiver (or transmitter) to keep up. Geostationary satellites are also used to monitor and track the weather from above.

There are downsides though — they are expensive and pose a small risk of something going wrong and the satellite falling back to Earth.

Figure 3: *Geostationary satellites remain in a fixed position relative to a point on the Earth's surface.*

Example — Maths Skills

Calculate the height above the Earth's surface that a geostationary satellite orbits. The mass of the Earth is 5.97×10^{24} kg and the radius is 6370 km.

You know that $T = 24 \times 60 \times 60 = 86\,400$ s and $M = 5.97 \times 10^{24}$ kg,

so rearrange $T^2 = \left(\frac{4\pi^2}{GM}\right)r^3$ to find r:

$$r = \sqrt[3]{\frac{T^2GM}{4\pi^2}} = \sqrt[3]{\frac{(86\,400)^2 \times 6.67 \times 10^{-11} \times 5.97 \times 10^{24}}{4\pi^2}} = 4.22... \times 10^7\,\text{m}$$

r is the orbital radius from the Earth's centre, so subtract the Earth's radius:

Height of orbit $= 4.22... \times 10^7 - 6370 \times 10^3$

$= 3.585... \times 10^7 = 3.59 \times 10^7$ m (to 3 s.f.)

Tip: All geostationary satellites orbiting Earth are at the same height — regardless of their mass.

Kepler's three laws of planetary motion

Kepler came up with these three laws around 1600, about 80 years before Newton developed his law of gravitation. They're usually used to describe the planets in our solar system, but can be used for any object and its satellite.

HOW SCIENCE WORKS

- **Kepler's first law**: Each planet moves in an ellipse around the Sun, with the Sun at one focus (a circle is just a special kind of ellipse).
- **Kepler's second law**: A line joining the Sun to a planet will sweep out equal areas in equal times. So in Figure 4, if moving from A to B takes the same amount of time as moving from C to D, the two shaded sections will have equal areas.

Tip: An ellipse is a curve determined by two focal points (foci) inside it.

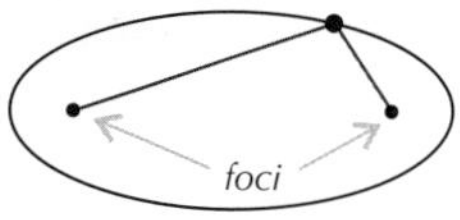

At any point on the edge of the ellipse, the total length of the red line shown is constant. In a circle, the foci are both at its centre.

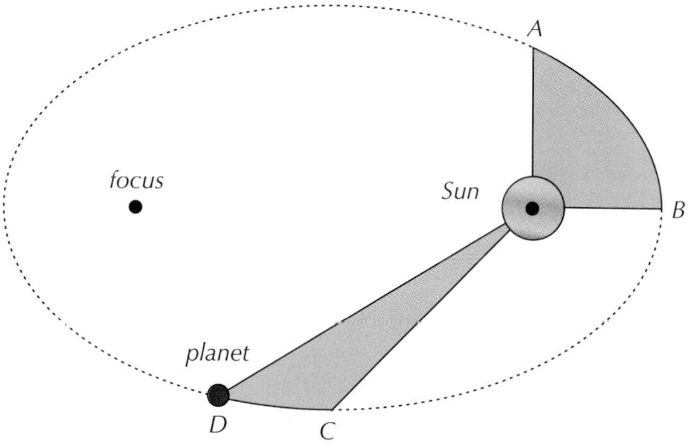

Figure 4: A diagram showing Kepler's second law. The planet takes the same amount of time to move from A to B, as from C to D, and the shaded areas are equal.

- **Kepler's third law**: The period of the orbit and the mean distance between the Sun and the planet are related by Kepler's third law:

$$T^2 \propto r^3$$

So the greater the radius of a satellite's orbit, the slower it will travel and the longer it will take to complete one orbit. You can see this relationship in the expression for the orbital period derived on the previous page.

Tip: You'll usually assume that orbits are circular when using this relation.

Tip: An exoplanet is just a name for a planet that orbits a different star to our own Sun.

Tip: You can leave the orbital period of *A* in hours in your calculation, as you're asked to give your answer in hours too.

Example — Maths Skills

The diagram shows the orbits of two exoplanets around a star in a nearby solar system. Exoplanet A completes one orbit of the star in 42.5 hours. Find the orbital period of exoplanet B to the nearest hour, assuming both orbits are circular.

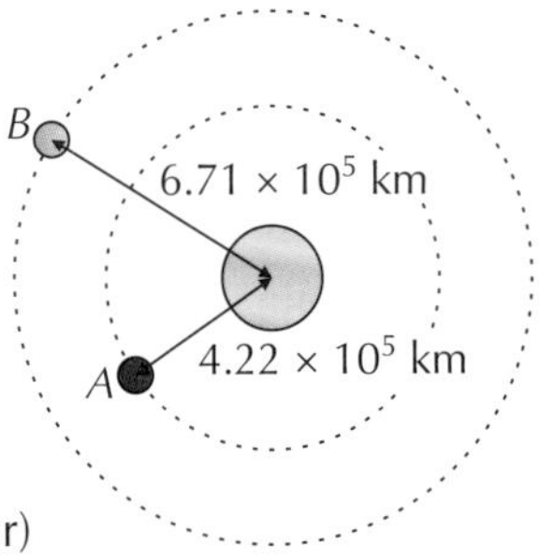

Using Kepler's third law:

$$\frac{T_A^2}{r_A^3} = \frac{T_B^2}{r_B^3}$$

Rearranging:

$$T_B = \sqrt{\frac{T_A^2 r_B^3}{r_A^3}} = \sqrt{\frac{(42.5)^2 \times (6.71 \times 10^5)^3}{(4.22 \times 10^5)^3}}$$

$$= 85.212... = 85 \text{ hours (to the nearest hour)}$$

Tip: An atmosphere is the layer of gases that surround a planet or other large body.

Figure 5: The thickness of the Earth's atmosphere is relatively tiny compared to the Earth's radius.

Atmospheric thickness

As well as predicting the motion of satellites, Newton's law of gravitation can also help to explain how thick a planet's atmosphere is. The planet's gravitational field exerts a force on everything around it, including the particles which make up its atmosphere. Otherwise, the particles would float off into space.

The thickness of a planet's atmosphere is dependent on the gravitational force on the particles of the atmosphere around the planet's surface. You've already seen that this force is dependent on the mass of the planet and the distance from the planet (see p.63). So, for example, if two planets have the same volume but different masses, the planet with the larger mass will have a thicker atmosphere. This is because, at any distance above its surface, the gravitational force will be greater than at the same distance above the planet with the lower mass. So the larger mass planet can stop more atmosphere particles escaping into space, leading to a thicker atmosphere.

Practice Questions — Application

Q1 Using the expressions $F = \frac{GMm}{r^2}$ and $F = \frac{mv^2}{r}$, derive an expression for the speed of a planet orbiting a star.

Q2 A satellite orbiting a planet has orbital period T = 42 hours and orbital radius $r = 3.95 \times 10^5$ km. Calculate the mass of the planet.

Q3 A satellite with orbital radius r has orbital period T. The satellite is moved so that its orbital radius is halved. What is its new orbital period in terms of T?

Practice Questions — Fact Recall

Q1 Define a satellite.

Q2 How is the orbital speed of a satellite related to the radius of its orbit?

Q3 How is the orbital period of a satellite related to the radius of its orbit?

Q4 What is a geostationary satellite and why are they useful for transmitting TV and telephone signals?

Q5 State Kepler's three laws of planetary motion.

Section Summary

Make sure you know...

- That a force field is a region where an object will experience a non-contact force, and that force fields cause interactions between objects or particles.
- That gravitational fields are a type of force field that are caused by objects having mass.
- How to use gravitational field lines to map gravitational fields, including for a point mass and for a uniform field.
- That spherical objects can be treated as point masses, where all of their mass is concentrated at their centre.
- How to use Newton's law of gravitation, $\boldsymbol{F} = -\frac{GMm}{r^2}$, to calculate the force between two point masses.
- That gravitational field strength, $\boldsymbol{g}$, tells you the force per unit mass due to gravity at a point in a gravitational field.
- How to use $\boldsymbol{g} = \frac{\boldsymbol{F}}{m}$ to calculate gravitational field strength.
- That gravitational field strength is uniform close to the surface of the Earth and equal to 9.81 Nkg^{-1}.
- That $\boldsymbol{g}$ is also numerically equal to acceleration due to gravity in ms^{-2}.
- How to use $\boldsymbol{g} = -\frac{GM}{r^2}$ to calculate gravitational field strength for a point mass.
- That gravitational potential at a point is the work done in bringing a unit mass from infinity to the point, and that it is always negative, and is zero at infinity.
- How to calculate gravitational potential using $V_g = -\frac{GM}{r}$.
- That two points at different distances from a mass will have different gravitational potentials.
- That work has to be done to move an object to change its gravitational potential.
- How to sketch a force-distance graph for a point mass or spherical mass.
- That the area under the force-distance graph for a point mass or spherical mass between two distances gives the work done moving the object between those two distances.
- How to use $E = mV_g = -\frac{GMm}{r}$ to calculate the gravitational potential energy of a mass m in the gravitational field of a mass M.
- That escape velocity is the velocity needed so an object has just enough kinetic energy to escape a gravitational field, and that it can be calculated using $\boldsymbol{v} = \sqrt{\frac{2GM}{r}}$.
- That the centripetal force on a planet is provided by the gravitational force between it and the Sun.
- How to use and derive the equation $T^2 = \left(\frac{4\pi^2}{GM}\right)r^3$, where T is the period of an orbit and r is the distance from the centre of the object being orbited to the centre of the orbiting satellite.
- That each planet moves in an ellipse around the Sun, with the Sun at one focus (Kepler's first law).
- That a line joining the Sun to a planet will sweep out equal areas in equal times (Kepler's second law).
- That the period of the orbit and the mean distance between the Sun and the planet are related by $T^2 \propto r^3$ (Kepler's third law).
- That the relationship for Kepler's third law, $T^2 \propto r^3$, can be applied to systems other than our solar system.
- That a geostationary orbit of the Earth takes exactly 24 hours, follows the same direction as the rotation of the Earth and is located directly over the equator.
- Some uses of geostationary satellites, such as for sending TV and telephone signals.

Exam-style Questions

1 Which of the following graphs shows how the magnitude of the gravitational field strength varies with distance from a spherical mass with a radius R?

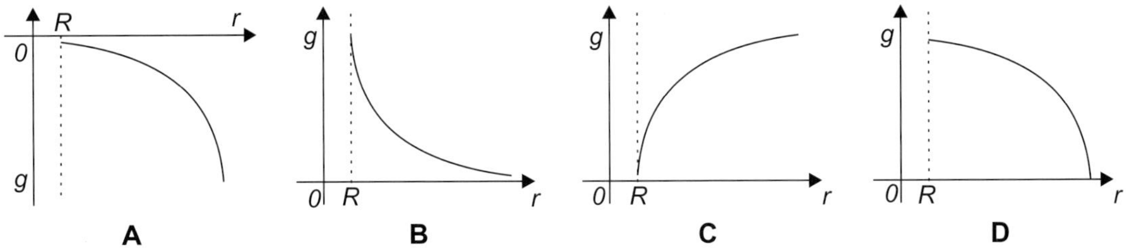

(1 mark)

2 Two identical objects of mass m, are a distance of r away from each other.
Which of the following is an expression for the magnitude of the force F acting on each of them?

A $\frac{Gm^2}{r^2}$ **B** $\frac{Gm}{r^2}$ **C** $\frac{Gm}{r}$ **D** $\frac{Gm^2}{r}$

(1 mark)

3 The four largest moons of Jupiter are known as the Galilean moons. They can be considered to be uniformly spherical in shape and to have regular circular orbits.

(a) The orbital speed, v, of a satellite undergoing circular motion can be calculated using:

$$v = \sqrt{\frac{GM}{r}}$$

where G is the gravitational constant, M is the mass of the object being orbited, and r is the distance from the centre of the object being orbited to the centre of the orbiting satellite.

Use the orbital speed equation to derive this expression for the orbital period:

$$T^2 = \left(\frac{4\pi^2}{GM}\right)r^3$$

(2 marks)

(b) The closest of the Galilean moons, Io, takes 1.8 days to orbit Jupiter at an average orbital radius of 420 000 km. Europa is another Galilean moon. Europa takes 3.6 days to orbit Jupiter.
Calculate Europa's average orbital radius.

(2 marks)

(c) The gravitational field strength on the surface of Io is 1.8 Nkg^{-1}.
Calculate the gravitational force that would act on a 65 kg person stood on the surface of Io, assuming the gravitational force from Jupiter is negligible.

(1 mark)

(d) Carpo is a smaller moon of Jupiter with an elliptical orbit. State what Kepler's second law of planetary motion says about the orbit of the moon Carpo.

(1 mark)

4 (a) Define a force field.

(1 mark)

(b) Sketch the gravitational field lines for the gravitational field around Earth.

(1 mark)

A student is conducting an experiment to determine the local value of $\boldsymbol{g}$.
He finds the magnitude of $\boldsymbol{g}$ to be 9.83 Nkg^{-1}, to 3 significant figures.

Another student conducts the same experiment and finds the value to be 6.31 Nkg^{-1}.
He claims his value is about 1/3 lower because his work bench is about 1/3 taller than the other student's.

(c) Explain whether the student's claim is true or not.

(1 mark)

(d) If the Earth has radius 6370 km and mass 5.97×10^{24} kg, calculate the altitude at which the local value of $\boldsymbol{g}$ is -6.31 Nkg^{-1}. (Altitude is measured from sea level.)

(2 marks)

A satellite is in orbit around Earth at a gravitational potential of -20.6 $MJkg^{-1}$.

(e) Calculate the value of $\boldsymbol{g}$ at this point.

(3 marks)

(f) Calculate the speed that the satellite would need in order to escape the gravitational field of Earth from its current position.

(2 marks)

(g) Another satellite orbits the Earth once every 24 hours in the same plane as the equator, and in the same direction as the Earth's rotation. Name this type of orbit.

(1 mark)

5 An asteroid passes close to a planet of mass 2.14×10^{24} kg then continues travelling away. **Fig 5.1** shows the gravitational force acting on the asteroid against its distance from the centre of the planet.

Fig 5.1

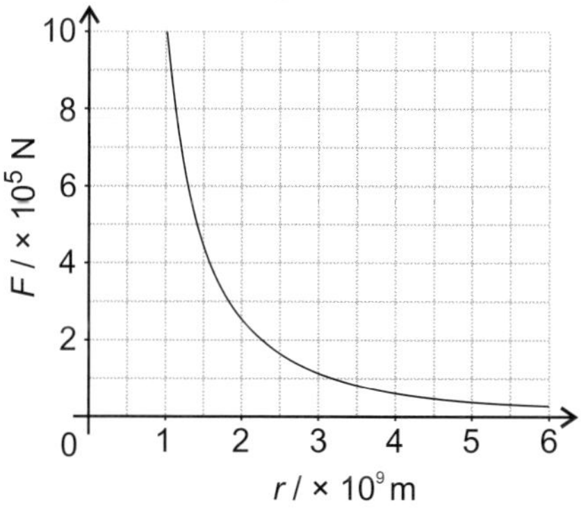

(a) Use the graph to calculate the mass of the asteroid.

(2 marks)

(b) Explain why the gravitational potential energy of the asteroid, E, always has a negative value at a distance, r, from the planet.

(2 marks)

(c) Sketch a graph of the gravitational potential energy of the asteroid as it gets further away from the planet.

(1 mark)

1. The Solar System

A solar system is made up of a star (e.g. the Sun) at the centre, all the objects that orbit around it — and everything else which orbits around them.

Learning Objectives:
- Understand the terms 'universe', 'galaxy', 'solar system', 'planet', 'planetary satellite' and 'comet'.
- Understand distances measured in astronomical units, AU.

Specification References 5.5.1 and 5.5.3

Our place in the universe

The **universe** is everything that exists — this includes plenty you can see, like **stars** and **galaxies**, and plenty that you can't see, like microwave radiation (page 101), **dark energy** and **dark matter** (pages 104-105). Galaxies, like our Milky Way galaxy, are clusters of stars and planets that are held together by gravity.

Inside the Milky Way is our **solar system**. A solar system consists of a star and all the objects that orbit it. In our solar system, the star is the Sun. The Sun is orbited by:

- **Planets** — these are large objects which orbit a star. There are eight of them orbiting the Sun. They have to be large enough to have "cleared their neighbourhood". This means that their gravity is strong enough to have pulled in any nearby objects apart from their satellites.
- **Dwarf planets** — e.g. Pluto. These are planet-like objects that orbit stars, but are too small to meet all of the rules for being a planet.
- **Planetary satellites** — these are objects that orbit a planet. For example:
  1. Moons — these orbit planets. They're natural satellites.
  2. Artificial satellites are satellites that humans have built. There are lots orbiting the Earth and some orbiting other planets.

Tip: There are a lot of other conditions that an object has to meet to be considered a planet. Size is just one of them.

Figure 1: *A photograph of the planet Jupiter. The two black spots are caused by two of Jupiter's moons passing in front of the planet.*

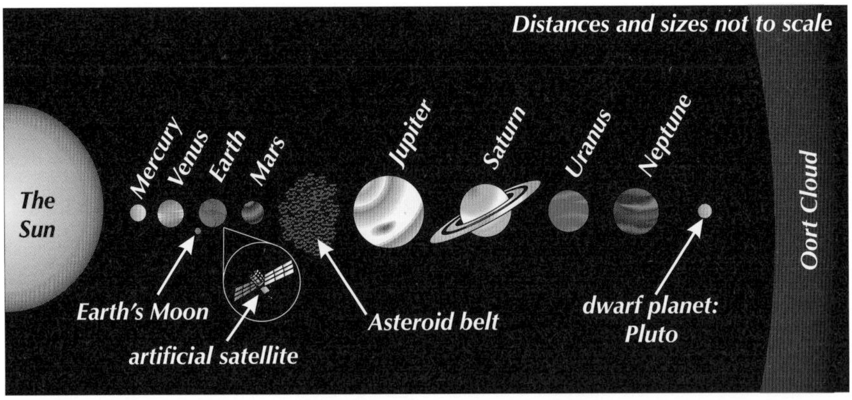

Figure 2: *A diagram showing some of the contents of our solar system, including the eight planets in order.*

- **Asteroids** — these are chunks of rock and mineral that orbit the Sun. They range in size, but can be anywhere from just over 1 m in diameter to the size of a dwarf planet. The asteroid belt is a disc of many asteroids that all sit in orbits between Mars and Jupiter.

- **Comets** — these are "dirty snowballs", made of ice, dust and rock. We think most comets usually orbit the Sun about 1000 times further away than Pluto does (in the "Oort cloud"). Occasionally one gets dislodged and heads towards the Sun. It then follows a new elliptical orbit, which can take millions of years to complete. The orbits of the comets we see are highly elliptical. Some comets (from closer in than the Oort cloud) follow a smaller orbit and swing round the Sun more often. The most famous is Halley's comet, which orbits in 76 years.

Figure 3: *A comet seen travelling across the night sky, as it passes the Earth.*

The scale of the solar system

HOW SCIENCE WORKS

From Copernicus onwards, astronomers have tried to work out the distances of the planets from the Sun. They could not work out the actual distances in standard units, but they could work out relative distances by comparing the orbits of other planets to the orbit of the Earth. This led to the creation of the **astronomical unit** (AU).

> One astronomical unit (AU) is defined as the mean distance between the Earth and the Sun.

The size of the AU wasn't known accurately until 1769 — when it was carefully measured during a transit of Venus (when Venus passed between the Earth and the Sun). We now know that 1 AU is equal to about 150 million km (1.5×10^{11} m). Neptune, the outermost planet in our solar system, is about 30 AU from the Sun.

Practice Questions — Fact Recall

Q1 What is a galaxy?

Q2 What is a solar system?

Q3 Where do most comets usually orbit the Sun?

Q4 State what is meant by an astronomical unit.

2. Astronomical Distances

The distances to stars can be difficult to calculate. Astronomers have several methods to work them out — but not all of the methods work for all distances. Astronomers also use a few different units to measure distance.

Learning Objectives:

- Understand stellar parallax, and how it can be used to find distances in parsecs.
- Understand distances measured in parsecs (pc).
- Be able to use the equation $p = \frac{1}{d}$ to make parallax calculations.
- Understand distances measured in light-years (ly).

Specification Reference 5.5.3

Parallax

Imagine you're in a moving car. You see that (stationary) objects in the foreground seem to be moving faster than objects in the distance. This is an example of **parallax**.

The distance to nearby stars can be calculated by observing how they appear to move relative to stars that are so distant that they appear not to move at all — background stars. This is done by comparing the position of the nearby star in relation to the background stars at different parts of the Earth's orbit.

Parallax is measured in terms of the angle of parallax. If you observe the position of the star from opposite points of the Earth's orbit (6 months apart), the angle of parallax is half the angle that the star appears to move in relation to the background stars. The greater the angle, the nearer the object is to you.

Tip: You can see parallax at work by holding your hand out in front of your face and closing just your right eye, followed by just your left eye. Your hand will move in relation to the background. The closer your hand is to your face, the more it moves in relation to the background.

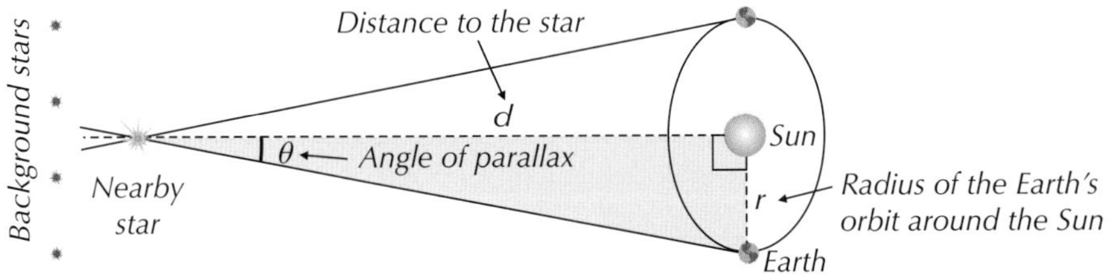

Figure 1: *The distance of a nearby star can be calculated by measuring the angle of parallax and using the diameter of the Earth's orbit.*

Using the shaded triangle in Figure 1, you can calculate the distance to the nearby star, d, using trigonometry, if you know the angle of parallax and the radius of the Earth's orbit:

$$\tan\theta = \frac{r}{d}$$

$$\Rightarrow d = \frac{r}{\tan\theta}$$

Tip: Remember, trigonometry says that $\tan\theta = \frac{\text{opposite side}}{\text{adjacent side}}$.

For small angles $\tan\theta \approx \theta$, where θ is in radians. Because the angles used in astronomy are so tiny, you can use this assumption for calculations of parallax (as long as you're working in radians). So:

$$d \approx \frac{r}{\theta}$$

d *= distance to the star*

r *= radius of the Earth's orbit*

θ *= angle of parallax in radians*

Remember, the angle in radians = angle in degrees × $\frac{\pi}{180}$ (see page 245).

Tip: You can use any units for d and r, but you must use the same unit for both.

Example — Maths Skills

A nearby star has an angle of parallax of 1.39×10^{-4} °. Calculate the distance to the star. The average radius of the Earth's orbit is 1.50×10^{11} m.

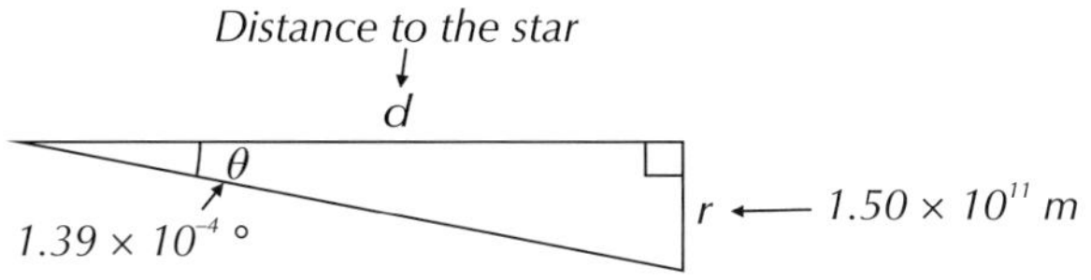

Convert the angle into radians:

$$1.39 \times 10^{-4}\,° \times \frac{\pi}{180} = 2.42... \times 10^{-6} \text{ rad}$$

$$d \approx \frac{r}{\theta} = \frac{1.50 \times 10^{11}}{2.42... \times 10^{-6}} = 6.182... \times 10^{16} \text{ m}$$
$$= 6.18 \times 10^{16} \text{ m (to 3 s.f.)}$$

Parsecs

The angle of parallax is used to define a unit of distance called a **parsec** (pc).

> A star is exactly one parsec (pc) away if the angle of parallax,
> $\theta = 1$ arcsecond $= \left(\frac{1}{3600}\right)^{\circ}$

The distances measured in astronomy are usually huge — even the nearest large galaxy to the Milky Way is 780 000 parsecs away. Astronomers often use parsecs (pc) or megaparsecs (Mpc) to measure these large distances. You need to be able to use this conversion:

> 1 pc = 3.1×10^{16} m

When you're calculating distances in parsecs, you can use a special form of the parallax equation from the previous page:

$$p = \frac{1}{d}$$

p *= angle of parallax in arcseconds*

d *= distance to the star in parsecs*

Tip: An arcsecond is just a measure of angle like degrees and radians. It's equivalent to $\left(\frac{1}{3600}\right)^{\circ}$, so it's useful when measuring the really small angles that astronomers use.

Exam Tip
This conversion between parsecs and metres will be in the exam data and formulae booklet, so don't worry about learning it.

Example — Maths Skills

Proxima Centauri has an angle of parallax of 0.77 arcseconds. Calculate the distance to Proxima Centauri in parsecs.

$$p = \frac{1}{d}$$

Rearrange the parsec parallax equation for distance, d:

$$d = \frac{1}{p} = \frac{1}{0.77} = 1.298... \text{ pc}$$
$$= 1.3 \text{ pc (to 2 s.f.)}$$

Light years

All electromagnetic waves travel at the speed of light, c, in a vacuum (where $c = 3.00 \times 10^8\ ms^{-1}$). The distance that electromagnetic waves travel through a vacuum in one year is called a **light year** (ly).

$$1 \text{ light year} = 9.5 \times 10^{15} \text{ m}$$

If we see the light from a star that is, say, 10 light years away then we are actually seeing it as it was 10 years ago. The further away the object is, the further back in time we are actually seeing it.

Examples

- Light from the Sun takes around 8 minutes to reach Earth, so the light that we see from the Sun actually left the Sun 8 minutes earlier.
- The star Proxima Centauri is 1.3 pc away from Earth, which is approximately 4.3 light years. So the light from Proxima Centauri will take around 4 years and 4 months to reach us.

The light from very distant galaxies has taken billions of years to reach us. Astronomers search for distant galaxies so that they can 'look into the past' at what the Universe was like billions of years ago.

Practice Questions — Application

Q1 The star Alpha Tauri in the constellation Taurus is 20 pc away. Calculate this distance in light years.

Q2 Scientists at an observatory are studying the star Sirius from Earth. One image of Sirius is captured, and another is captured 6 months later. The angle of parallax is recorded to be 0.37 arcseconds.

a) The average radius of the Earth's orbit is 1.50×10^{11} m. Calculate the distance to Sirius in metres.

b) How long will it take for light from Sirius to reach Earth in years and months?

Practice Questions — Fact Recall

Q1 What is meant by the angle of parallax?

Q2 What is the definition of a parsec?

Q3 What is the definition of a light year?

3. Stellar Evolution

'Stellar' means 'to do with stars', so this topic's all about how stars change. Stars go through several different stages in their lives. The Sun is a main sequence star right now, but it's still got a lot of life left in it.

Learning Objectives:

- Know that a star is formed from interstellar dust and gas.
- Understand how gravitational collapse, nuclear fusion and a balance between gravitational forces and radiation and gas pressure lead to the creation of a stable main sequence star.
- Understand how low-mass stars, like our Sun, evolve into red giants.
- Know that a low-mass red giant will collapse into a white dwarf and a planetary nebula.
- Understand that a star core with mass less than the Chandrasekhar limit becomes a white dwarf because electron degeneracy pressure prevents further collapse.
- Know that stars much more massive than the Sun evolve into red super giants, which will eventually explode in a supernova leaving behind either a neutron star or black hole.
- Know the characteristics of neutron stars and black holes.

Specification Reference 5.5.1

Formation

All stars are born in a cloud of dust and gas, most of which was left when previous stars blew themselves apart in supernovae (see page 87). The denser clumps of the cloud contract (very slowly) under the force of gravity.

When these clumps get dense enough, the cloud fragments into regions called **protostars** that continue to contract, and heat up as they do. Eventually the temperature at the centre of the protostar reaches a few million degrees, and hydrogen nuclei start to fuse together to form helium (see page 210).

As the star's temperature increases and its volume decreases (remember, it's contracting), the gas pressure increases (p.27). There is also radiation pressure in the star — a pressure exerted by electromagnetic radiation on any surface it hits. Radiation pressure is usually too tiny to notice, but becomes significant in stars because of the enormous amount of electromagnetic radiation released by fusion. The combination of gas pressure and radiation pressure counteract the force of gravity, preventing the star from contracting further. The star has now reached the **main sequence** and will stay there, relatively unchanged, while it fuses hydrogen into helium.

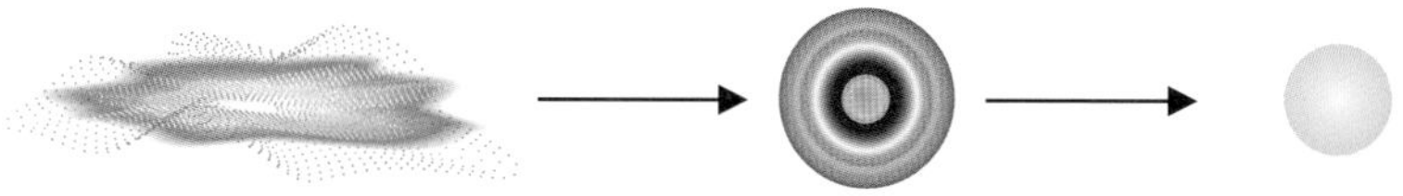

Figure 1: *A star in the early stages of its stellar evolution.*

Core and shell burning sequence

Stars spend most of their lives as main sequence stars. The pressure produced from hydrogen fusion in their core balances the gravitational force trying to compress them. This stage is called core hydrogen burning.

When all the hydrogen in the core has fused into helium, fusion stops, meaning the outward pressure also stops. The helium core contracts and heats up under the weight of the star. As a result, the outer layers expand and cool, and the star becomes a **red giant**.

The material surrounding the core still has plenty of hydrogen. The heat from the contracting helium core raises the temperature of this material enough for the hydrogen to fuse. This is called shell hydrogen burning. (Very low-mass stars stop at this point. They use up their fuel and slowly fade away...)

The helium core continues to contract until, eventually, it gets hot enough and dense enough for helium to fuse into carbon and oxygen. This is called core helium burning. This releases a huge amount of energy, which pushes the outer layers of the star further outwards.

When the core helium runs out, the carbon-oxygen core contracts again and heats a shell around it so that helium can fuse in this region — shell helium burning.

Tip: As a star contracts, the temperature increases due to conservation of energy — gravitational potential energy is converted to thermal energy.

Figure 3 shows a summary of the stages a star goes through during the main sequence and red giant phases.

Tip: This process is just a load of contracting, heating and fusing of successively bigger nuclei.

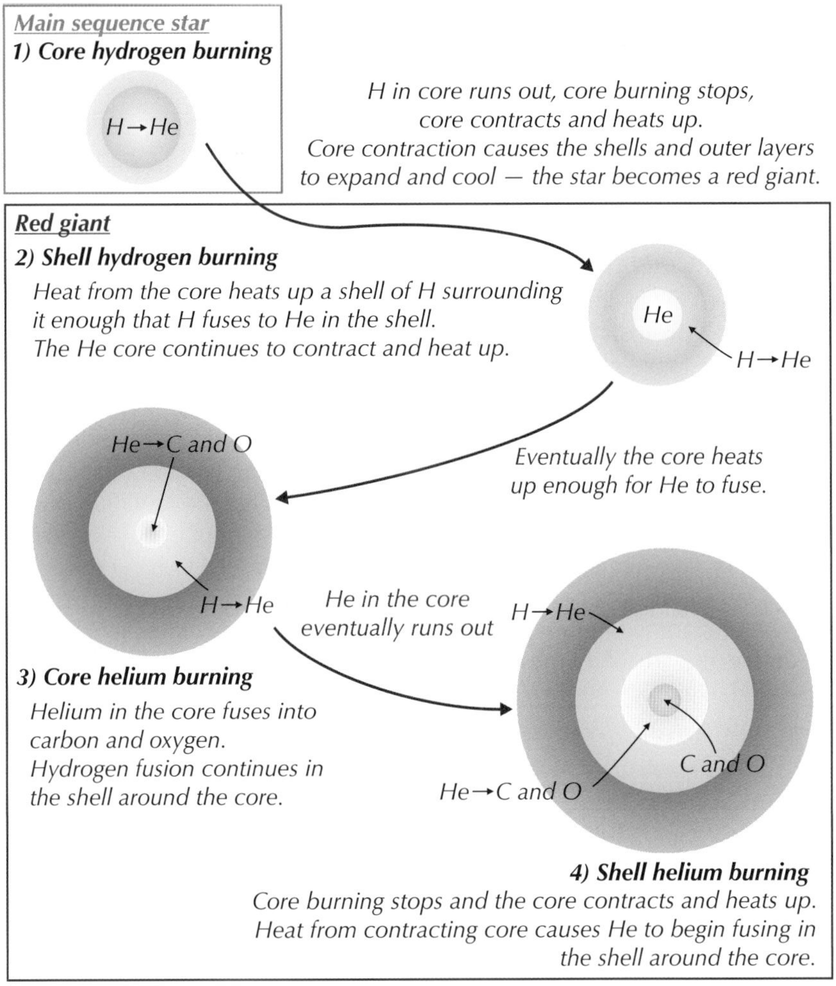

Figure 2: A red giant with a radius of around 1 billion km.

Tip: The cooling of the outer layers of the star makes the star's colour change to become redder — this is why we call them red giants.

Main sequence star
1) Core hydrogen burning

H→He

H in core runs out, core burning stops, core contracts and heats up. Core contraction causes the shells and outer layers to expand and cool — the star becomes a red giant.

Red giant
2) Shell hydrogen burning

Heat from the core heats up a shell of H surrounding it enough that H fuses to He in the shell. The He core continues to contract and heat up.

He

H→He

Eventually the core heats up enough for He to fuse.

He→C and O

H→He

He in the core eventually runs out

H→He

3) Core helium burning

Helium in the core fuses into carbon and oxygen. Hydrogen fusion continues in the shell around the core.

He→C and O

C and O

4) Shell helium burning

Core burning stops and the core contracts and heats up. Heat from contracting core causes He to begin fusing in the shell around the core.

Figure 3: The helium and hydrogen burning sequence of a star as it transitions from a main sequence star to a red giant.

White dwarfs

In low-mass stars, like our Sun, the carbon-oxygen core won't get hot enough for any further fusion and so it continues to contract under its own weight. Once the core has shrunk to about the size of the Earth, electrons exert enough pressure (**electron degeneracy pressure**) to stop it collapsing any more (don't worry — you don't have to know how).

The above is only true for stars with a core mass under about 1.4 times the mass of the Sun though — in bigger stars the electron degeneracy pressure isn't high enough to counteract the gravitational force and the star collapses (see next page). The maximum mass for which the electron degeneracy pressure can counteract the gravitational force is called the **Chandrasekhar limit**.

For stars below the Chandrasekhar limit, the helium shell becomes more and more unstable as the core contracts. The star pulsates and ejects its outer layers into space as a **planetary nebula**, leaving behind the dense core. The star is now a very hot, dense solid called a **white dwarf**, which will simply cool down and fade away.

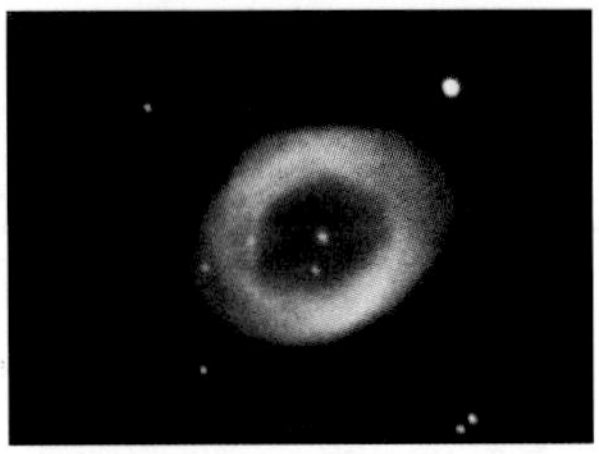

Figure 4: When a low-mass red giant cools down, the shells are ejected and create a beautiful planetary nebula, like this one, leaving a white dwarf at the centre.

Tip: The term 'planetary nebula' doesn't actually have anything to do with planets. They were called that because they tend to be roughly spherical, and so look like enormous planets.

Supernovae

High-mass stars have a shorter life and a more exciting death than lower-mass stars like the Sun. Even though stars with a large mass have a lot of fuel, they use it up more quickly and don't spend so long as main-sequence stars.

When they are red giants the 'core burning to shell burning' process can continue beyond the fusion of helium, building up layers of different fusing elements in an onion-like structure, to become **red super giants** (or super red giants). For really massive stars this can go all the way up to iron. Nuclear fusion beyond iron isn't energetically favourable (p.212), so once an iron core is formed then very quickly it's goodbye star...

When the core of a star runs out of fuel, it starts to contract. If this core's mass is larger than the Chandrasekhar limit, electron degeneracy can't stop the core contracting. This happens when the mass of the core is more than 1.4 times the mass of the Sun.

The core of the star continues to contract, and as it does, the outer layers of the star fall in and rebound off the core, setting up huge shockwaves. These shockwaves cause the star to explode cataclysmically in a **supernova**, leaving behind the core, which will either be a **neutron star** or (if the star was massive enough) a **black hole**.

Tip: Supernovae release a huge amount of energy — a supernova even as far as 1000 ly away could seriously damage the Earth's atmosphere.

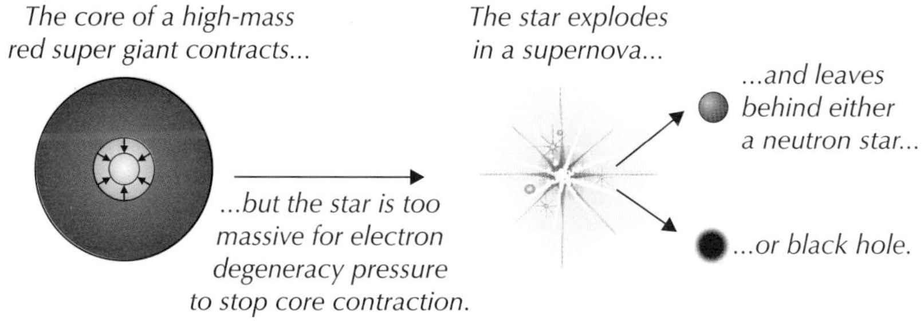

The core of a high-mass red super giant contracts...
...but the star is too massive for electron degeneracy pressure to stop core contraction.
The star explodes in a supernova...
...and leaves behind either a neutron star...
...or black hole.

Figure 5: *The possible evolution paths of a high-mass star from its red super giant phase.*

Figure 6: *An optical image of the Tarantula Nebula (top left) and a nearby supernova (bottom right). The Tarantula Nebula is an incredibly bright region, where many stars are being formed, but the supernova is able to reach a comparable level of brightness.*

When the star explodes in a supernova, it will experience a brief and rapid increase in brightness. The light from a supernova can briefly outshine an entire galaxy, before fading over the next few weeks or months.

Neutron stars

As the core of a massive star contracts, the electrons in the core material get squashed onto the atomic nuclei and combine with protons to form neutrons and neutrinos (hence the name neutron star).

If the star's core is between 1.4 and 3 solar masses, this is as far as the star can contract — the core suddenly collapses to become a neutron star made mostly of neutrons. The outer layers of the star fall onto the neutron star, which causes shockwaves in these layers and leads to a supernova. After the supernova, the neutron star is left behind.

Neutron stars are incredibly dense (about 4×10^{17} kg m^{-3}). They're also very small, typically about 20 km across, and they can rotate very fast (up to 600 times a second).

Neutron starts emit two beams of radio waves as they rotate. These beams sometimes sweep past the Earth and can be observed as radio pulses rather like the flashes of a lighthouse. These pulsing neutron stars are called pulsars.

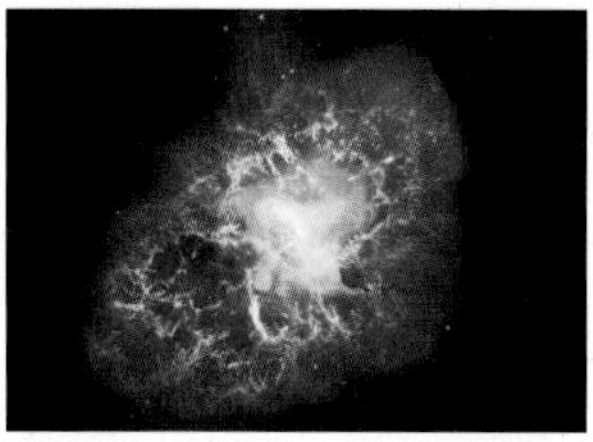

Figure 7: *The Crab Nebula — a supernova remnant with a pulsar (rotating neutron star) in the centre.*

Black holes

If the core of the star is more than 3 times the Sun's mass, the core will contract until neutrons are formed, but now the gravitational force on the core is greater. The neutrons can't withstand this gravitational force, so the star continues to collapse. For something of this size, there are no known mechanisms left to stop the core collapsing to an infinitely dense point, called a singularity. At that point, the laws of physics break down completely.

Tip: Remember, the kinetic energy of an object is given by $E_k = \frac{1}{2}mv^2$.

Tip: Nothing can travel faster than the speed of light, c.

The **escape velocity** (p.71) is the velocity that an object would need to travel at to have enough kinetic energy to escape a gravitational field. When a massive star collapses into an infinitely dense point, a region around it has such a strong gravitational field that it becomes a black hole — an object whose escape velocity is greater than the speed of light, c. If you enter this region, there's absolutely no escape — not even light can escape it.

The boundary of this region is called the **event horizon**. At the event horizon, the escape velocity is equal to c, so light has just enough kinetic energy to escape the black hole's gravitational pull. The radius of a black hole is considered to be the radius of the event horizon. Past that point, everything, including light, can do nothing but travel further into the black hole.

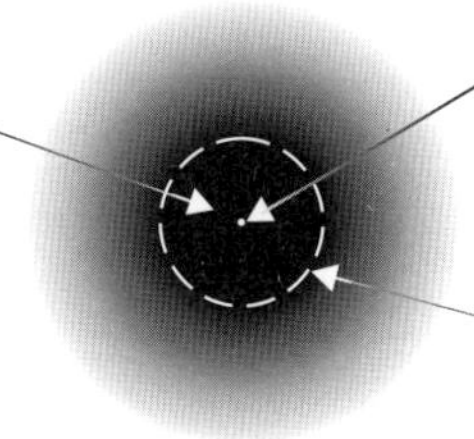

Inside the event horizon, nothing escapes as the escape velocity > c
Infinitely dense point
At the event horizon, escape velocity = c

Figure 8: *The structure of a black hole.*

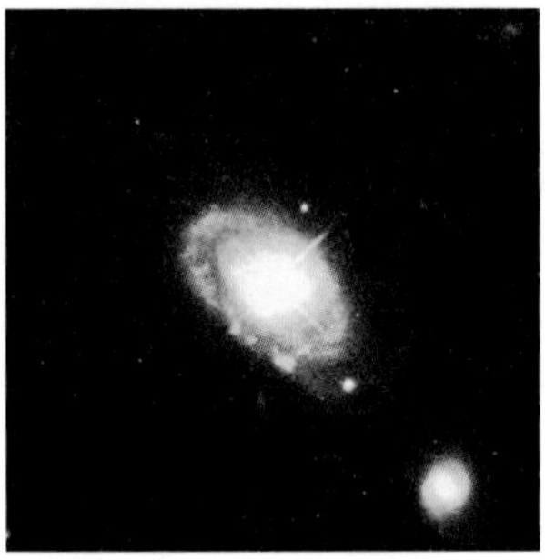

Figure 9: *Astronomers think the intense radiation seen at the centre of galaxies is caused by matter falling into a supermassive black hole.*

Astronomers now believe that there is a supermassive black hole (more than 10^6 times more massive than our Sun) at the centre of every galaxy. As stars get closer to the event horizon of the black hole, and begin to be consumed, they get very hot and produce intense radiation, making the centre of galaxies very bright.

Practice Questions — Fact Recall

Q1 What causes a cloud of dust and gas to contract and form a star?

Q2 In what stage of evolution does a star spend most of its life?

Q3 Explain how the process of core hydrogen burning stops a main sequence star from compressing under the gravitational force.

Q4 What happens to the core of a star when it runs out of hydrogen? What effect does this have on the hydrogen layer surrounding the core of the star?

Q5 What causes a main-sequence star to become a red giant?

Q6 How is a white dwarf formed?

Q7 What is a planetary nebula?

Q8 Why can stars with a core mass greater than 1.4 solar masses not form white dwarfs?

Q9 Describe how the contracting core of a star creates a supernova.

Q10 What is left after a supernova if the core of the star is not massive enough to form a black hole? What are they mostly made of?

Q11 What is a black hole? How is one formed?

4. Stellar Radiation and Luminosity

Astronomers can estimate a star's temperature based on the radiation it emits. They can then use this information to help identify what type of star it is.

Learning Objectives:

- Know how Wien's displacement law can be used to estimate the peak surface temperature of a star.
- Know what is meant by the luminosity of a star.
- Be able to use Stefan's law to calculate luminosity.
- Be able to use Wien's displacement law and Stefan's law to estimate the radius of a star.
- Know that the Hertzsprung-Russell (HR) diagram is a plot of luminosity against temperature for stars.
- Be able to identify the regions of the HR diagram that correspond to main sequence stars, red giants, super red giants and white dwarfs.

Specification References 5.5.1 and 5.5.2

Wien's displacement law

Objects emit electromagnetic radiation due to their temperature. At everyday temperatures this is mostly in the infrared part of the spectrum (which we can't see). But heat something up enough and it will start to glow.

Stars can be assumed to emit radiation in a continuous spectrum. The relationship between intensity ('power per unit area') and wavelength for this radiation varies with temperature. As shown in Figure 1, a hotter star will emit more radiation than a cooler one, and so have a higher total power output (assuming it has the same surface area). The most common wavelength of EM wave emitted becomes shorter as the surface temperature of the star increases. This is called the peak wavelength, λ_{max}, shown by the dotted lines on Figure 1.

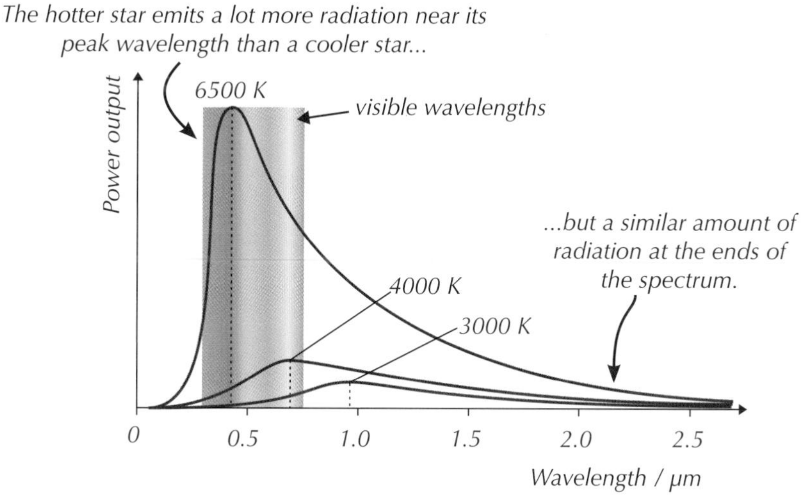

Figure 1: *Three curves showing the different range and intensity of wavelengths of radiation emitted by stars of different temperatures.*

You can use λ_{max} to estimate a star's peak surface temperature using **Wien's displacement law**:

$$\lambda_{max} \propto \frac{1}{T}$$

λ_{max} = *peak wavelength in m*

T = *temperature in K*

We can use this to estimate the temperature of stars from their colour. A star that appears red will emit light in the red part of the visible spectrum with the highest intensity. From the shape of the intensity-temperature curve, this means its peak wavelength can be no smaller than the red part of the visible spectrum. A star that appears white will emit all frequencies of visible light with a similar power output. This means they must have a shorter peak wavelength, around the middle of the EM spectrum. So a white star must have a higher temperature than a red star.

Figure 2: *Wilhelm Wien was a German physicist who was awarded the 1911 Nobel Prize in physics for his work on radiation emission.*

Stefan's law

The **luminosity** of a star is a measure of its brightness. It is the total energy it emits per second, i.e. its power output. The luminosity of a star is directly proportional to the fourth power of the star's surface temperature and is also directly proportional to the surface area. This is **Stefan's law**:

$$L = 4\pi r^2 \sigma T^4$$

L *= luminosity of the star in W*

T *= surface temperature in K*

r *= radius of the star in m*

σ = the Stefan constant = 5.67×10^{-8} $Wm^{-2}K^{-4}$

Tip: The surface area of a sphere is given by $4\pi r^2$. Since stars are spherical, this is the surface area of a star.

Exam Tip
Stefan's law and the Stefan constant are given in the data and formulae booklet.

Example — Maths Skills

The Sun has a surface temperature of 5800 K, and a radius of 6.96×10^8 m. What surface temperature would a star have if it had a radius of 9.42×10^7 m but the same luminosity as the Sun?

$L = 4\pi r^2 \sigma T^4$. The luminosities of the star and the Sun are the same, so:

$$4\pi (r_{Sun})^2 \sigma (T_{Sun})^4 = 4\pi (r_{star})^2 \sigma (T_{star})^4$$

$$\Rightarrow \quad T_{star} = \sqrt[4]{\frac{(r_{Sun})^2 (T_{Sun})^4}{(r_{star})^2}} \quad \text{(cancelling } 4\pi\sigma \text{ from both sides and rearranging)}$$

$$\Rightarrow \quad T_{star} = \sqrt[4]{\frac{(6.96 \times 10^8)^2 \times (5800)^4}{(9.42 \times 10^7)^2}}$$

$$= 15\,765.47... = 16\,000 \text{ K (to 2 s.f.)}$$

Tip: A smaller star needs to be a lot hotter to have the same power output.

You can combine Stefan's law and Wien's displacement law (see previous page) to estimate a star's radius:

Example — Maths Skills

Sirius A is a main sequence star, and Sirius B is a white dwarf. Sirius A has a surface temperature of 9800 K and produces electromagnetic radiation with a peak wavelength of 300 nm (to 2 s.f.). Sirius B has a luminosity of 9.4×10^{24} W and produces electromagnetic radiation with a peak wavelength of 115 nm. Estimate the radius of Sirius B.

First, find the temperature of Sirius B. $\lambda_{max} \propto \frac{1}{T}$, so $T\lambda_{max}$ = constant.

This means $T_A \lambda_{max\,A} = T_B \lambda_{max\,B}$ so $9800 \times 300 \times 10^{-9} = (115 \times 10^{-9}) T_B$

$$\text{So } T_B = \frac{9800 \times (300 \times 10^{-9})}{115 \times 10^{-9}} = 25\,565.2... \text{ K}$$

Then use Stefan's law to find the star's radius:

$$L = 4\pi r^2 \sigma T^4 \quad \text{so} \quad r = \sqrt{\frac{L}{4\pi\sigma T^4}}$$

$$= \sqrt{\frac{9.4 \times 10^{24}}{4\pi \times (5.67 \times 10^{-8}) \times (25\,565.2...)^4}}$$

$$= 5.557... \times 10^6 \text{ m} = 5.6 \times 10^6 \text{ m (to 2 s.f.)}$$

The Hertzsprung-Russell diagram

You can plot a graph of luminosity against temperature for stars. In the 1910s, this was done with data from a survey of a large number of stars. Independently, two scientists, Hertzsprung and Russell, noticed that the graph didn't just throw up a random scattering of points, but showed points clustered in distinct areas. These distinct areas turned out to correspond to different stages of a star's life cycle.

This diagram ended up being really important for studying how stars evolve (p.85-88) and became known as the **Hertzsprung-Russell (HR) diagram** (Figure 3).

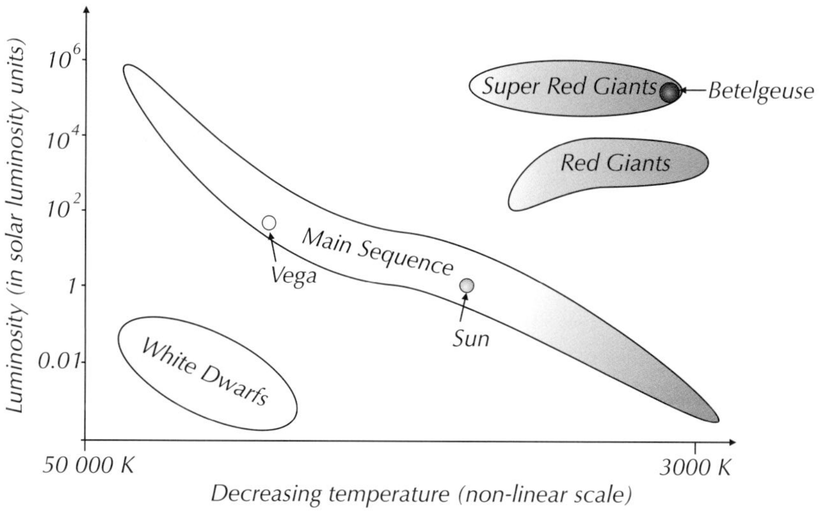

Figure 3: *The Hertzsprung-Russell diagram with the positions of the Sun and the stars Vega and Betelgeuse labelled.*

Tip: A 'solar luminosity unit' is just the luminosity of the Sun. So, a star with a luminosity of 10^2 luminosity units will be 100 times more luminous than the Sun.

Tip: Be careful with the axes on an HR diagram. The horizontal axis goes from high temperatures to low temperatures from left to right.

Distinct areas of the HR diagram

The distinct areas in which stars fall on the HR diagram correspond to four main types of stars:

- There is a long, diagonal band along the middle of the graph, where luminosity increases with temperature. This corresponds to main sequence stars.
- There is a section with high luminosity (10^2-10^4), but relatively low surface temperature. From Wien's displacement law, we can tell these stars should appear reddish. And from Stefan's law we know they must be very large to produce the same luminosity as much hotter main sequence stars. Therefore, this section must correspond to red giants.
- There is another section, at similar temperatures to the red giants, but even higher luminosity ($\sim 10^6$). These stars must be even bigger than red giants, and so this section corresponds to super red giants.
- There is a fourth section with very high temperatures, but low luminosity. From Stefan's law, we know they must be small to be so hot while only being as luminous as low temperature main sequence stars. And from Wien's displacement law we can tell these stars should appear white. So this section corresponds to white dwarfs.

Tip: In several billion years, the Sun will become a red giant (see pages 85-86). It will expand to be around 20% larger than the Earth's current orbit and shine 3000 times brighter than it does now.

Practice Questions — Application

Q1 The diagram below shows two stars, A and B, marked on a luminosity-temperature graph.

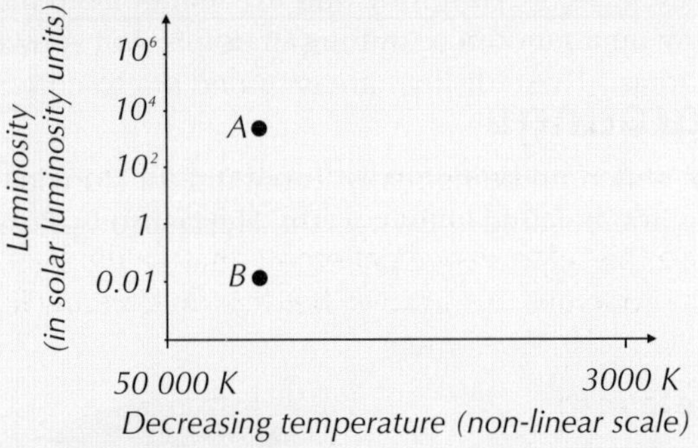

a) Which star you would expect to be bigger, and why?

b) What type of star is Star A?

c) What type of star is Star B?

d) Which star is further along in its evolution sequence?

Q2 The star Rigel, in the constellation Orion, is measured to have a surface temperature of 12 000 K. The Sun has a surface temperature of 5800 K.

a) Suggest how Rigel's peak wavelength will differ to that of the Sun.

b) The luminosity of Rigel is measured as 4.5×10^{31} W. Calculate the radius of Rigel.

Practice Questions — Fact Recall

Q1 Which would you expect to be hotter, a white star or a red star?

Q2 Write down Stefan's law. Define all the symbols you use.

Q3 Sketch the HR diagram. Mark on your diagram the regions corresponding to the main sequence, red giants, super red giants and white dwarfs. Make sure you label both your axes.

5. Stellar Spectra

One way astronomers analyse stars is by splitting the light received from them using a diffraction grating. By comparing the spectrum produced to the emission spectra of elements, they can work out what a star is made of.

Learning Objectives:

- Understand how a transmission diffraction grating can be used to determine the wavelength of light (PAG5).
- Know and understand that $d\sin\theta = n\lambda$ is the condition for maxima in an interference pattern produced by a diffraction grating with spacing d.
- Know that electrons in isolated gas atoms orbit the nucleus in distinct energy levels.
- Know that electron energy levels are given negative values by convention.
- Be able to use the equations $\Delta E = hf$ and $\Delta E = \frac{hc}{\lambda}$ to calculate the energy of emitted photons.
- Understand how the emission of photons by electrons moving between energy levels leads to the creation of emission spectra.
- Understand what is meant by a continuous spectrum and an absorption line spectrum.
- Understand how the spectral lines of different atoms can be used to identify elements within stars.

Specification Reference 5.5.2

Diffraction gratings

PRACTICAL ACTIVITY GROUP 5

You can pass light through a **transmission diffraction grating** to form an interference pattern. Since different wavelengths of light diffract differently, you can use this interference pattern to analyse the light, and determine some of its properties. As always, before you start any investigation, make sure you carry out a risk assessment.

Determining wavelength

If you shine monochromatic light (light with a single wavelength or frequency) through a diffraction grating, you'll get a pattern of bright lines (maxima) on a dark background. This is a result of the light interfering with itself constructively and destructively (you met interference in year 1 of A-level).

The line of maximum brightness at the centre is called the zero order line, the next lines on each side are called first order lines, and so on.

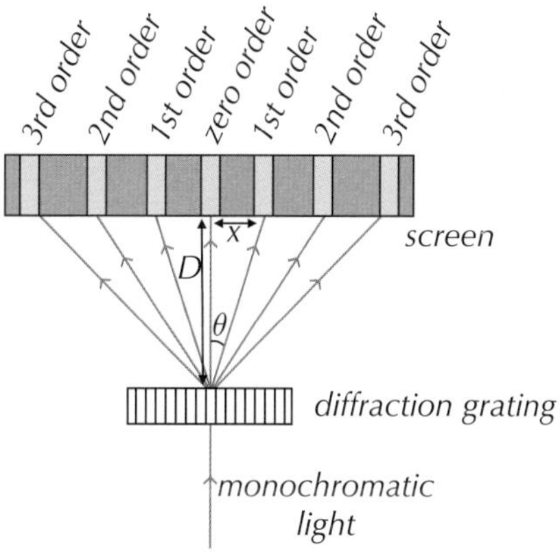

Figure 1: *A diagram of light passing through a diffraction grating and forming an interference pattern.*

Using the fringe width (the distance between the maxima), x, and the distance to the screen, D, the angle the first order line makes with the zero order line can be calculated using the small angle approximation:

$$\theta \approx \tan\theta = \frac{x}{D}$$

- $\boldsymbol{\theta}$ = *angle between the zero and first order fringes in radians*
- $\boldsymbol{x}$ = *fringe width in m*
- $\boldsymbol{D}$ = *distance to the screen in m*

If you know the slit separation, d, the order of the maximum you're observing, n, and the angle between this maximum and the incident light, θ, you can find the wavelength of the incident light:

$$d\sin\theta = n\lambda$$

- $\boldsymbol{d}$ = *slit spacing in m*
- $\boldsymbol{\theta}$ = *angle between the zero and nth order fringes in radians*
- $\boldsymbol{\lambda}$ = *wavelength of incident light in m*
- $\boldsymbol{n}$ = *number of fringe order*

Example — Maths Skills

A laser beam is passed through a diffraction grating with grating spacing of 2.0 × 10⁻⁵ m. The interference pattern it produces is displayed on a screen 1.6 m away. The first order fringe is located 0.05 m away from the zero order fringe. Calculate the wavelength of the laser light.

First, calculate θ: $\tan\theta = \frac{0.05}{1.6} \Rightarrow \theta \approx \frac{0.05}{1.6} = 0.03125$

Then find λ, when $n = 1$: $\lambda = d\sin\theta = 2.0 \times 10^{-5} \times \sin(0.03125)$

$= 6.24898... \times 10^{-7}$ m $= 6.2 \times 10^{-7}$ m (to 2 s.f.)

White light and diffraction gratings

White light is a continuous spectrum of all the wavelengths of visible light. If you pass white light through a diffraction grating, it will be split up, as the different wavelengths within the white light are diffracted by different amounts.

Each order in the pattern becomes a spectrum, with red on the outside and violet on the inside, as shown in Figure 2. The zero order maximum stays white because all the wavelengths produce a maximum when $\theta = 0$.

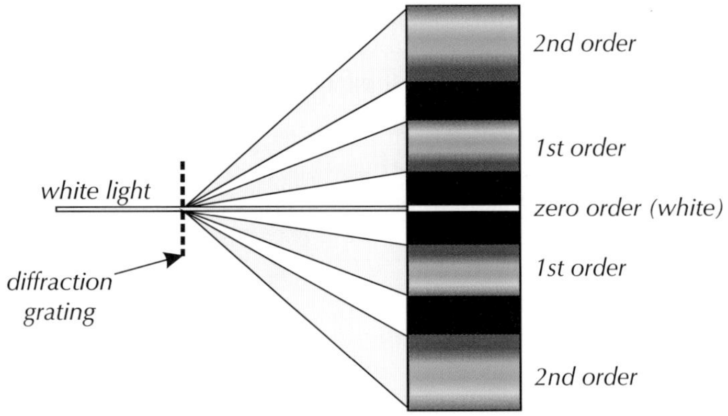

Figure 2: *White light being split into spectra as it passes through a diffraction grating.*

Electron energy levels and photons

Tip: When we say 'isolated gas atom', we mean an atom of a gas that is not interacting in any way with the other atoms of the gas.

Electrons in an isolated gas atom can only exist in certain well-defined energy levels. Each level is given a number, with $n = 1$ representing the ground state (i.e. the lowest energy level).

When energy is transferred to a substance (e.g. by heating), electrons in its atoms can be 'excited' — they move to higher energy levels. Electrons can move back down energy levels by emitting a photon. Since these transitions are between discrete energy levels, the energy of each photon emitted can only take a certain allowed value.

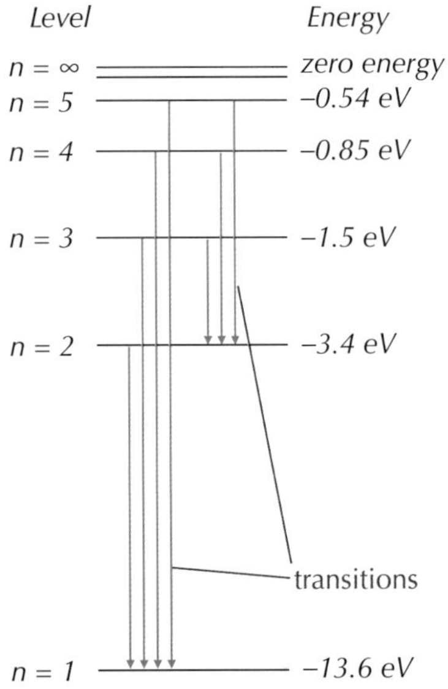

Figure 3: *The electron energy levels in atomic hydrogen.*

Tip: Remember, an electron volt, eV, is equal to 1.60×10^{-19} J.

Figure 3 shows the energy levels for atomic hydrogen. The energies are all negative because of how 'zero energy' is defined. The energy carried by each photon emitted is equal to the difference in energies between the two levels. For example, a photon produced by an electron moving from the $n = 3$ to the $n = 2$ level would have energy equal to $\Delta E = -1.5 - (-3.4) = 1.9$ eV.

Electrons can also move up energy levels (be excited) by absorbing photons. For an electron to absorb a photon and become excited, the photon must be carrying an amount of energy that exactly matches the difference between the electron's current energy level and a higher energy level.

You can relate the energy change of an electron in an atom to the frequency of the photon it absorbs or emits using the following equation:

$$\Delta E = hf$$

ΔE = energy carried by photon (or change in electron energy) in J

f = frequency of the photon in Hz

h = Planck's constant in Js $= 6.63 \times 10^{-34}$ Js

You can also write this in terms of wavelength:

$$\Delta E = \frac{hc}{\lambda}$$

c = speed of light in ms^{-1} $= 3 \times 10^{8}\ ms^{-1}$

λ = wavelength of the photon in m

Tip: This uses the fact that $c = f\lambda$.

Examples — Maths Skills

An electron in the ground state of a hydrogen atom (E = –13.6 eV) absorbs a photon and is excited to a higher energy level with an energy of –0.850 eV. Calculate the frequency of the absorbed photon.

First, calculate ΔE, and convert it to J.

$\Delta E = (-0.850 - (-13.6)) = 12.75\text{ eV} = (12.75 \times 1.6 \times 10^{-19}) = 2.04 \times 10^{-18}\text{ J}$

Rearranging the equation for frequency:

$$f = \frac{\Delta E}{h} = \frac{2.04 \times 10^{-18}}{6.63 \times 10^{-34}} = 3.07692... \times 10^{15}\text{ Hz} = 3.08 \times 10^{15}\text{ Hz (to 3 s.f.)}$$

The electron emits a photon, and drops from the –0.850 eV energy level to an energy level with energy –3.40 eV. Calculate the wavelength of the photon emitted in this transition.

$\Delta E = (-0.850 - (-3.40)) = 2.55\text{ eV} = (2.55 \times 1.6 \times 10^{-19}) = 4.08 \times 10^{-19}\text{ J}$

Rearranging the energy equation for wavelength:

$$\lambda = \frac{hc}{\Delta E} = \frac{6.63 \times 10^{-34} \times 3 \times 10^{8}}{4.08 \times 10^{-19}} = 4.875 \times 10^{-7}\text{ m} = 4.88 \times 10^{-7}\text{ m (to 3 s.f.)}$$

Tip: You'll usually be given electron energy level values in eV. Make sure that you convert them to J before you try and use any energy values in these equations.

Emission line spectra

If you heat a gas to a high temperature, many of its electrons move to higher energy levels — the gas is said to be 'excited'. As they fall back to the ground state, these electrons emit energy as photons. If you split the light from a hot gas with a diffraction grating, you get an **emission line spectrum** — see Figure 4.

Tip: You can assume that all the atoms in these gases are isolated.

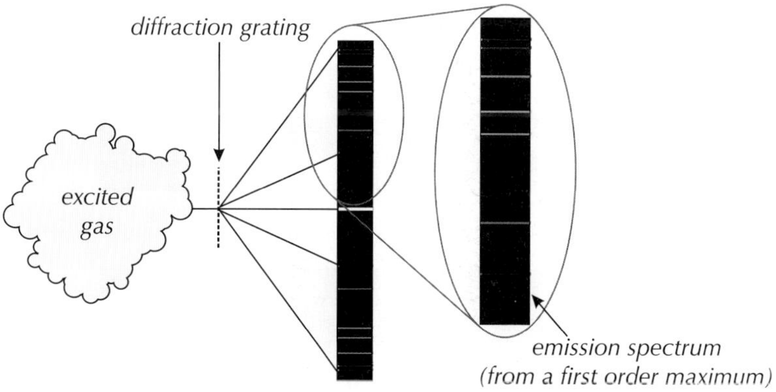

Figure 4: *A line emission spectrum, produced by passing light emitted by a hot gas through a diffraction grating.*

Each line on the spectrum corresponds to a particular wavelength of light emitted by electrons in the gas as they drop down energy levels. Since only certain photon energies are allowed, you only see the corresponding wavelengths. You can calculate the wavelength, λ, of each line in a line emission spectrum using $d\sin\theta = n\lambda$ (see page 93), if you know which order maximum the spectrum is.

Different atoms have different electron energy levels and so different sets of emission spectra. This means you can identify a gas from its emission spectrum.

Absorption line spectra

Tip: Remember, the change in energy of an electron that absorbs a photon is the same change as if it had emitted a photon of the same frequency, but in reverse.

Tip: You can use $d\sin\theta = n\lambda$ to find the wavelength of absorption lines too.

You get an **absorption line spectrum** when light with a continuous spectrum of energy (white light) passes through a cool gas:

- At low temperatures, most of the electrons in the gas atoms will be in their ground states.
- Photons of the correct wavelength are absorbed by the electrons to excite them to higher energy levels.
- These wavelengths are missing from the continuous spectrum when it comes out on the other side.
- You see a continuous spectrum with black lines in it corresponding to the absorbed wavelengths — see Figure 5.

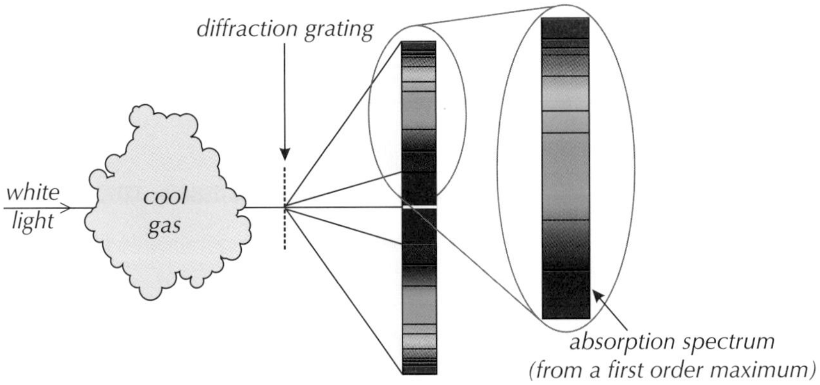

Figure 5: An absorption spectrum, produced by passing white light through a cold gas and then through a diffraction grating.

If you compare the absorption and emission spectra of a particular gas, the black lines in the absorption spectrum match up to the bright lines in the emission spectrum — as seen in Figure 6.

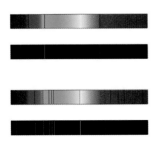

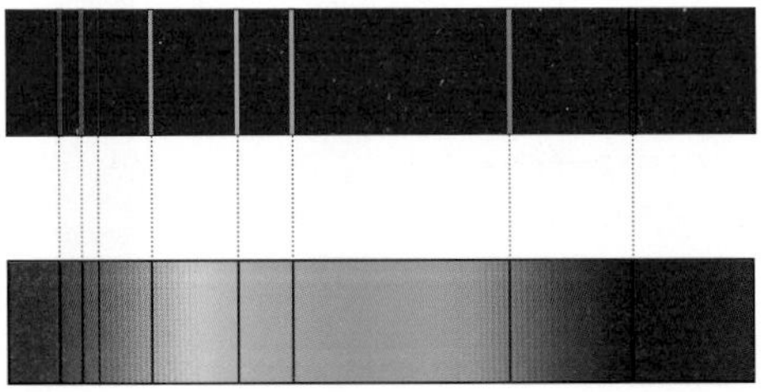

Figure 6: An emission spectrum and absorption spectrum of the same substance. The lines in the emission spectrum correspond to the missing parts of the full spectrum in the absorption spectrum.

Figure 7: The absorption and emission spectra of hydrogen (top) and helium (bottom).

You get absorption lines in the spectra of light from stars. Stars can be assumed to emit radiation in a continuous spectrum. This radiation has to pass through a large amount of gas at the surface of the star (the star's 'atmosphere') before travelling to Earth. This gas absorbs particular wavelengths of light depending on the elements it consists of.

Comparing the absorption spectra of stars to sets of emission spectra from the lab therefore allows you identify elements within a star. The most common element in most stars is hydrogen, so the spectral lines for hydrogen are usually the clearest. This makes these lines the easiest to identify and measure.

Practice Questions — Application

Q1 The diagram below shows three line emission spectra produced by elements in a lab, and an absorption spectrum of a star.

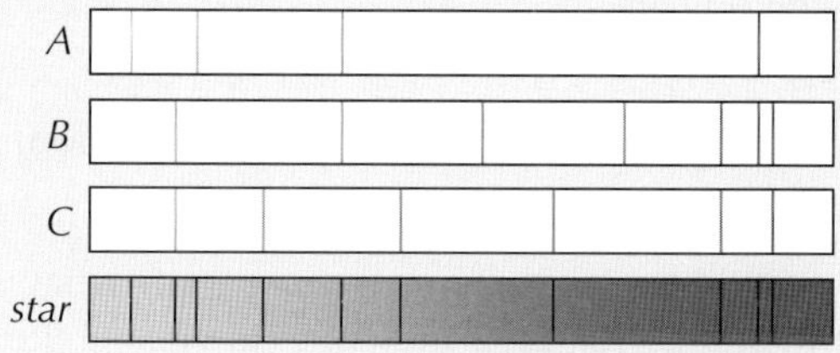

Which element, A, B or C, is not present in the star?

Q2 An electron in a hydrogen atom emits a photon as it drops from the –3.4 eV energy level to the –13.6 eV level. Calculate the frequency of the emitted photon.

Q3 An electron in an atom absorbs a photon with a wavelength of 3.60×10^{-7} m, and moves to a –2.00 eV energy level. Find the energy value of the original energy level of the electron.

Q4 Laser light with wavelength of 4.7×10^{-7} m is passed through a diffraction grating. It produces an interference pattern on a screen. The third order fringe of the diffraction pattern is at an angle of 0.020 radians to the incident light. Calculate the grating spacing of the diffraction grating.

Practice Questions — Fact Recall

Q1 What is the zero order maximum of an interference pattern produced by a diffraction grating?

Q2 Why are spectra formed when white light passes through a diffraction grating?

Q3 What is emitted by an electron when it moves from a higher energy level to a lower energy level? How does its energy relate to the energy change of the electron?

Q4 Why do absorption lines only appear at particular wavelengths in the spectra of stars?

6. The Big Bang Theory

Learning Objectives:

- Know that the cosmological principle states that the universe is homogeneous and isotropic, and that the laws of physics are universal.
- Understand what is meant by the Doppler effect.
- Understand the effects of Doppler shift on electromagnetic radiation.
- Be able to use the Doppler equation $\frac{\Delta\lambda}{\lambda} \approx \frac{\Delta f}{f} \approx \frac{v}{c}$ to quantify the effect of Doppler shift.
- Understand how galactic red shift supports the idea of an expanding universe.
- Understand Hubble's law, $v \approx H_0 d$, for receding galaxies.
- Know that the Hubble constant, H_0, can be measured in both $\text{kms}^{-1}\text{Mpc}^{-1}$ and s^{-1}.
- Know the Big Bang theory, and understand that the Big Bang gave rise to the expansion of space-time.
- Understand how the existence of microwave background radiation at a temperature of 2.7 K provides experimental evidence for the Big Bang theory.

Specification Reference 5.5.3

While astronomers examine space as it is now, cosmologists try to use all the evidence we've got to discover more about the origin, evolution and fate of the universe. And right now, the best idea they've got is the Big Bang theory.

The cosmological principle

It's easy to imagine that the Earth is at the centre of the universe, or that there's something really special about it. Earth is special to us because we live here — but on a universal scale, it's just like any other lump of rock.

The idea that no part of the universe is any more special than any other is summarised by the **cosmological principle**.

The Cosmological Principle:
The laws of physics are universal (they're the same everywhere), and on a large scale the universe is:
- **homogeneous** (every part is the same as every other part) and
- **isotropic** (everything looks the same in every direction)

This is a powerful idea — it means we can apply what we know about physics on Earth and in our Solar System to the rest of the universe.

The Doppler effect

You'll have experienced the **Doppler effect** loads of times with sound waves. Imagine a police car driving past you. As it moves towards you its siren sounds higher-pitched, but as it moves away, its pitch is lower. This change in frequency and wavelength is called the Doppler effect.

The frequency and the wavelength change because the waves bunch together in front of the source and stretch out behind it. The amount of stretching or bunching together depends on the speed of the source.

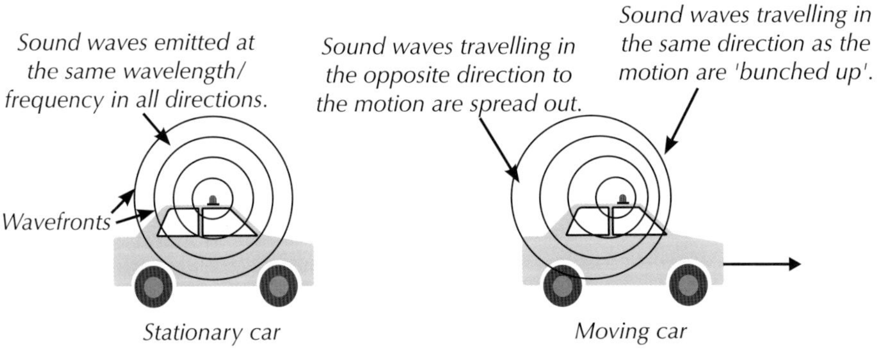

Sound waves emitted at the same wavelength/ frequency in all directions.
Sound waves travelling in the opposite direction to the motion are spread out.
Sound waves travelling in the same direction as the motion are 'bunched up'.
Wavefronts
Stationary car
Moving car

Figure 1: *The Doppler effect of sound waves from a moving police car siren.*

Doppler shift

The Doppler effect happens with all waves, including electromagnetic radiation. When a light source moves away from us, the wavelength of the light reaching us becomes longer and the frequencies become lower. This shifts the light that we receive towards the red end of the electromagnetic spectrum and is called a **red shift**.

The light that we receive from a star moving away from us is redder than the actual light emitted by the star.

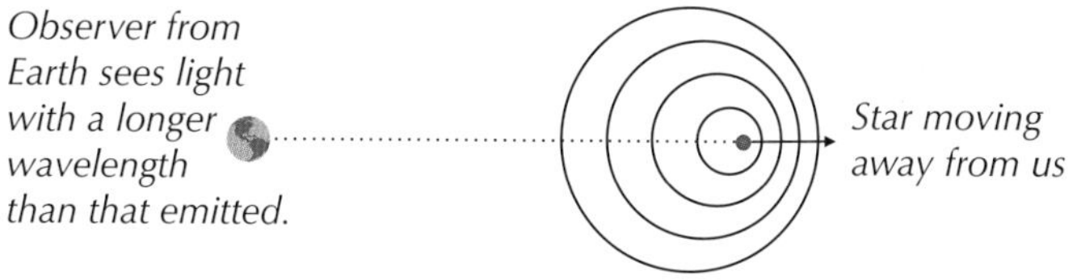
Observer from Earth sees light with a longer wavelength than that emitted.
Star moving away from us

Figure 2: The red shift of light emitted by a star moving away from us.

When a light source moves towards us, the opposite happens. The light from the star undergoes a **blue shift** and looks bluer than it actually is.

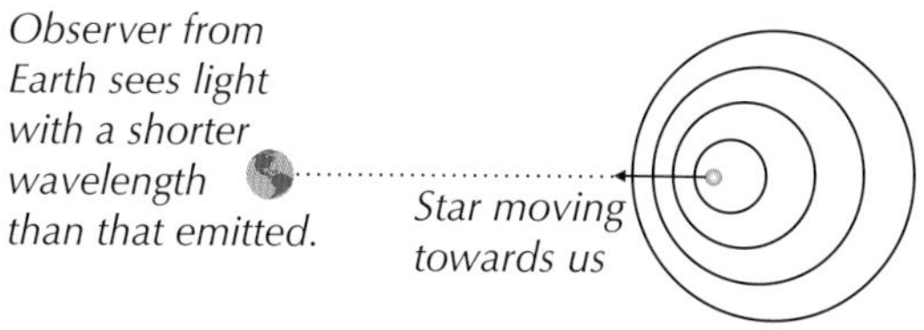
Observer from Earth sees light with a shorter wavelength than that emitted.
Star moving towards us

Figure 3: The blue shift of light emitted by a star moving towards us.

The amount of Doppler shift (red or blue) depends on how fast the star is moving relative to us (either away from or towards us). The higher the speed, the more the waves are shifted.

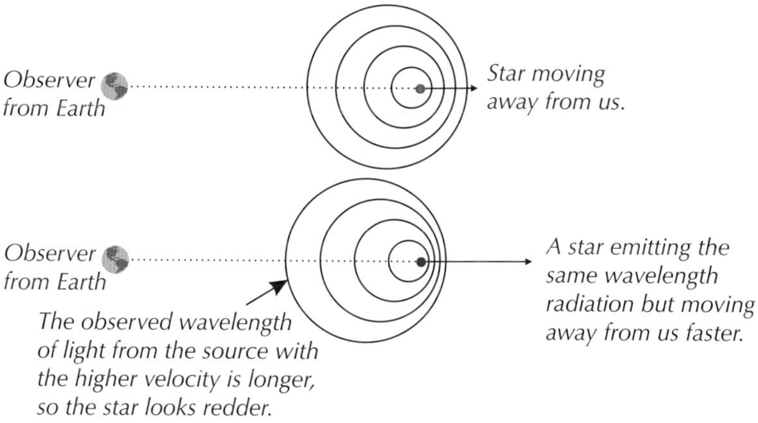
Observer from Earth
Star moving away from us.
Observer from Earth
A star emitting the same wavelength radiation but moving away from us faster.
The observed wavelength of light from the source with the higher velocity is longer, so the star looks redder.

Figure 4: Red (or blue) shift increases with the velocity of the source.

You can quantify the amount of Doppler shift using the equation:

$$\frac{\Delta\lambda}{\lambda} \approx \frac{\Delta f}{f} \approx \frac{v}{c}$$

Where:

- $\Delta\lambda$ is the difference between the observed and emitted wavelengths, and λ is the emitted wavelength,
- Δf is the difference between the observed and emitted frequencies, and f is the emitted frequency,
- v is the speed of the source relative to the observer and c is the speed of light.

Tip: Remember, red light is the lowest frequency of light in the visible spectrum.

Tip: It's important to realise that the frequency of the source doesn't change, just the frequency of the radiation reaching us.

Tip: Blue light is at the higher frequency end of the visible spectrum. In blue shift, light gets shunted towards (or beyond) the blue end of the visible spectrum.

Tip: You can use any units for v and c in this equation, as long as they are the same.

Tip: This equation only works when the speed of the source is much less than the speed of light.

Examples — Maths Skills

A star emitting light with a peak wavelength of 617 nm is moving away from the Earth. When it is detected in the lab, this wavelength of light is Doppler shifted by 5 nm. Calculate the speed at which the star is receding from Earth.

$\frac{v}{c} = \frac{\Delta\lambda}{\lambda}$ so $v = \frac{\Delta\lambda}{\lambda} \times c$

$v = \frac{5}{617} \times (3.00 \times 10^{8}) = 2.431... \times 10^{6}\ \text{ms}^{-1}$
$= 2.43 \times 10^{6}\ \text{ms}^{-1}$ (to 3 s.f.)

The frequency of this wavelength of light is 4.86 × 10¹⁴ Hz. Calculate the size of the Doppler shift in light of this frequency from the star.

$\frac{\Delta f}{f} = \frac{\Delta\lambda}{\lambda}$ so $\Delta f = \frac{\Delta\lambda}{\lambda} \times f$

$\Delta f = \frac{5}{617} \times (4.86 \times 10^{14}) = 3.938... \times 10^{12}$ Hz
$= 3.94 \times 10^{12}$ Hz (to 3 s.f.)

The expansion of the universe

Until the early 20th century, cosmologists believed that the Universe was infinite in both space and time (that is, it had always existed) and static. This seemed the only way it could be stable. This changed when Edwin Hubble realised that the spectra from galaxies (apart from a few very close ones) all show red shift — so they're all moving away from us. The amount of galactic red shift gives the recessional velocity — how fast the galaxy is moving away.

Tip: Some nearby galaxies are moving towards us due to gravitational attraction. The light from these galaxies shows blue shift.

Plotting recessional velocity against the distance of the object from the Earth shows that they're proportional — i.e. the speed that galaxies move away from us depends on how far away they are. This suggests that the universe is expanding, and gives rise to Hubble's law:

$$v \approx H_0 d$$

v = recessional velocity in kms^{-1}

H_0 = Hubble constant in $\text{kms}^{-1}\text{Mpc}^{-1}$

d = distance in Mpc

Since distance is very difficult to measure, astronomers used to disagree greatly on the value of H_0, with measurements ranging from 50 to 100 $\text{kms}^{-1}\text{Mpc}^{-1}$. It's now generally accepted that H_0 lies between 65 and 80 $\text{kms}^{-1}\text{Mpc}^{-1}$ and most agree it's around the low to mid 70s. You'll be given a value to use in the exam. The SI unit for H_0 is s^{-1}. To get H_0 in SI units, you need v in ms^{-1} and d in m.

Tip: To convert a value in $\text{kms}^{-1}\text{Mpc}^{-1}$ to s^{-1}, first multiply it by 1000 to get it in units of $\text{ms}^{-1}\text{Mpc}^{-1}$. Then divide it by the number of metres in an Mpc, 3.1×10^{22}. This will give you units of $\text{ms}^{-1}\text{m}^{-1}$, which is the same as s^{-1}.

Example — Maths Skills

The Perseus Cluster is located at a distance of around 2.28 × 10²⁴ m from Earth. Assuming that the Hubble constant is equal to 71.9 kms⁻¹Mpc⁻¹, estimate the recessional velocity of the Perseus Cluster.

First convert H_0 to s^{-1}:

$H_0 = 71.9 \times 1000 = 71\,900\ \text{ms}^{-1}\text{Mpc}^{-1}$
$= 71\,900 \div 3.1 \times 10^{22} = 2.319... \times 10^{-18}\ \text{s}^{-1}$

Then plug the values you're given into Hubble's Law:

$v \approx H_0 d$
$= 2.319... \times 10^{-18} \times 2.28 \times 10^{24}$
$= 5.2881... \times 10^{6}\ \text{ms}^{-1} = 5.29 \times 10^{6}\ \text{ms}^{-1}$ (to 3 s.f.)

The red shift of light from other galaxies isn't caused by them flying away from us through space — they are moving away from us because space itself is expanding. Since the universe is expanding uniformly away from us it seems as though we're at the centre of the universe, but this is an illusion. You would observe the same thing at any point in the universe.

A good way to visualise this is to think of the universe as the surface of a balloon with lots of dots on it, representing galaxies. As you blow up the balloon, the space between all the galaxies (dots) gets bigger. Each galaxy sees all the other galaxies moving away from it, even though it's not at the centre of the motion.

The Big Bang theory

So the universe is expanding and cooling down (because it's a closed system). So further back in time it must have been smaller, denser and hotter. If you trace time back far enough, you get a Big Bang:

The Big Bang theory:
The universe started off very hot and very dense (perhaps as an infinitely hot, infinitely dense singularity) and has been expanding ever since.

Tip: A closed system just means that energy and momentum are conserved. If there's only so much energy in the universe, as it expands, it must also be cooling.

According to the Big Bang theory, before the Big Bang, there was no space or time — space-time began with the Big Bang, (when time = 0 and the radius of the universe = 0) and has been expanding ever since.

There is a lot of evidence to support the Big Bang theory for the creation of the universe, such as the red shift of galaxies, and the existence of the cosmic microwave background radiation.

Tip: If the universe began at a specific point in time, i.e. with the Big Bang, then it has a finite age — see p.103.

Cosmic microwave background radiation

The Big Bang model predicts that loads of gamma radiation was produced in the very early universe. This radiation should still be observed today (it hasn't had anywhere else to go). Because the universe has expanded, the wavelengths of this cosmic background radiation have been stretched and are now in the microwave region.

Cosmic microwave background radiation (CMBR) was discovered accidentally by Penzias and Wilson in the 1960s. In the late 1980s a satellite called the Cosmic Background Explorer (COBE) was sent up to have a detailed look at the radiation. It found a continuous spectrum corresponding to a temperature of about 2.7 K. The radiation is largely the same everywhere (homogeneous) and in all directions (isotropic), in line with the cosmological principle.

Figure 5: *A WMAP image of the CMBR. You can see that, except for the red horizontal band, which shows Milky Way emissions, the CMBR is mostly homogeneous and isotropic, but shows small variations.*

There are very tiny fluctuations in temperature, which were at the limit of COBE's detection. These are due to tiny energy-density variations in the early universe, and are needed for the initial 'seeding' of a star or galaxy formation.

The background radiation also shows a Doppler shift, indicating the Earth's motion through space. It turns out that the Milky Way is rushing towards an unknown mass (the Great Attractor) at over a million miles an hour.

Practice Questions — Application

Q1 The Antennae galaxies are a pair of colliding galaxies at a distance of 1.4×10^7 pc from Earth. Calculate the recessional velocity of the Antennae galaxies using $H_0 = 70$ kms^{-1} Mpc^{-1}.
Give your answer in ms^{-1}.

Q2 The Whirlpool galaxy is receding from us at a speed of 463 kms^{-1}.
A frequency of radiation emitted by atomic hydrogen has a wavelength of 0.211 m in the lab. The same radiation, when emitted from the Whirlpool galaxy, is observed to have been red shifted when it is observed from Earth. Calculate the change in the wavelength of this radiation due to red shift.

Practice Questions — Fact Recall

Q1 State the cosmological principle.

Q2 Explain how the Doppler effect makes police car sirens sound higher pitched as they travel towards us.

Q3 What is red shift? What is blue shift?

Q4 Write down Hubble's law. Define all symbols used, and give the units of each quantity.

Q5 What is the Big Bang theory?

Q6 Explain what cosmic microwave background radiation is.

7. The Evolution of the Universe

So we have a pretty good idea of how the universe started. But how we got from there to our current state is another question. And there are still a lot of things we don't have a concrete answer for.

Learning Objectives:

- Be able to estimate the age of the universe from Hubble's law, using $t \approx H_0^{-1}$.
- Know how the universe has evolved from the Big Bang to its present state.
- Know some current ideas about cosmology, including dark matter and dark energy.
- Know that the universe is made up of dark energy, dark matter and a small percentage of ordinary matter.

Specification Reference 5.5.3

The age and size of the universe

If the universe has been expanding at the same rate for its whole life, the age of the universe can be estimated by plugging Hubble's Law into velocity = distance ÷ time. This gives us the equation:

$$t \approx \frac{\text{distance}}{\text{velocity}} \approx H_0^{-1}$$

This is only an estimate, since the universe probably hasn't always been expanding at the same rate. And, unfortunately, since no one knows the exact value of H_0 we can only guess the universe's age.

Example — Maths Skills

If H_0 = 72 $\text{kms}^{-1}\text{Mpc}^{-1}$, calculate the age of the universe.

The Hubble constant is in $\text{kms}^{-1}\text{Mpc}^{-1}$, but you need it in s^{-1} to get the time in s.

Multiplying by 10^3 gives it in $\text{ms}^{-1}\text{Mpc}^{-1}$. Then, since 1 Mpc = 3.1×10^{22} m, dividing by 3.1×10^{22} gives it in $\text{ms}^{-1}\text{m}^{-1}$, which is just s^{-1}.

$72\ \text{kms}^{-1}\text{Mpc}^{-1} = (72 \times 10^3) \div (3.1 \times 10^{22})\ \text{s}^{-1} = 2.3... \times 10^{-18}\ \text{s}^{-1}$

$$\text{So } t \approx H_0^{-1} \approx \frac{1}{2.3... \times 10^{-18}} = 4.305... \times 10^{17}\ \text{s}$$
$$= 4.31 \times 10^{17}\ \text{s}$$
$$= 14 \text{ billion years (to 2 s.f.).}$$

The absolute size of the universe is unknown but there is a limit on the size of the observable universe. This is simply a sphere (with the Earth at its centre) with a radius equal to the maximum distance that light can travel during the universe's existence. So if H_0 = 72 $\text{kms}^{-1}\text{Mpc}^{-1}$ then this sphere will have a radius of 14 billion light years. However, when scientists take into account the expansion of the universe (p.100), they find that it is likely to be more like 46-47 billion light years.

Tip: The universe is bigger than this sphere, but it isn't observable to us, because we can only see light that has had time to travel to us.

The history of the universe

Before 10^{-4} seconds after the Big Bang, this is mainly guesswork. There are plenty of theories around, but not much experimental evidence to back them up. The general consensus at the moment goes something like this:

- **Big Bang to 10^{-43} seconds** — Well, it's anybody's guess, really. At this sort of size and energy, even general relativity stops working properly. This is the "infinitely hot, infinitely small, infinitely dense" bit.
- **From 10^{-43} seconds to 10^{-4} seconds after the Big Bang** — At the start of this period, there's no distinction between different types of force — there's just one grand unified force. Then the universe expands and cools, and the unified force splits into gravity, strong nuclear, weak nuclear and electromagnetic forces. Many cosmologists believe the universe went through a rapid period of expansion called inflation at about 10^{-34} s.

During this period, the universe is a sea of quarks, antiquarks, leptons and photons. The quarks aren't bound up in particles like protons and neutrons, because there's too much energy around. At some point, matter-antimatter symmetry gets broken, so slightly more matter is made than antimatter. Nobody knows exactly how or when this happened, but most cosmologists like to put it as early as possible in the history of the universe.

Once we reach 10^{-4} s after the Big Bang, we're a bit more confident with what went on.

- **10^{-4} seconds after the Big Bang** — This corresponds to a temperature of about 10^{12} K. The universe is cool enough for quarks to join up to form particles like protons and neutrons. They can never exist separately again. Matter and antimatter annihilate each other, leaving a small excess of matter and huge numbers of photons (resulting in the cosmic background radiation that we observe today).
- **About 100 seconds after the Big Bang** — Temperature has cooled to 10^{9} K. The universe is similar to the interior of a star. Protons are cool enough to fuse to form helium nuclei.
- **About 300 000 years after the Big Bang** — Temperature has cooled to about 3000 K. The universe is cool enough for electrons (that were produced in the first millisecond) to combine with helium and hydrogen nuclei to form atoms. The universe becomes transparent since there are no free charges for the photons to interact with. This process is called recombination.
- **About 14 billion years after the Big Bang (now)** — Temperature has cooled to about 2.7 K. Slight density fluctuations in the universe mean that, over time, clumps of matter have been condensed by gravity into galactic clusters, galaxies and individual stars.

This covers the basic milestones that cosmologists think the universe must have gone through in order to become the universe we see today. But a lot of it is still a mystery. We can't be fully certain about the timing or mechanism of any of this yet, as there are still parts of the universe whose behaviour we don't understand. Chief among these unknown parts of the universe are dark matter and dark energy.

Dark matter

HOW SCIENCE WORKS

In the 1930s, the Swiss astronomer Fritz Zwicky calculated the mass of a cluster of galaxies (the COMA cluster) based on the velocity of its outer galaxies and compared this figure to the mass of the cluster as estimated from its luminosity. The mass calculated from the velocity was much bigger, suggesting there was 'extra' mass in the cluster that couldn't be seen.

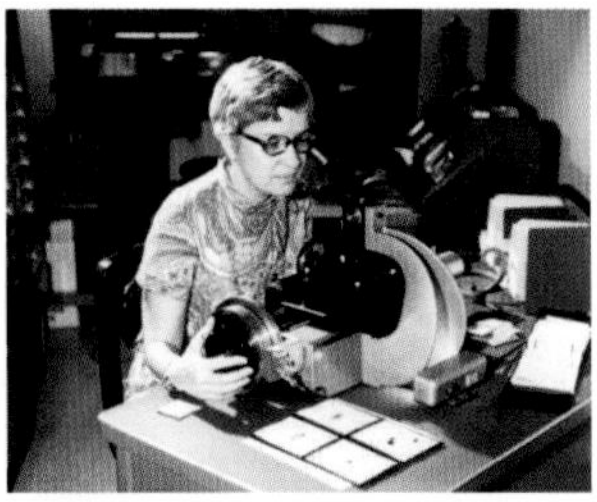

Figure 1: *The astronomer Vera Rubin, whose discovery of the 'galaxy rotation problem' revolutionised modern astronomy and cosmology.*

In the 1970s, Vera Rubin observed that stars at the edges of galaxies were moving faster than they should given the mass and distribution of stars in the galaxy. For Newton's laws to hold, there needed to be extra matter in the galaxies that hadn't been accounted for.

These observations suggest there is something extra in the universe, giving mass to galaxies, that we can't see.

This theoretical substance has been called 'dark matter'. Astrophysicists now estimate that there is about five times as much dark matter as ordinary matter in the universe, and that dark matter makes up about 25% of the universe in total.

What is dark matter?

Physicists don't know for sure what dark matter is yet. But there are a number of theories about what it could be, and experiments are constantly being run to test them.

One explanation is that dark matter is made up of MACHOs (Massive Compact Halo Objects). These are objects made of normal matter in a very dense form that don't give off light and so are hard to detect, e.g. black holes (p.88) and brown dwarfs (stars that aren't massive enough for nuclear fusion to take place). Astronomers looking for evidence of these kinds of objects have had some success, but it's unlikely that MACHOs made of normal matter account for all the dark matter in the universe, as this would require more protons and neutrons to exist than is compatible with our current understanding of the Big Bang.

Another idea is that dark matter is made of WIMPs (Weakly Interacting Massive Particles). These are exotic particles that don't interact with the electromagnetic force, but do interact with gravity. As yet, though, no particle like this has ever been detected, and WIMPs are currently purely theoretical.

There's also the possibility that dark matter doesn't really exist at all, and is an illusion caused by mistakes in other theories. But most scientists agree that it is there, even if we don't know what it is yet.

Figure 2: *Some equipment being used in an experiment that aims to detect signs of WIMP dark matter.*

Dark energy

HOW SCIENCE WORKS

Everything in the universe is attracted to everything else by gravity. This means the expansion of the universe should be slowing down. Historically, astronomers debated whether this would slow the expansion of the universe enough to cause it to contract back in on itself (in a so called 'Big Crunch'), or if the universe would go on expanding forever.

In the late 1990s, astronomers discovered something entirely unexpected. Rather than slowing down, the expansion of the universe appears to be accelerating. Astronomers are trying to explain this acceleration using dark energy — a type of energy that fills the whole of space.

There are various theories of what this dark energy is, but it's really hard to test them. So like dark matter, it's currently a mystery.

Based on current observations, dark energy makes up about 70% of the universe. As dark matter makes up another 25%, this means that only about 5% of the universe is made up of ordinary matter. Or to put it another way, we have very little idea what 95% of the universe is made up of.

Tip: There has been a lot of controversy surrounding dark energy in the astrophysics community — nobody really knows what it is, or if it's even there — but it is now generally accepted that it does exist.

Practice Question — Application

Q1 A student uses the equation $t \approx H_0^{-1}$ to calculate that the universe is approximately 13.7 billion years old.

a) Calculate the value of H_0 used to calculate this age of the universe. Give your answer in $\text{kms}^{-1}\text{Mpc}^{-1}$.

b) State the assumption that the student must have made to make this calculation.

Practice Questions — Fact Recall

Q1 How can we estimate the age of the universe from the Hubble constant?

Q2 Why can't we measure the absolute size of the universe?

Q3 State the temperature of the universe and describe one change happening in it at each of the following times after the Big Bang:

a) 10^{-4} s

b) 300 000 years

c) 14 billion years

Q4 State two observations which could be explained by the existence of dark matter.

Q5 What does the acronym MACHO stand for?

Q6 What does the acronym WIMP stand for?

Q7 What observations might dark energy explain?

Section Summary

Make sure you know...

- What is meant by the universe, a galaxy, a solar system, a planet, a planetary satellite and a comet.
- How to use distances measured in astronomical units (AU), parsecs (pc) and light years (ly).
- That stellar parallax is used to measure distances to nearby stars due to their apparent motion in the night sky against the distant background as the Earth orbits the Sun.
- How to use the equation $p = \frac{1}{d}$ to make calculations of parallax angle and distance.
- How stars are formed from interstellar dust and gas drawn together by gravity.
- That stars are kept stable by a balance between outwards pressure (from radiation and gas) and gravitational attraction.
- The evolutionary stages of a low-mass star like our Sun, including the terms red giant star, white dwarf star and planetary nebula.
- That a star may only form a white dwarf star if the mass of the star's core is less than the Chandrasekhar limit (approximately 1.4 solar masses).
- That a white dwarf star is prevented from collapsing in on itself by the electron degeneracy pressure within the star.
- The evolutionary stages of stars much more massive than the Sun, including the terms red super giant and supernova, and that the star will become either a neutron star or a black hole.
- That a neutron star is an incredibly dense, very small star which rotates very fast, and emits beams of radio waves.
- That a black hole is an infinitely dense point, with such a strong gravitational field that past a certain point in the field the escape velocity exceeds the speed of light.
- That Wien's displacement law states that $\lambda_{max} \propto \frac{1}{T}$, and how to use this to estimate the peak surface temperature of a star.
- That the luminosity of a star is a measure of the total energy it emits per second and is given by Stefan's law, $L = 4\pi r^2 \sigma T^4$, where σ is Stefan's constant.

cont...

- How to use Wien's displacement law and Stefan's law to estimate the radius of a star.
- That the Hertzsprung-Russell (HR) diagram is a luminosity-temperature plot of a range of stars.
- That there are distinct areas on the HR diagram which correspond to main sequence stars, red giant stars, super red giant stars and white dwarf stars.
- How a transmission diffraction grating can be used to determine the wavelength of light.
- That the fringes of the interference pattern produced by a diffraction grating are defined by the equation $d\sin\theta = n\lambda$.
- That electrons in isolated gas atoms exist in distinct energy levels.
- That electron energy levels are considered to have negative energy values.
- That electrons can emit and absorb photons with specific energies by moving between discrete energy levels and how this gives rise to line emission spectra and line absorption spectra.
- That the energy change in an atomic electron is related to the frequency and wavelength of the emitted/absorbed photon by the equations $\Delta E = hf$ and $\Delta E = \frac{hc}{\lambda}$.
- That the fact that different elements produce unique spectral lines in emission and absorption spectra allows us to use analysis of spectra to identify elements present in stars.
- That the cosmological principle states that the universe is homogeneous and isotropic, and that the laws of physics are the same everywhere in the universe.
- That the Doppler effect is a change in perceived frequency (and so wavelength) of a wave due to the relative motion between the source and the observer.
- That the Doppler effect of electromagnetic waves can result in red shift or blue shift.
- That the observed Doppler shift of an electromagnetic wave can be quantified using the Doppler equation $\frac{\Delta\lambda}{\lambda} \approx \frac{\Delta f}{f} \approx \frac{v}{c}$, where v is the speed of the source relative to the observer.
- That light from nearly all other galaxies is observed to have been red shifted, indicating that galaxies are moving away from each other, and that this supports the idea of an expanding universe.
- That Hubble's law states that the further away a galaxy is, the faster it is receding, and is expressed as $\mathbf{v} \approx H_0 d$, where H_0 is the Hubble constant.
- That the Hubble constant is a measure of the rate of expansion of the universe, and can be measured in either $\text{kms}^{-1}\text{Mpc}^{-1}$ or s^{-1}.
- That the Big Bang theory states that the universe started off very hot and very dense and has been expanding ever since.
- That the Big Bang gave rise to the expansion of space-time.
- That the cosmic microwave background radiation is a continuous spectrum of EM radiation corresponding to a temperature of 2.7 K that pervades throughout the universe and how this is evidence in support of the Big Bang theory.
- That the age of the universe can be estimated using the Hubble constant with the equation $t \approx H_0^{-1}$.
- How the universe has evolved from the Big Bang to its present state.
- That dark matter is a theoretical form of matter that cannot be seen directly, but which could explain inconsistencies between the velocities of galaxies in clusters, or of stars in galaxies, and their apparent masses.
- That dark energy is a theoretical type of energy that fills the universe, and drives the accelerated expansion of the universe.
- That the universe appears to be made up of dark energy, dark matter, and a small percentage of ordinary, observable matter.

Exam-style Questions

1 A star is reaching the end of the red giant stage of its life. The core of the star has a mass of 2.17×10^{30} kg. Given that the mass of the Sun is 1.99×10^{30} kg, what is the next stage in this star's life cycle?

A neutron star

B super red giant

C white dwarf

D black hole

(1 mark)

2 As a main sequence star evolves into a red giant star, it goes through four phases of nuclear fusion:

1. shell helium fusion
2. core hydrogen fusion
3. core helium fusion
4. shell hydrogen fusion

Which is the correct order in which these phases of fusion begin?

A 4, 2, 3, 1

B 3, 1, 2, 4

C 2, 3, 4, 1

D 2, 4, 3, 1

(1 mark)

3 An electron in a hydrogen atom is in the –13.6 eV energy level. It absorbs a photon with frequency 3.151×10^{15} Hz, and moves to a higher energy level. What is the energy value of this energy level?

A 13.1 eV

B –0.54 eV

C –26.6 eV

D 13.6 eV

(1 mark)

4 **Fig. 4.1** shows a set of axes for a Hertzsprung-Russell (HR) diagram.

Fig. 4.1

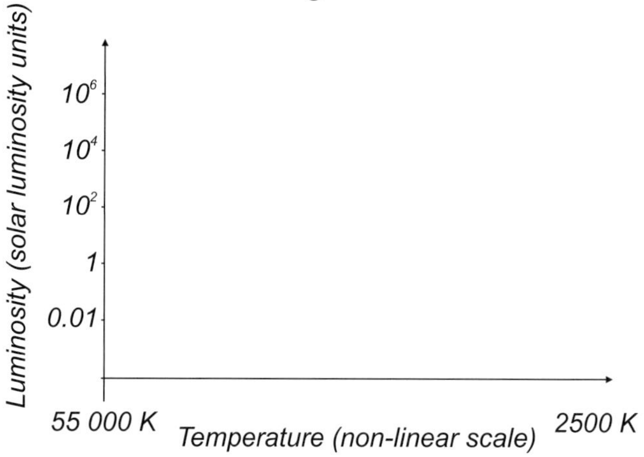

(a) Copy and complete the diagram, marking the main sequence stars, white dwarfs, red giants and super red giants.

(3 marks)

(b) Star X has a surface temperature of around 39 000 K and a luminosity of 200 000 solar luminosity units. Mark the position of Star X on your HR diagram.

(1 mark)

The mass of Star X is around 35 times that of the Sun.
It will eventually become a black hole.

(c) Explain how Star X will evolve from its red super giant phase to become a black hole.

(3 marks)

5 Mu Cephei is a red super giant star in the constellation Cepheus. Analysis of the radiation emitted by Mu Cephei gives the value of its peak wavelength as λ_{max} = 828.6 nm.

(a) Given that the Sun has a surface temperature of 5800 K and a peak wavelength of approximately 500.0 nm, estimate the surface temperature of Mu Cephei in kelvin.

(2 marks)

A scientist measures the luminosity of Mu Cephei as 1.083×10^{32} W.

(b) Estimate the radius of Mu Cephei in metres.

(2 marks)

(c) The scientist measures the angle of parallax of Mu Cephei as 5.5×10^{-4} arcseconds to the Earth.
Calculate the distance to Mu Cephei in light years.

(3 marks)

(d) Mu Cephei is near the end of its life. It will soon explode in a supernova.
Describe what occurs when a star explodes in a supernova.

(2 marks)

6 The static theory of the universe states that the universe is infinite in both space and time and it has always been that way. This theory is now widely discredited in favour of the Big Bang theory.

(a) Describe how the universe began according to the Big Bang theory.

(1 mark)

Hubble observed that the recessional velocity of distant objects was proportional to their distance from Earth. This observation led to Hubble's law:

$$v \approx H_0 d$$

(b) Explain how Hubble's law supports the Big Bang theory.

(1 mark)

(c) Assuming that H_0 = 65 $\text{kms}^{-1}\,\text{Mpc}^{-1}$, estimate the age of the universe in years using Hubble's law. (1 Mpc = 3.1×10^{22} m)

(3 marks)

(d) Red-shift data is only one piece of evidence supporting the Big Bang theory. Describe another piece of evidence that supports the Big Bang theory and explain how it supports the theory.

(2 marks)

7 A scientist is analysing the light from stars using diffraction gratings.

(a) The scientist passes the light produced by a heated sample of hydrogen gas through a diffraction grating with a slit separation of 5.4×10^{-6} m. She uses the second order fringe to produce an emission line spectrum of hydrogen. One of the lines from the emission spectrum makes an angle of 0.18 radians to the incident light. Calculate the wavelength of the light which produced this line.

(2 marks)

(b) Calculate the energy carried by a photon of light with this wavelength. Give your answer in eV.

(3 marks)

(c) The scientist passes light from the Sun through the same diffraction grating. When she analyses the spectrum produced by this, she notices dark lines in the spectrum that match the emission lines produced by hydrogen. Explain why this indicates that hydrogen is present in the Sun.

(3 marks)

(d) The scientist compares this spectrum from the Sun to one from a distant galaxy. Parts of these spectra are shown in **Fig. 7.1**. The dark line corresponding to the hydrogen line from part (a) in the galaxy's spectrum is displaced towards the red end of the spectrum by an amount equal to a change in wavelength of 0.019 nm. Using your answer to part (a), calculate the recessional velocity of the galaxy.

Fig. 7.1

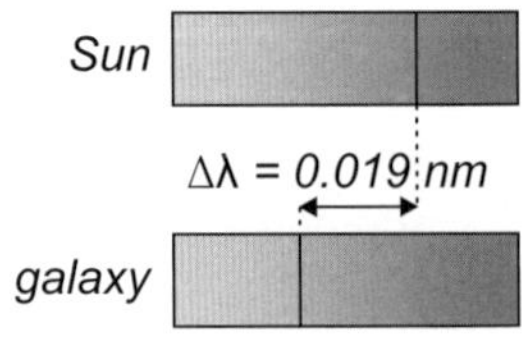

(2 marks)

Module 6 Section 1: Capacitors

1. Capacitors

Capacitors are devices that can store up charge and energy. The capacitance of a capacitor tells you how much charge it can hold. Almost all electrical products will contain one somewhere...

What is a capacitor?

Capacitors are electrical components that can store electrical charge. They are made up of two electrical conducting plates separated by a gap or a dielectric (an electrically insulating material). The circuit symbol is two parallel lines — see Figure 1.

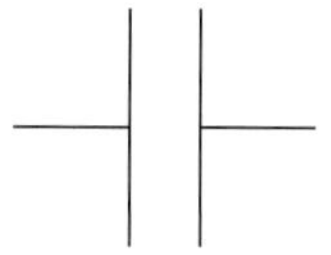

Figure 1: *Circuit symbol for a capacitor.*

When a capacitor is connected to a direct current (d.c.) power source, charge builds up on its plates — one plate becomes negatively charged and one becomes positively charged (there's more on this on the next page). The plates are separated by an electrical insulator (which could just be air), so no charge can move between them. This means that a potential difference builds up between the plates of the capacitor. This creates a uniform electric field (p.137) between the two plates — see Figure 2.

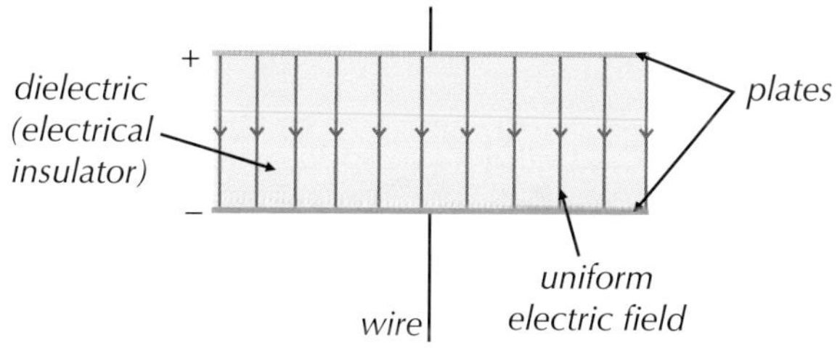

Figure 2: *The uniform electric field formed between the plates of a capacitor.*

The ability of a capacitor to store charge is measured by its capacitance.

> The **capacitance** of a capacitor is the amount of charge it is able to store per unit potential difference (p.d.) across it.

$$C = \frac{Q}{V}$$

C *= capacitance (in F)*

Q *= charge (in C)*

V *= potential difference (in V)*

Learning Objectives:

- Know what is meant by the term 'capacitance'.
- Be able to calculate capacitance using the equation $C = \frac{Q}{V}$.
- Know that the unit of capacitance is the farad, F.
- Understand how capacitors are charged and discharged, in terms of electron flow.
- Understand a p.d.-charge graph of a capacitor.
- Know that the area under a p.d.-charge graph gives the energy stored by a capacitor.
- Be able to calculate the energy stored by a capacitor using the equations: $W = \frac{1}{2}QV$, $W = \frac{1}{2}\frac{Q^2}{C}$ and $W = \frac{1}{2}V^2C$
- Know some uses of capacitors for storing energy.

Specification References 6.1.1 and 6.1.2

Tip: In the capacitance equation, Q is the charge stored on each capacitor plate — the charge is just negative on one plate and positive on the other.

Capacitance is measured in farads. 1 farad (F) is equal to 1 coulomb per volt (CV^{-1}). This is a huge amount of capacitance, so you'll usually see capacitances given in microfarads, μF (1×10^{-6} F), nanofarads, nF (1×10^{-9} F), or picofarads, pf (1×10^{-12} F).

Example — Maths Skills

A 100 μF capacitor is charged to a potential difference of 12 V. How much charge is stored by the capacitor?

$C = \frac{Q}{V}$ so by rearranging, $Q = C \times V = (100 \times 10^{-6}) \times 12 = 1.2 \times 10^{-3}$ C

Charging and discharging

Figure 3: *A disassembled capacitor, showing two sheets of aluminium foil with paper between them. The foil and paper are rolled up like a big Swiss roll.*

Tip: You can make your own simple capacitor using two pieces of aluminium separated by a piece of paper.

When a capacitor is connected to a d.c. power supply (e.g. a battery), a current flows in the circuit until the capacitor is fully charged, then stops.

The electrons flow from the negative terminal of the supply onto the plate connected to it, so a negative charge builds up on that plate.

At the same time, electrons flow from the other plate to the positive terminal of the supply, making that plate positive. These electrons are repelled by the negative charge on the negative plate and attracted to the positive terminal of the supply.

The same number of electrons are repelled from the positive plate as are built up on the negative plate. This means an equal but opposite charge builds up on each plate, causing the potential difference between the plates.

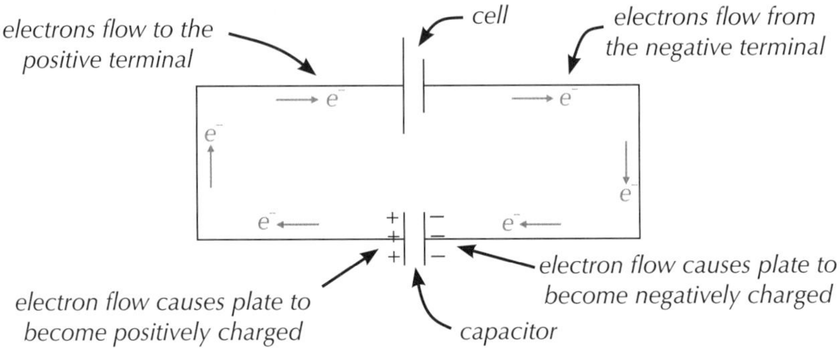

Figure 4: *Flow of electrons as a capacitor charges.*

Tip: After removing the power supply, you need to reconnect the circuit to discharge a capacitor. If the capacitor isn't connected in a full circuit it will hold its charge.

Initially the current through the circuit is high. But, as charge builds up on the plates, electrostatic repulsion makes it harder and harder for more electrons to be deposited. When the p.d. across the capacitor is equal to the p.d. across the supply, the current falls to zero. The capacitor is fully charged.

If the source of e.m.f. (in this case, a cell) is removed, and the circuit is reconnected without it, the capacitor will begin to discharge. The capacitor then becomes the circuit's source of e.m.f. A current flows around the circuit in the opposite direction to the charging current, as the electrons from the negatively charged plate drift towards the positively charged plate through the circuit.

Initially, the current is high, due to the strong forces between charges. There is a strong force of repulsion between the electrons on the negative plate, which pushes them away from the negative plate. There is also a strong force of attraction from the positive plate, pulling electrons through the circuit towards the positive plate. But as the electrons reach the positive plate, the potential difference decreases, and so the current decreases. The capacitor is completely discharged when the potential difference between the plates, and so the charge on each plate, becomes zero.

Energy stored by capacitors

Remember that when a capacitor charges, one plate becomes negatively charged while the other becomes positively charged. Like charges repel, so when each plate of a capacitor becomes charged, the charges on that plate are being forced together 'against their will'. This requires energy, which is supplied by the power source and stored as electric potential energy for as long as the charges are held. When the charges are released, the electric potential energy is released.

Deriving the energy-stored equations

You can find the energy stored by a capacitor by using the graph of potential difference against charge for the capacitor (see Figure 6).

The p.d. across a capacitor, of capacitance C, is directly proportional to the charge stored on it, so the graph is a straight line through the origin (since $C = \frac{Q}{V}$, page 111).

Consider a tiny increase in the charge on the plates during the charging process. The electric potential energy stored is the work done to move the extra charge onto the plates against the potential difference across the plates, given by $W = VQ$ (from year 1). Let the small charge being moved be q. The average p.d. over that step is v. So in that small step, the energy stored is $W = qv$, which is given by the area of the red rectangle in Figure 6. The area of the green trapezium in Figure 6 is the area under the graph over the charge difference q, and it is the same as the area of the red rectangle.

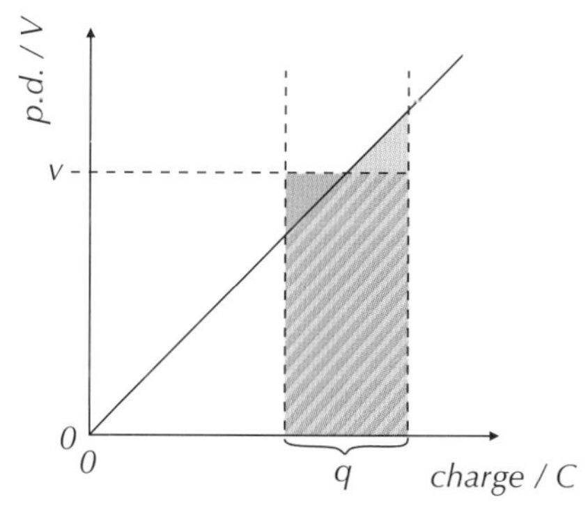

Figure 6: *A potential difference-charge graph showing the energy stored by a small increase in charge on the plates of a capacitor.*

The total energy stored by the capacitor is the sum of all the energies stored in each small step increase in charge, until the capacitor is fully charged. So it's just the area under the graph of p.d. against charge.

Figure 5: *Storm clouds store massive amounts of energy — forming huge natural capacitor plates. Particle collisions cause electrons to be knocked off atoms, giving the bottom of the cloud a charge. This induces a charge in the surface of the Earth. The air in between acts as the dielectric, and if the potential difference gets big enough, the dielectric breaks down and lightning bolts travel between the cloud and the Earth.*

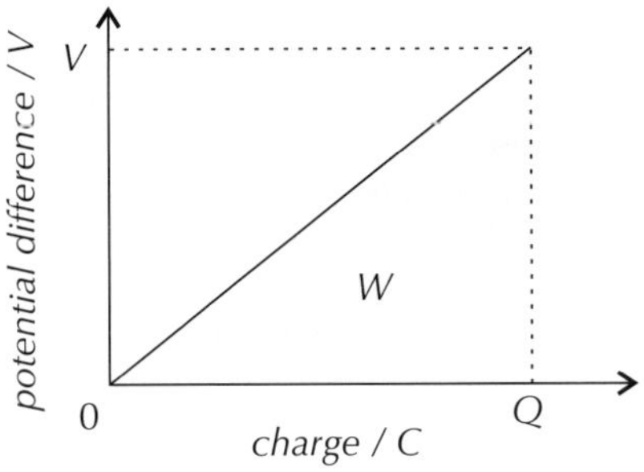

Figure 7: A plot of potential difference against charge stored for a capacitor. The shaded area gives the energy stored by the capacitor.

Exam Tip
This formula is given in the data and formulae booklet, but you need to understand where it comes from — remember it's the area under a V-Q graph.

The energy stored is given by the yellow triangle. The area of a triangle is equal to ½ × base × height, so the energy stored by the capacitor is:

$$W = \frac{1}{2}QV$$

W = energy stored by capacitor (in J)

V = potential difference across capacitor (in V)

Q = charge on capacitor (in C)

Example — Maths Skills

A capacitor is fully charged to a potential difference of 12 V. When it is fully charged, it stores 0.15 C of charge. Calculate the energy stored by it.

$W = \frac{1}{2}QV = \frac{1}{2} \times 0.15 \times 12 = 0.9$ J

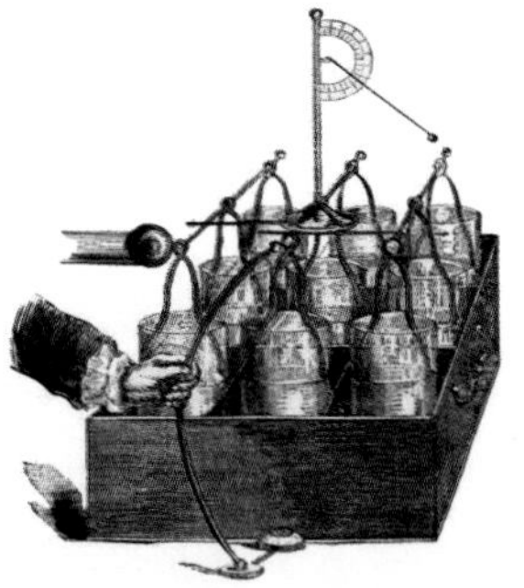

Figure 8: Benjamin Franklin was one of the first people to store electrical energy, in the mid 1700s. He invented what he called a 'battery' by grouping together many Leyden jars (simple capacitors).

Using the capacitance equation $C = \frac{Q}{V}$, you can derive two more equations for the energy stored by a capacitor. Simply rearrange the capacitance equation for Q and V, and substitute them into the energy equation:

Using $Q = CV$:

$$W = \frac{1}{2}V^2C$$

Using $V = \frac{Q}{C}$:

$$W = \frac{1}{2}\frac{Q^2}{C}$$

Examples — Maths Skills

A 900 μF capacitor is charged up to a potential difference of 240 V. Calculate the energy stored by the capacitor.

First, choose the best equation to use — you've been given V and C, so you need $W = \frac{1}{2}V^2C$.

Substitute the values in:

$W = \frac{1}{2}V^2C = \frac{1}{2} \times 240^2 \times (900 \times 10^{-6}) = 25.9$ J (to 3 s.f.)

When capacitor X is fully charged, it stores 63 mC of charge and 15 J of electrical energy. Calculate the capacitance of capacitor X.

You've been given W and Q, and need to find C, so use the equation:

$W = \frac{1}{2}\frac{Q^2}{C}$

Rearrange for C, and substitute in the values given:

$C = \frac{Q^2}{2W} = \frac{(63 \times 10^{-3})^2}{2 \times 15} = 1.323 \times 10^{-4}$ F $= 130$ μF (to 2 s.f.)

Uses of capacitors

HOW SCIENCE WORKS

Capacitors are found in loads of electronic devices. They don't store much charge, so can't replace batteries, but they can discharge quicker than batteries, which makes them very useful. What's more, the amount of charge that can be stored and the rate at which it's released can be controlled by selecting different types of capacitor. Some uses for them are:

- Flash photography — when you take a picture, the capacitor has to discharge really quickly to give a short pulse of high current to create a brief, bright flash.
- Back-up power supplies — these often use lots of large capacitors that can release charge for a short period if the power supply goes off — e.g. for keeping computer systems running if there's a brief power outage.
- Smoothing out p.d. — when converting an a.c. power supply to d.c. power, capacitors charge up during the peaks and discharge during the troughs, helping to maintain a constant output.

Figure 9: *Some touch screens use capacitors. The screen contains a layer of capacitive material that holds an electrical charge. Since your body is an electrical conductor, when your finger touches the capacitive layer, it changes the charge at that specific point, so the touch screen can work out exactly where you made contact.*

Practice Questions — Application

Q1 A 0.10 F capacitor is used in a circuit as a back-up in case of a short interruption in the mains power supply. The p.d. supplied to the circuit is 230 V. How much charge can the capacitor store?

Q2 A 40 mF capacitor is connected to a 230 V power source. When fully charged, how much energy will be stored by the capacitor?

Q3 Explain why a capacitor would not be a good source of power for a portable media player.

Q4 A capacitor is charged with 2.25 mC of charge. It stores 1.30 J of energy while holding this charge. Calculate its capacitance.

Practice Questions — Fact Recall

Q1 Write down the definition of capacitance.

Q2 Explain how a charge builds up on each plate of a capacitor when it is connected in a circuit.

Q3 How would you find the energy stored in a capacitor from a graph of potential difference against charge?

Q4 Write down three equations that can be used to calculate the energy stored in a capacitor. Define all symbols used.

Learning Objectives:

- Know that the total capacitance of two or more capacitors in parallel is given by: $C = C_1 + C_2 + ...$
- Know that the total capacitance of two or more capacitors in series is given by: $\frac{1}{C} = \frac{1}{C_1} + \frac{1}{C_2} + ...$
- Be able to analyse circuits containing capacitors.
- Understand techniques and procedures used to investigate capacitors in both series and parallel combinations using ammeters and voltmeters (PAG9).

Specification Reference 6.1.1

2. Capacitors in Circuits

Capacitors are really useful in circuits, but the effect they have in the circuit depends on how they are connected. You'll get a larger total capacitance when they're connected in parallel than when they're in series.

Capacitors in parallel

If you connect two or more capacitors in parallel with each other, the potential difference across each one is the same. So each capacitor can store the same amount of charge as it would if it was the only capacitor in the circuit. The total charge stored in the circuit is split between the two capacitors, since the charges can take either path.

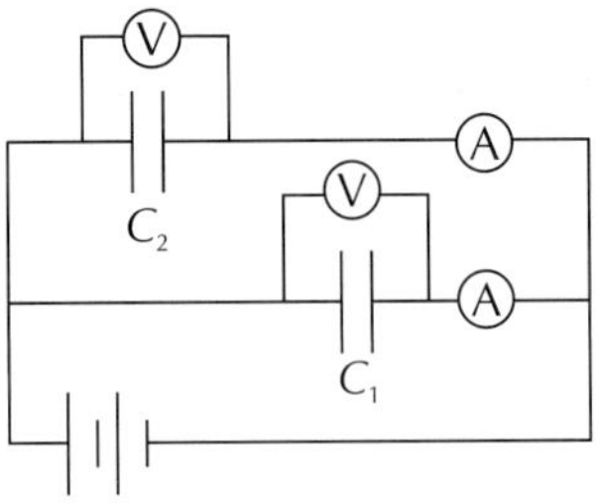

Figure 1: *An example of a circuit containing capacitors connected in parallel*

So the total capacitance of the circuit, C_{total} is equal to:

$$C_{total} = \frac{Q_{total}}{V}$$

$$\Rightarrow C_{total} = \frac{Q_1 + Q_2}{V}$$

$$\Rightarrow C_{total} = \frac{Q_1}{V} + \frac{Q_2}{V}$$

$$\Rightarrow C_{total} = C_1 + C_2$$

So the total capacitance is just the sum of the individual capacitances:

$$C_{total} = C_1 + C_2 + ...$$

Examples — **Maths Skills**

The total capacitance of the circuit shown below is 27 μF. Calculate the capacitance of capacitor X.

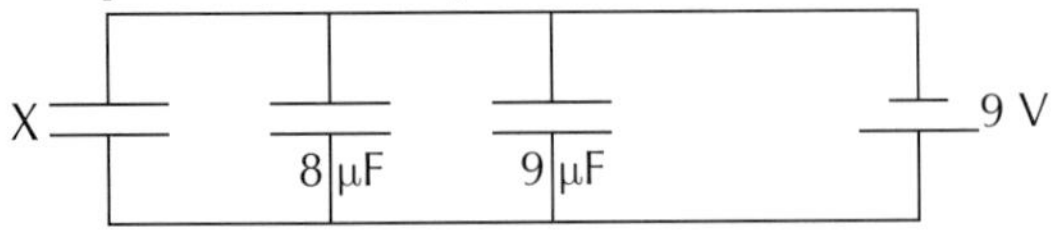

$C_{total} = C_1 + C_2 + C_3$, so $C_3 = C_{total} - (C_1 + C_2)$.

So, $C_X = 27 - (8 + 9) = 10$ μF

A 12 μF capacitor is connected in parallel with the other capacitors in the circuit above. Calculate the new total capacitance.

$C_{total} = C_1 + C_2 + C_3 + C_4 = 27 + 12 = 39$ μF

Capacitors in series

When you connect capacitors in series, like in Figure 2, the potential difference is shared between them, but each capacitor stores the same charge.

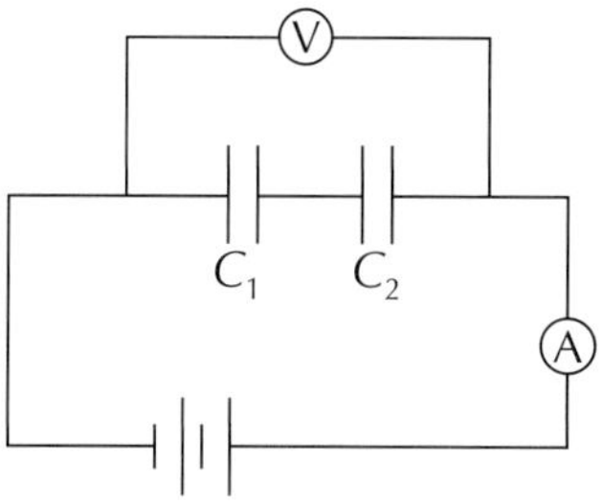

***Figure 2:** An example of a circuit containing capacitors connected in series.*

This is because, as a negative charge builds up on the left plate of capacitor C_1, an equal positive charge builds up on the right plate of C_1. This causes a negative charge to build up on the left plate of C_2 that is equal in size to the charge on the right plate of C_1. This happens because these two plates are disconnected from the rest of the circuit (see Figure 3), so the electrons repelled from the right plate of C_1 have to end up on the left plate of C_2. This then produces an equal positive charge on the right plate of C_2. All the plates have the same charge, so each capacitor stores the same charge.

So the total capacitance of the circuit is:

$$C_{total} = \frac{Q}{V_{total}} = \frac{Q}{V_1 + V_2}$$

$$\text{So, } \frac{1}{C_{total}} = \frac{V_1 + V_2}{Q}$$

$$\Rightarrow \frac{1}{C_{total}} = \frac{V_1}{Q} + \frac{V_2}{Q}$$

$$\Rightarrow \frac{1}{C_{total}} = \frac{1}{C_1} + \frac{1}{C_2}$$

This means that you can find the total capacitance of capacitors connected in series using:

$$\frac{1}{C_{total}} = \frac{1}{C_1} + \frac{1}{C_2} + \ldots$$

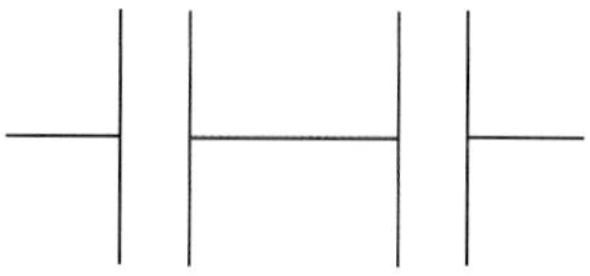

***Figure 3:** The red part of the circuit is disconnected from the rest of the circuit by the dielectrics, so the charge can't leave it.*

Tip: You'll have met series and parallel relations like this in your first year of A-level Physics with resistors. Don't get them confused though — total capacitance is bigger when components are connected in parallel than in series, while total resistance is the opposite.

Example — Maths Skills

The diagram on the right shows a circuit containing capacitors. Calculate the total capacitance of the circuit.

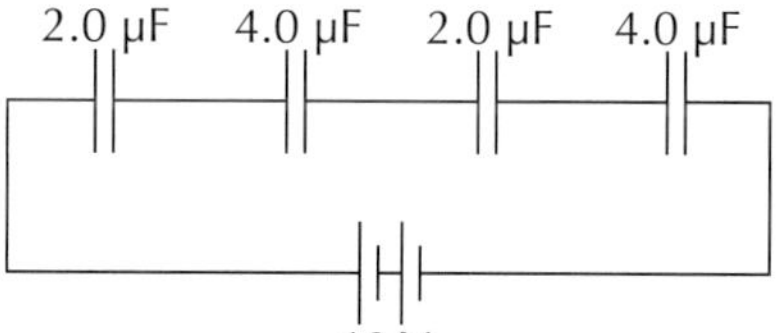

$\frac{1}{C_{total}} = \frac{1}{C_1} + \frac{1}{C_2} + \frac{1}{C_3} + \ldots$, so:

$\frac{1}{C_{total}} = \frac{1}{2} + \frac{1}{4} + \frac{1}{2} + \frac{1}{4} = \frac{3}{2}$

so $C_{total} = \frac{2}{3} = 0.666... = 0.67$ μF (to 2 s.f.)

Tip: You could be asked to solve circuit problems using the equations for both series and parallel capacitors. They'll work for any number of capacitors.

Tip: Remember to carry out a risk assessment before you start any experiment.

Investigating capacitors in series and in parallel

PRACTICAL ACTIVITY GROUP 9

You can investigate the properties of capacitors in series and in parallel for yourself. Start by setting up the circuit shown in Figure 4, using three identical capacitors.

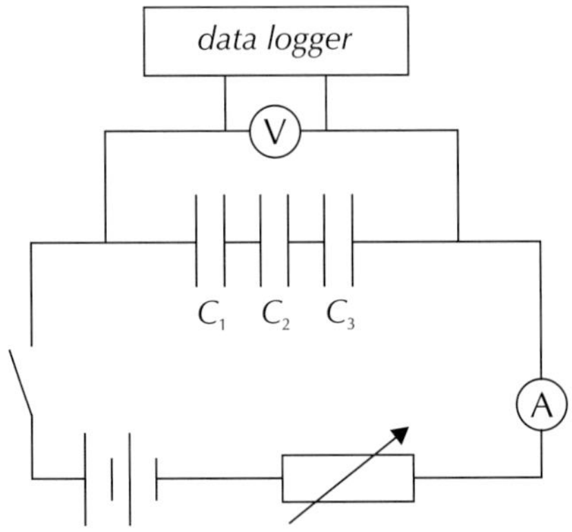

Figure 4: *An example of a circuit that can be used to investigate capacitance in series.*

1. Set the variable resistor to a fairly high resistance value and record it, making sure there is plenty of room for you to decrease the resistance during the experiment.
2. Close the switch. The capacitors will begin charging. Record the initial value of the current in the circuit.
3. The data logger connected to the voltmeter can be used to record the potential difference over time. Constantly adjust the variable resistor to keep the charging current constant for as long as you can (it'll be almost impossible when the capacitor is nearly fully charged).
4. Once the capacitors are fully charged (when the current drops to zero), open the switch.

Tip: You could use the same three capacitors from the first experiment, but if you do, make sure they have all been fully discharged before you start this experiment.

Next, set up the circuit shown in Figure 5, using another three capacitors, identical to the first three.

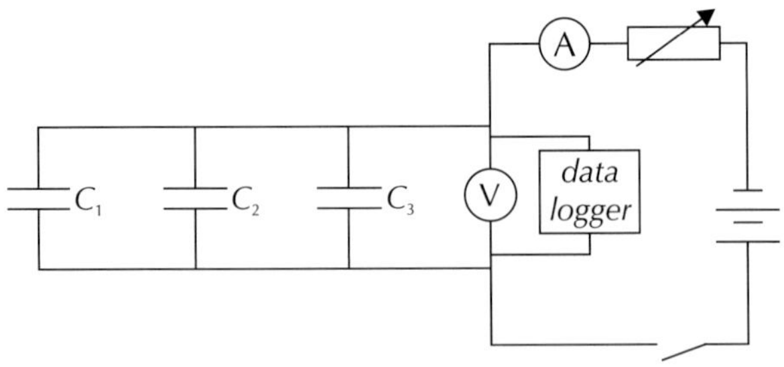

Figure 5: *An example of a circuit that can be used to investigate capacitance in parallel.*

Repeat steps 1-4 for this circuit (making sure the variable resistor starts at the same resistance as it did for the first experiment).

Once you have results for both circuits, plot a graph of current against time for each circuit. Use the constant value of current you kept the circuit at, and use the time value measured by the data logger when the current dropped to zero (the capacitors reached full charge).

Assuming each of your identical capacitors has a capacitance C and could store a maximum charge of Q_c if they were the only capacitor in the circuit, your results should look something like those in Figure 6.

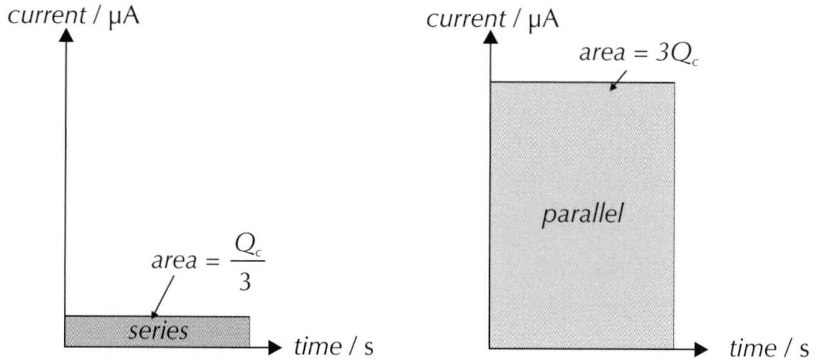

Figure 6: *The current-time graphs of three capacitors connected in series (left) and in parallel (right).*

The area beneath a current-time graph is equal to $I \times \Delta t = \Delta Q$, so the area gives the total charge stored by the capacitors.

You can also plot a graph of charge against potential difference for each circuit. Calculate the charge stored by the capacitors at each time reading taken by the data logger, using the equation $\Delta Q = I\Delta t$. Your graphs should look something like those in Figure 8.

The graphs are straight lines through the origin, so the gradient of the graphs is equal to charge divided by potential difference at any point on the line. So the gradient gives the total capacitance of the capacitors in the circuit. You should see that the gradient of your graph gives the same answer for the total capacitance of your circuit as the equations on pages 116-117.

Figure 7: *Some digital multimeters can be used to measure capacitance directly.*

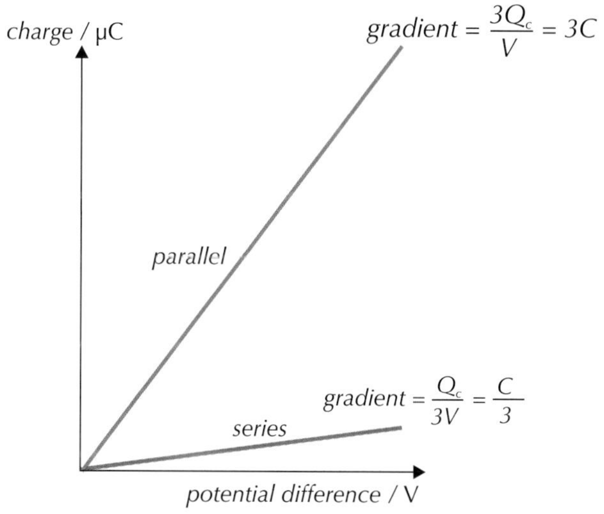

Figure 8: *A graph for charge stored against p.d. for three capacitors connected in series and in parallel. The gradient of the graph gives the total capacitance.*

Practice Questions — Application

Q1 Two identical capacitors are connected in series in a circuit. It takes 2 minutes for them to become fully charged when they are supplied with a constant current. Given that one of the capacitors connected on its own in the same circuit can store a charge of 1.2×10^{-3} C, calculate the current used to charge the two capacitors.

Q2 The diagram below shows the graph of charge against p.d. for a number of identical capacitors connected in parallel in a circuit. The capacitance of one of these capacitors is 1.5×10^{-3} F.

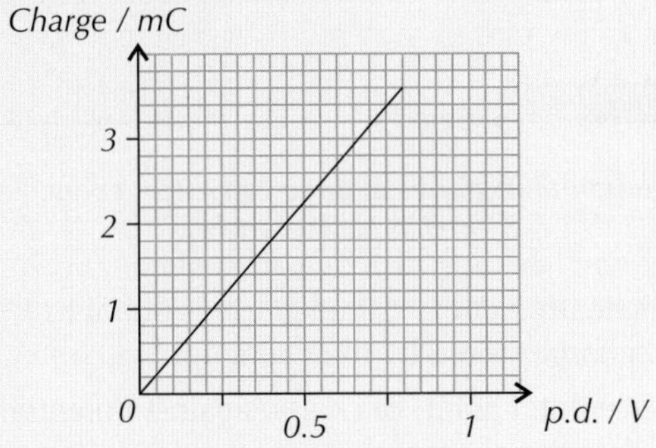

a) Calculate the total capacitance of the circuit.

b) How many capacitors were connected in the circuit?

Practice Questions — Fact Recall

Q1 Write down the formula you would use to find the total capacitance of a circuit containing a number of capacitors connected in series.

Q2 What value is given by the gradient of a straight line graph of charge against potential difference across a set of capacitors?

3. Investigating Charging and Discharging Capacitors

Charging or discharging a capacitor 'through' a fixed resistor makes it easier to investigate.

Investigating charging a capacitor

PRACTICAL ACTIVITY GROUP 9

As you'll have seen when trying to keep the current constant in the experiment on page 118, capacitors have an effect on the properties of a circuit (e.g. current in the circuit, potential difference across other components, etc). You need to know how to measure these effects, and explain how the capacitor causes them.

A capacitor can charge (and discharge) fairly quickly, making the process difficult to investigate. Including a fixed resistor in your test circuit and charging the capacitor 'through' the resistor is a good idea. The resistance in the circuit will slow down the rate at which the capacitor charges or discharges, which makes the process much easier to investigate.

To investigate a charging capacitor, set up the test circuit in Figure 1.

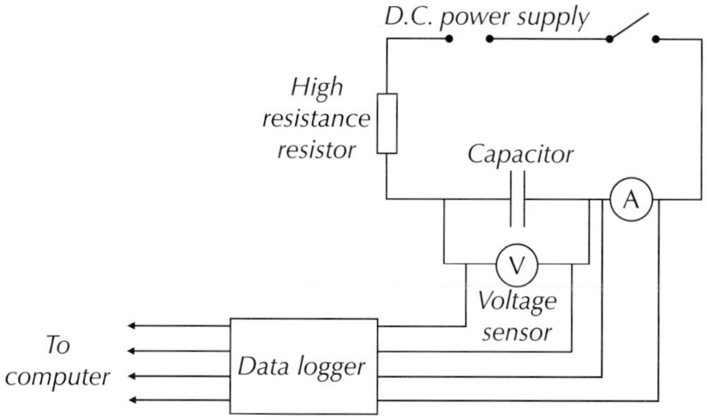

Figure 1: *A test circuit which can be used to investigate a circuit containing a charging capacitor.*

- Close the switch to allow the capacitor to begin charging.
- Let the capacitor charge whilst the data logger records both the p.d. (from the voltmeter) and the current (from the ammeter) over time.
- When the current through the ammeter is 0, the capacitor is fully charged.
- You can then use a computer to plot a graph of current, p.d. or charge against time (remember $\Delta Q = I\Delta t$) — see Figure 2.

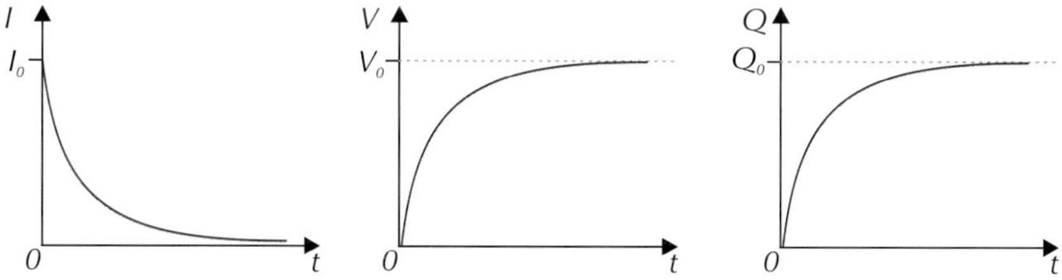

Figure 2: *Graphs to show charging current against time (left), potential difference against time (middle) and charge against time (right) for a capacitor being charged through a fixed resistor.*

Learning Objectives:

- Understand the effect of charging and discharging a capacitor through a resistor.
- Understand techniques and procedures to investigate the charge and discharge of a capacitor using both meters and data loggers (PAG9).
- Understand graphical methods and spreadsheet modelling of the equation $\frac{\Delta Q}{\Delta t} = -\frac{Q}{CR}$ for a discharging capacitor.

Specification Reference 6.1.3

Tip: As always, be sure to carry out a risk assessment before you start any of the investigations in this section.

Tip: Data loggers can be connected to a computer which will collect all your data for you and plot all sorts of graphs of it.

Tip: In Figure 2, I_0 is the initial current in the circuit. V_0 and Q_0 are the p.d. across and charge stored by the capacitor when it is fully charged.

Tip: Current is the odd one out here — while everything else increases, current decreases as time passes.

Tip: Remember, $V = IR$ so $I = V/R$.

Tip: You could also do these experiments without the data logger, by manually taking readings from the ammeter and voltmeter at fixed time intervals. But you'll need to use a fixed resistor with a very high resistance to make sure the capacitor charges and discharges slowly enough for you to take enough measurements.

As soon as the switch is closed, a current starts to flow. The potential difference across the capacitor is zero at first, so there is no p.d. opposing the current. The potential difference of the power supply causes an initial relatively high current of $\frac{V}{R}$ to flow (where V is the voltage of the power supply and R is the resistance of the resistor).

As the capacitor charges, the p.d. across the capacitor gets bigger, the p.d. across the resistor gets smaller, and the current drops. The charge (Q) on the capacitor is proportional to the potential difference across it, so the Q-t graph is the same shape as the V-t graph.

Investigating discharging a capacitor

You can carry out a similar experiment to investigate a discharging capacitor. Simply open the switch, remove the power supply, and then reconnect the circuit as shown in Figure 3.

PRACTICAL ACTIVITY GROUP 9

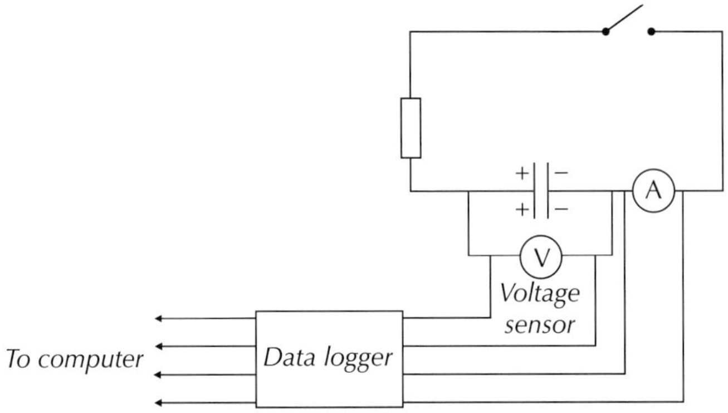

Figure 3: *A test circuit which can be used to investigate a circuit containing a discharging capacitor.*

Figure 4: *Some capacitors must be connected the right way round in the circuit. In this photo, the strip of minus signs indicates which pin should be connected to the negative terminal.*

- Close the switch to allow the capacitor to begin discharging.
- Let the capacitor discharge whilst the data logger records potential difference and current over time.
- When the current through the ammeter and the potential difference across the plates fall to zero, the capacitor is fully discharged.
- You can once more plot graphs of p.d., charge and current against time (see Figure 5).

Tip: V is proportional to Q (see page 111), so you always get the same shape graph whether you use V or Q on the vertical axis.

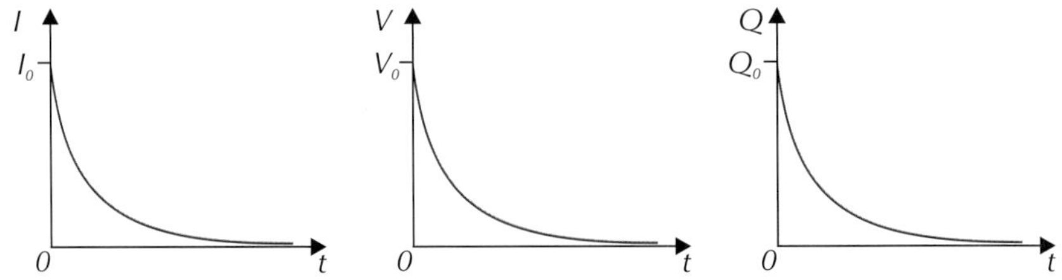

Figure 5: *Graphs to show discharging current against time (left), potential difference against time (middle) and charge against time (right) for a capacitor discharging through a fixed resistor.*

Tip: Depending on your apparatus, you may get graphs like this:

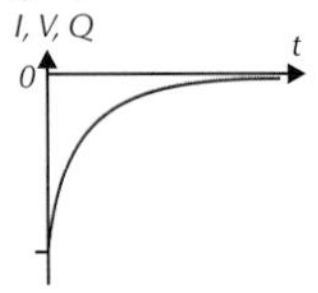

This is because the charge is flowing in the opposite direction to the charging current, so the reading on your meters will have changed sign.

The electrons (current) flow from the negative plate to the positive plate. Initially, the current is high, but as the charge leaves the plates, the potential difference across the plates decreases. So the electrostatic repulsion decreases, reducing the flow of current.

Modelling a discharging capacitor

HOW SCIENCE WORKS

You can predict roughly how the stored charge will change over time for any discharging capacitor with capacitance C and initial charge Q, in a circuit with resistance R. You need to find the change in charge over a tiny time interval and repeat this process for a long period of time. To do this, you'll need an equation which relates ΔQ and Δt.

You know that $I = \frac{\Delta Q}{\Delta t}$ and also that $I = \frac{V}{R}$, so:

$$\frac{\Delta Q}{\Delta t} = \frac{V}{R}$$

For capacitors, $Q = CV$ so $V = \frac{Q}{C}$. Combine these and you get:

$$\frac{\Delta Q}{\Delta t} = -\frac{Q}{CR}$$

Tip: There's a minus sign here because the charge is decreasing over time.

Finally, multiply both sides by Δt to get:

$$\Delta Q = -\frac{Q}{CR}\Delta t$$

You can use this formula to plot a graph of Q against t, by working out how Q changes for each small change in t. Graphical methods like this are really useful for modelling relationships, but they can take a long time to do by hand. So using a spreadsheet can be very helpful.

Using a spreadsheet to model discharge

You can use this equation in a spreadsheet to demonstrate how a capacitor gradually discharges over time.

Tip: There's another example of modelling exponential relationships with spreadsheets on p.199.

1. Create a new spreadsheet, with columns for the time elapsed since the capacitor began discharging, t, the change in charge, ΔQ, and the charge remaining on the capacitor, Q.
2. Choose an initial starting charge, Q_{init}, for the capacitor along with a value for the capacitance C and the resistance R of the resistor.
3. Choose a sensible time interval Δt that is significantly less than CR. This will let you plot more precise graphs showing the relationship between charge and time.
4. In the initial row, $t_0 = 0$ and $Q_0 = Q_{init}$. Leave ΔQ blank in this row.
5. In the next row:
   - The new time $t_1 = t_0 + \Delta t$.
   - You can write a formula for the change in charge, ΔQ, over the time interval by using the equation $\Delta Q = -\frac{Q}{CR}\Delta t$ above. Let's call it $(\Delta Q)_1$.
   - The new charge is $Q_1 = Q_0 + (\Delta Q)_1$ so write a formula for that too.
6. Repeat this process of calculating the values in each row from Q and t in the row above. If you write the formulas correctly, you can use the spreadsheet program to automatically fill in as many rows as you want.
7. Once you have enough points, plot a graph of charge against time — see Figure 8. It should be similar to the graph for a discharging capacitor shown on page 122.

Tip: Figure 6 on the next page shows how all this can be laid out.

Tip: This process is called iteration — where you apply an equation repeatedly, using the previous result as the input for the next calculation. A spreadsheet that does this is known as an iterative spreadsheet, and its data is known as iterative data.

Tip: The graph you would get from these initial values is plotted in Figure 8.

Data input cells		
	Δt	E.g. 1×10^{-12} s
	C	E.g. 3×10^{-12} F
	R	E.g. 12 Ω
	Q_{init}	E.g. 6×10^{-7} C

t (s)	ΔQ (C)	Q (C)
$t_0 = 0$		$Q_0 = Q_{init}$
$t_1 = t_0 + \Delta t$	$(\Delta Q)_1 = -\frac{Q_0}{CR}\Delta t$	$Q_1 = Q_0 + (\Delta Q)_1$
$t_2 = t_1 + \Delta t$	$(\Delta Q)_2 = -\frac{Q_1}{CR}\Delta t$	$Q_2 = Q_1 + (\Delta Q)_2$

Figure 6: An example of the formulas that can be used to create an iterative spreadsheet of charge over time for a discharging capacitor.

	A	B	C
1	Δt in s	1E-12	
2	C in F	3E-12	
3	R in Ω	12	
4	Q_{init} in C	6E-07	
5			
6	t in s	ΔQ in C	Q in C
7	0.00E+00		6.00E-07
8	1.00E-12	-1.67E-08	5.83E-07
9	2.00E-12	-1.62E-08	5.67E-07
10	3.00E-12	-1.58E-08	5.51E-07
11	4.00E-12	-1.53E-08	5.36E-07
12	5.00E-12	-1.49E-08	5.21E-07
13	6.00E-12	-1.45E-08	5.07E-07
14	7.00E-12	-1.41E-08	4.93E-07
15	8.00E-12	-1.37E-08	4.79E-07

Figure 7: A sample of the spreadsheet produced by the example values in Figure 6. Once you've filled in rows 7 and 8, you can automatically fill the rest of the rows using the formulas in row 8, if you've written the formulas correctly.

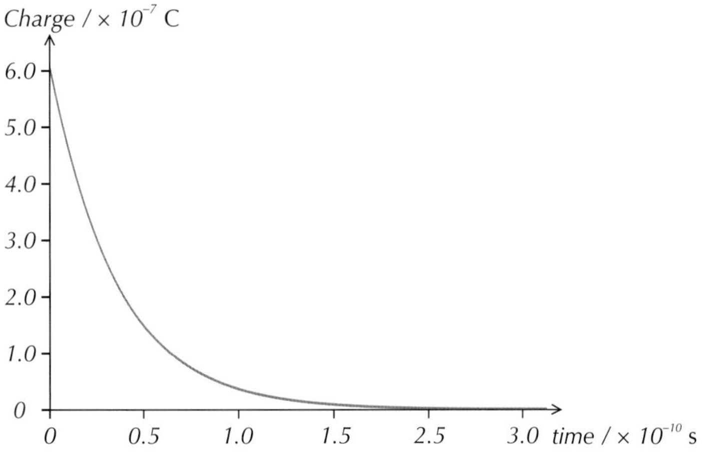

Figure 8: An example of a charge-time graph produced from a spreadsheet model of a discharging capacitor, using the values in Figures 6 and 7.

Tip: To produce the graph in Figure 8, from the data in Figure 7, over 200 rows of iterative data were used.

Practice Question — Application

Q1 A fully-charged 4.2 mF capacitor is connected in a circuit in series with a 5.5×10^3 Ω resistor. It discharges 85 nC of charge in the first 4.5 μs. Calculate the charge on the capacitor when it is fully charged.

Practice Questions — Fact Recall

Q1 Draw curves showing how the p.d. across the capacitor and current through the circuit vary with time when a capacitor is charged through a fixed resistor.

Q2 How can you alter the circuit used to charge a capacitor so that the capacitor discharges?

Q3 Describe an experiment you could do to investigate a discharging capacitor and plot graphs of current in the circuit and potential difference across the capacitor against time.

4. Charging and Discharging Calculations

The shapes of the graphs that you plotted in the previous section are actually exponential graphs. There are exponential relationships between V, Q or I and time for charging and discharging capacitors.

Equations for a charging capacitor

The charge on a charging capacitor in a capacitor-resistor circuit after a given time t is given by:

$$Q = Q_0(1 - e^{-\frac{t}{CR}})$$

- $\boldsymbol{Q}$ = *charge of the capacitor at time t, in C*
- $\boldsymbol{Q_0}$ = *charge of the capacitor in C when fully charged*
- $\boldsymbol{t}$ = *time since charging began in s*
- $\boldsymbol{R}$ = *resistance of fixed resistor in Ω*
- $\boldsymbol{C}$ = *capacitance of capacitor in F*

The formula for calculating the voltage across a charging capacitor is of the same form, but the formula for the charging current is different — it decreases exponentially. The formulas for both are:

$$V = V_0(1 - e^{-\frac{t}{CR}})$$

- $\boldsymbol{V_0}$ = *potential difference across the capacitor when fully charged, in V*
- $\boldsymbol{V}$ = *potential difference across the capacitor at time t, in V*

$$I = I_0 e^{-\frac{t}{CR}}$$

- $\boldsymbol{I_0}$ = *initial current, in A*
- $\boldsymbol{I}$ = *current at time t, in A*

Example — Maths Skills

When fully charged, a 3.00 μF capacitor holds 36.0 μC of charge. It is connected in series with a 56.0 kΩ resistor. Calculate the charge on the capacitor 0.500 seconds after it begins charging.

$$Q = Q_0(1 - e^{-\frac{t}{CR}}) = 36.0 \times 10^{-6}(1 - e^{\frac{-0.500}{3.00 \times 10^{-6} \times 56.0 \times 10^{3}}})$$
$$= 3.416... \times 10^{-5} = 34.2\,\mu\text{C (to 3 s.f.)}$$

Equations for a discharging capacitor

The charge left on the plates of a capacitor discharging from full in a capacitor-resistor circuit is given by the equation:

$$Q = Q_0 e^{-\frac{t}{CR}}$$

- $\boldsymbol{Q_0}$ = *charge of the capacitor in C when fully charged*
- $\boldsymbol{Q}$ = *charge of the capacitor at time t, in C*
- $\boldsymbol{t}$ = *time since discharging began in s*
- $\boldsymbol{R}$ = *resistance of fixed resistor in Ω*
- $\boldsymbol{C}$ = *capacitance of capacitor in F*

Learning Objectives:

- Be able to analyse capacitor-resistor circuits.
- Know and understand equations in the form of $x = x_0 e^{-\frac{t}{CR}}$ and $x = x_0(1 - e^{-\frac{t}{CR}})$ for capacitor-resistor circuits.
- Understand the constant-ratio property of exponential decay graphs.
- Know what is meant by the time constant of a capacitor-resistor circuit, $\tau = CR$.

Specification References 6.1.1 and 6.1.3

Tip: Make sure you don't miss a prefix — capacitance is often given in micro-, nano- or picofarads.

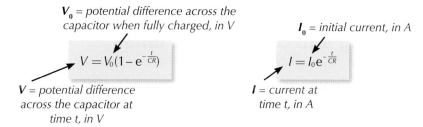

Tip: V_0 is the potential difference across the capacitor when it is fully charged, and I_0 is the initial current flowing through the circuit.

The same is true for the potential difference and current:

$$V = V_0 e^{-\frac{t}{CR}}$$

$$I = I_0 e^{-\frac{t}{CR}}$$

Example — Maths Skills

A 0.20 mF capacitor was charged to a potential difference of 12 V and then discharged through a fixed 50 kΩ resistor. Calculate the charge on the capacitor after 1 second.

The question tells you $C = 0.20 \times 10^{-3}$ F, $V = 12$ V, $R = 50 \times 10^3\ \Omega$ and $t = 1$.

First you need to calculate the initial charge.

$C = \frac{Q}{V}$ so $Q = CV$, so the initial charge is:

$Q_0 = CV = 0.20 \times 10^{-3} \times 12 = 2.4 \times 10^{-3}$ C

Then use the equation:

$$Q = Q_0 e^{-\frac{t}{CR}}$$
$$= 2.4 \times 10^{-3} \times e^{-\frac{1}{(0.20 \times 10^{-3})(50 \times 10^3)}}$$
$$= 2.4 \times 10^{-3} \times e^{-0.1}$$
$$= 2.171... \times 10^{-3} = 2 \times 10^{-3}\ \text{C (to 1 s.f.)}$$

Tip: There's some help with logarithms and exponentials on pages 236-237. There should be a button on your calculator for 'e' that you can use for this calculation.

Charging and discharging times

The time taken for a capacitor to charge and discharge depends on two factors:

- The capacitance of the capacitor (C). This affects the amount of charge that can be transferred at a given voltage.
- The resistance of the circuit (R). This affects the current in the circuit.

For given values of C and R, the quantity x in an exponential relationship of the form $x = x_0 e^{-\frac{t}{CR}}$ always takes the same time to halve, no matter what the initial value of x is.

This is because, when x is halved, $x = \frac{1}{2}x_0$.

So:

$$\frac{1}{2}x_0 = x_0 e^{-\frac{t}{CR}}$$
$$\Rightarrow \frac{1}{2} = e^{-\frac{t}{CR}}$$

Taking logs of both sides:

$$\ln\left(\frac{1}{2}\right) = \ln\left(e^{-\frac{t}{CR}}\right)$$

$\frac{1}{2} = 2^{-1}$, and $\ln(A^B) = B\ln(A)$, so:

$$-\ln(2) = \frac{-t}{CR}$$
$$\Rightarrow t = \ln(2)CR$$

ln(2) is just a number, and the capacitance, C, and resistance, R, are fixed for a given capacitor-resistor circuit, so:

$$t = \text{constant}$$

So the time taken to halve is always the same. In fact, for a given proportion it always takes the same time for that proportion of the charge to be lost — it's known as the constant-ratio property of exponential relationships.

Tip: The same applies to potential difference and current. In a discharging capacitor it will always take the same time for the p.d. or current to decrease by a given proportion.

Time constant

When the discharge time t is equal to CR the equation for the charge left on a discharging capacitor becomes:

$$Q = Q_0 e^{-1}.$$

So when $t = CR$:

$$\frac{Q}{Q_0} = \frac{1}{e}, \text{ where } \frac{1}{e} \approx \frac{1}{2.718} \approx 0.37$$

The time $t = CR$ is known as the **time constant**, τ, and is the time taken for the charge on a discharging capacitor (Q) to fall to about 37% of Q_0. It's also the time taken for the charge of a charging capacitor to rise to about 63% of Q_0 (Figure 1).

It doesn't matter when you start looking at the charge on the capacitor after it has begun discharging (i.e. what value you take as Q_0), it will always have dropped to 37% of that value after a period of time equal to the time constant because of the constant-ratio property of exponential decay.

The larger the resistance in series with the capacitor, the longer it takes to charge or discharge. In practice, the time taken for a capacitor to charge or discharge fully is taken to be about $5CR$ or 5τ.

Tip: τ is the Greek letter 'tau'.

Tip: For a charging capacitor, no matter what charge it starts at, after the time constant has elapsed the charge on the capacitor will be 63% closer to fully charged.

Tip: Q is proportional to V, so the time constant is also the time taken for the voltage to decrease to about 37% of the initial voltage (or increase to about 63% of the source voltage while charging).

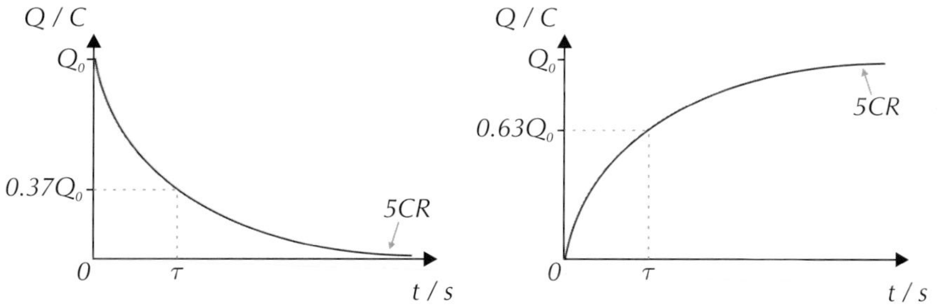

***Figure 1:** Q-t graphs showing the time constant for discharging (left) and charging (right) a capacitor through a fixed resistor.*

You can calculate the time constant directly using $\tau = CR$ or you can find it from a graph. These may be graphs of Q against t (as shown above and in the next example) or log-plots of $\ln(Q)$ against t (see next page).

Example — Maths Skills

A capacitor was discharged from full through a 10 kΩ resistor (to 2 s.f.). The graph below shows how the charge on the capacitor changed over the first 2.0 seconds. Calculate the time constant and the capacitance.

The initial charge is 12×10^{-4} C, and you want to find the time taken for the charge to decrease to 37% of that.

So $0.37 \times (12 \times 10^{-4}) = 4.4 \times 10^{-4}$ C.

Using the graph, the charge is 4.4×10^{-4} C when the time is 1.0 s, so the time constant, τ, is 1.0 s.

$\tau = CR$, so, $C = \frac{\tau}{R}$

$$C = \frac{1.0}{10 \times 10^3}$$

$= 1.0 \times 10^{-4}$ F $= 0.10$ mF (to 2 s.f.)

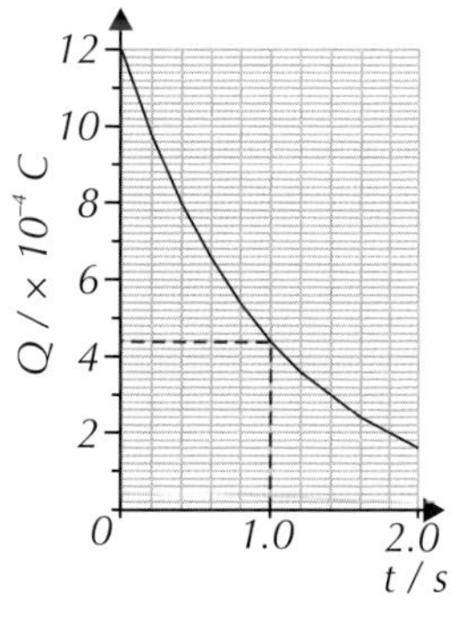

Tip: You might be asked to find the time it takes for a capacitor to discharge (or charge) to a certain percentage of its total charge.

Tip: Make sure that you remember that Q / Q_0 is the proportion of the charge on the plates, not the proportion of the charge that's been lost.

Log-linear graphs

Tip: Log-log plots are where both axes are logarithmic.

Tip: Log graphs come up a few times in physics — see p.244 for how to deal with them.

Log-linear graphs are plots where one of the axes is logarithmic. Log-linear and log-log plots are useful as they can often be used to produce a graph which is a straight line when linear axes would give a curve. So they're another good way of graphically displaying how charge, potential difference and current vary over time for a discharging capacitor.

Starting from the equation for charge on a discharging capacitor, the natural logarithm is taken of both sides:

$$Q = Q_0 e^{-\frac{t}{CR}}$$

$$\ln(Q) = \ln\left(Q_0 e^{-\frac{t}{CR}}\right)$$

$\ln(A \times B) = \ln(A) + \ln(B)$, so this can be written as:

$$\ln(Q) = \ln(Q_0) + \ln(e^{-\frac{t}{CR}})$$

Another log rule is $\ln(e^A) = A$, so:

$$\ln(Q) = \left(-\frac{1}{CR}\right)t + \ln(Q_0)$$

Tip: This equation is in the form $y = mx + c$ — the equation of a straight line.

If you then plot a graph of ln(Q) against *t*, you will see it is a straight line (Figure 2). By comparing the equation for ln(Q) to the equation for a straight line, you can see that the gradient is equal to $-\frac{1}{CR} = -\frac{1}{\tau}$.

So you can find the time constant from the log-linear graph by dividing –1 by the gradient of the line.

Tip: The graph's intercept with the vertical axis is $\ln(Q_0)$ — it's the *c* in $y = mx + c$.

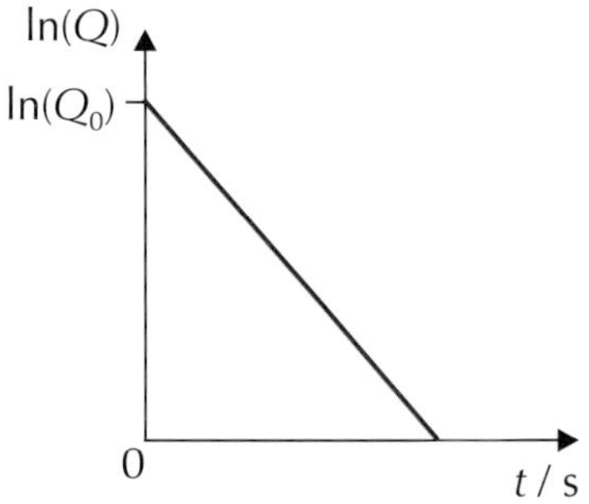

Figure 2: *A plot of ln(Q) against t.*

Tip: You can also plot a graph like this for the current in a charging capacitor. It's a bit more complicated for charge and potential difference though.

You could also plot ln(*V*) or ln(*I*) against time and find the time constant in the same way, as *V* and *I* both have the same dependence on *C* and *R*.

Example — Maths Skills

A student has done an experiment to find out how the charge on a discharging capacitor varies with time. She plots her results on a ln*Q*-*t* graph and draws a line of best fit. The graph is shown below. Use the graph to find the time constant of the circuit.

The gradient of a ln*Q*-*t* graph is equal to $-\frac{1}{CR}$.

To find the time constant, $\tau = CR$, you divide –1 by the gradient.

$$\text{Gradient} = -\frac{2.0 - 0}{0.06 - 0} = -33.333...$$

$$\tau = \frac{-1}{-33.333...} = 0.030 \text{ s (to 2 s.f.)}$$

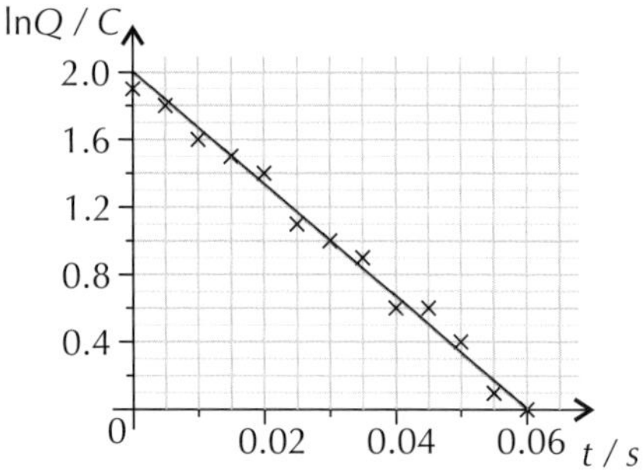

Practice Questions — Application

Q1 A capacitor with capacitance 0.60 μF is initially charged to 15.0 V. It is then connected across a resistor with resistance 2.6 kΩ. Calculate the potential difference across the capacitor 1.5 ms after it has begun to discharge.

Q2 Calculate the charge stored on a 15 nF capacitor, originally charged to a potential difference of 230 V and then discharged for 0.01 seconds through a 50 kΩ resistor.

Tip: You'll need $Q = CV$ for this, which is given in the data and formulae book.

Q3 A capacitor discharged in series with a resistor can be used to create a time delay function in electronics. In a particular burglar alarm, the alarm goes off when a 15 μF capacitor discharged through a 400 kΩ resistor has lost 63% of its initial charge.

a) Calculate the time delay on the alarm.

The manufacturer wants to increase the capacitance of the capacitor in order to make the time delay 60 seconds.

b) What capacitance would be needed for this time delay?

c) Explain why this might not be practical, and suggest what other component they could replace instead.

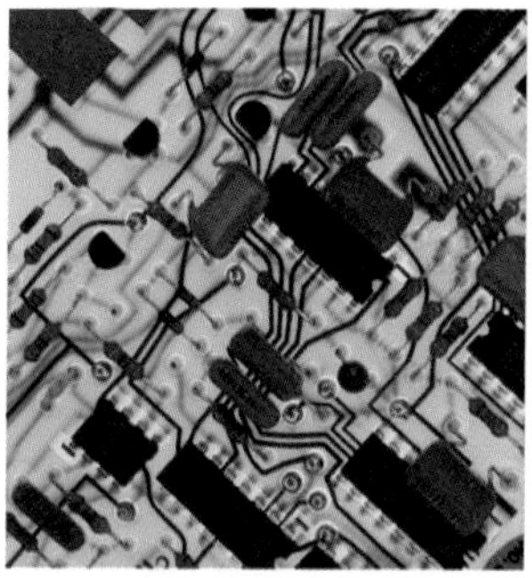

Figure 3: A circuit from a burglar alarm containing capacitors. They can be used with big resistors to create a time delay. When the door is opened the capacitor begins to discharge and when it has lost a certain amount of its charge the alarm goes off, giving you time beforehand to enter the code.

Q4 A discharging capacitor loses 70% of its initial charge, Q_0, in 20.0 seconds. Find the time constant of the capacitor-resistor circuit.

Q5 A plot of ln(V) against t is created for a discharging capacitor.

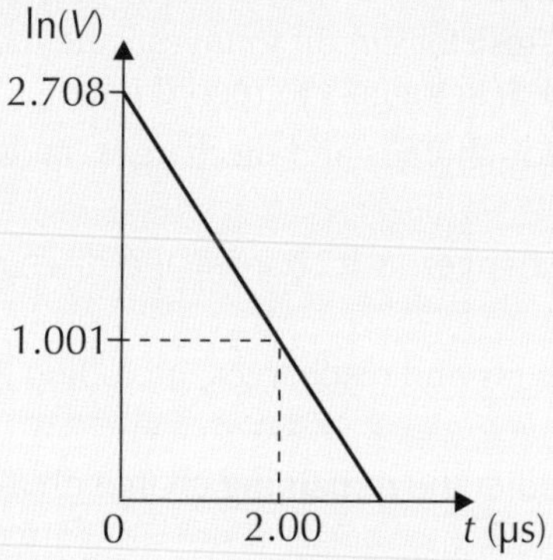

a) What does the vertical intercept of the graph represent?

b) Calculate the initial voltage.

c) Calculate the time constant for the capacitor.

Practice Questions — Fact Recall

Q1 What is the equation for calculating the charge on a capacitor charging through a fixed resistor R, after a certain time t?

Q2 Write down the equation for the p.d. across a discharging capacitor.

Q3 What two factors do the charging and discharging time of a capacitor depend on?

Q4 What is meant by the time constant when discharging a capacitor?

Section Summary

Make sure you know...

- That a capacitor is an electrical component that can store electric charge.
- That capacitance is defined as the amount of charge stored per unit potential difference, and it can be calculated using the equation $C = \frac{Q}{V}$.
- That capacitance is measured in farads, F.
- How the flow of electrons causes an equal and opposite charge to build up on each plate of a capacitor when the capacitor is being charged.
- How the flow of electrons causes the charge on each plate of a capacitor to decrease when it is being discharged.
- That the area underneath a graph of charge against p.d. is the energy stored by the capacitor.
- The equations for energy stored by a capacitor: $W = \frac{1}{2}QV = \frac{1}{2}V^2C = \frac{1}{2}\frac{Q^2}{C}$.
- Some uses of capacitors in the storage of electrical energy.
- That the total capacitance of two or more capacitors in parallel is given by: $C = C_1 + C_2 + ...$
- That the total capacitance of two or more capacitors in series is given by: $\frac{1}{C} = \frac{1}{C_1} + \frac{1}{C_2} + ...$
- How to perform an experiment to investigate the effect of connecting capacitors in series and in parallel.
- Understand that the rate at which a capacitor charges or discharges in a circuit can be slowed by including a resistor in the circuit.
- How to perform an experiment using a capacitor-resistor circuit to investigate the charging and discharging of a capacitor, using both meters and data loggers to record data.
- How the current in a circuit changes as a capacitor charges and discharges, including a graph of current against time.
- How the potential difference across a capacitor changes as it charges and discharges, including a graph of p.d. against time.
- How the charge on a capacitor changes as it charges and discharges, including a graph of charge against time.
- That the equation $\frac{\Delta Q}{\Delta t} = -\frac{Q}{CR}$ can be used to predict the remaining charge on a discharging capacitor after a given small period of time.
- How to use this equation to model a discharging capacitor using a spreadsheet.
- For a charging capacitor in a capacitor-resistance circuit, the charge on the capacitor at time t is given by $Q = Q_0(1 - e^{-\frac{t}{CR}})$ and the potential difference across the capacitor at time t is given by $V = V_0(1 - e^{-\frac{t}{CR}})$.
- For a discharging capacitor in a capacitor-resistance circuit, the charge on the capacitor at time t is given by $Q = Q_0 e^{-\frac{t}{CR}}$ and the potential difference across the capacitor at time t is given by $V = V_0 e^{-\frac{t}{CR}}$.
- For a charging or discharging capacitor in a capacitor-resistance circuit, the current at time t is given by $I = I_0 e^{-\frac{t}{CR}}$.
- The constant-ratio property of exponential functions, and how this applies to charging and discharging capacitors.
- That the time constant is given by the product of the capacitance and resistance of the circuit, $\tau = CR$.
- That the time constant, τ, of a capacitor-resistor circuit is the time taken for a capacitor to discharge to 37% of its initial value.

Exam-style Questions

1 When the switch S in the circuit shown below is in position 1, the capacitor C is fully charged by the battery through resistor R. The switch is then moved to position 2 and the capacitor is allowed to discharge fully through the resistor.

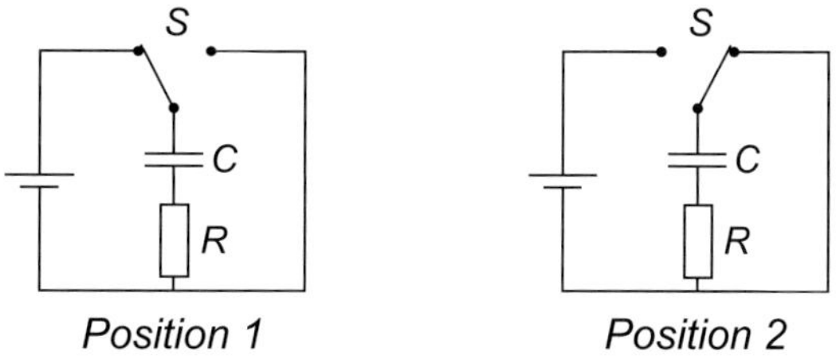

Which graph correctly shows how the charge, Q, on the capacitor varies with time, t, during this process?

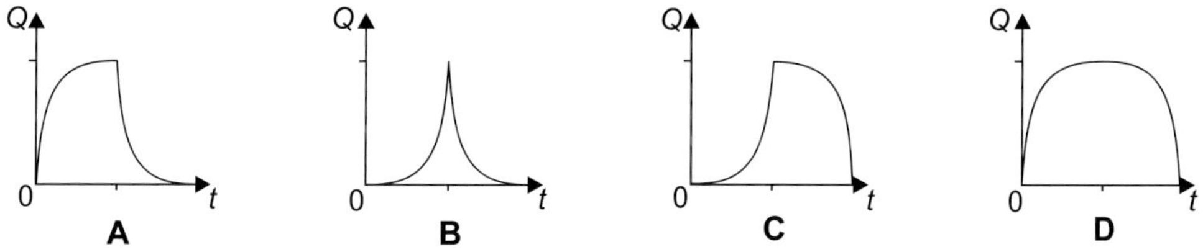

(1 mark)

2 A capacitor discharging through a fixed 100 kΩ resistor loses 35% of its charge in 1 s. What is the capacitance of the capacitor?

A 9.5×10^{-6} F

B 5.3×10^{-5} F

C 2.3×10^{-5} F

D 2.2×10^{-2} F

(1 mark)

3 The diagram on the right shows a circuit containing capacitors. Three 20 μF capacitors are connected in series in the circuit. A 12 μF is connected in parallel across these three capacitors. What is the total capacitance of the circuit?

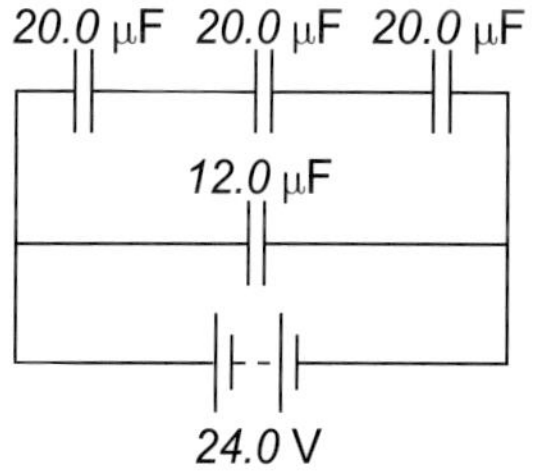

A 10.0 μF

B 18.7 μF

C 12.2 μF

D 72.0 μF.

(1 mark)

4 A student investigating capacitance charges a capacitor from a 12 V d.c. supply and uses a voltage sensor and a charge sensor connected to a data logger to measure the potential difference across the capacitor at regular intervals of charge stored by the capacitor. A computer is used to plot a graph of p.d. against charge.

(a) (i) Copy the axes below and sketch the graph obtained by the computer on the axes.

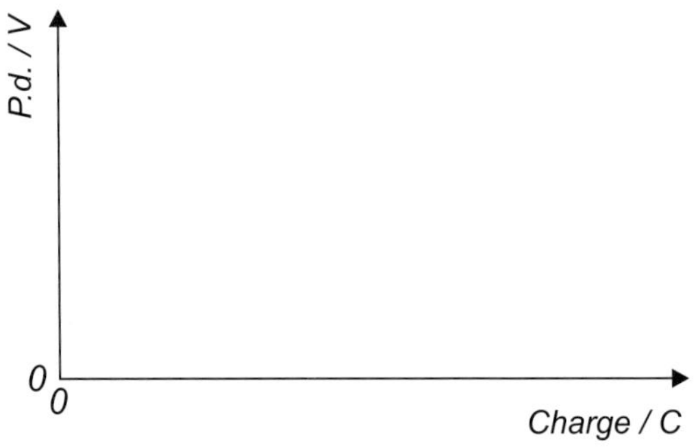

(1 mark)

(ii) State what quantity the area enclosed by the line and the horizontal axis represents.

(1 mark)

(b) When the capacitor is fully charged, it has a charge of 18 μC. Calculate the capacitance of the capacitor.

(2 marks)

The student then discharges the capacitor through a 29 kΩ resistor.

(c) What is the potential difference across the capacitor plates after 0.030 s?

(2 marks)

(d) Sketch a graph showing how the current supplied to the resistor varies over time as the capacitor discharges.

(1 mark)

(e) Calculate the time taken for the current in the circuit to fall to 10% of its initial value.

(2 marks)

5 A 3.0×10^{-3} F capacitor was fully charged to 50.0 V through a 2.0 kΩ resistor.

(a) Calculate how much energy was stored by the capacitor.

(2 marks)

(b) Show that the time constant of the circuit is 6 seconds.

(1 mark)

(c) Calculate the charge gained by the capacitor in 14 seconds.

(2 marks)

(d) Sketch a graph of charge against time for the first 14 seconds of charging.

(1 mark)

Module 6 Section 2: Electric Fields

1. Electric Fields

Electric fields are a lot like gravitational fields. But although the concepts are similar, there are still some subtle differences, so make sure you follow carefully.

What are electric fields?

There are several fields that can cause forces — for example you met gravitational fields on p.62. Electric fields can also give rise to a force, but electric fields can be attractive or repulsive, unlike gravitational fields (which are always attractive). It's all to do with charge. Any object with charge has an electric field around it — the region where it can attract or repel other charges.

Electric charge is measured in coulombs (C) and can be positive or negative. Oppositely charged particles attract each other, and like charges repel each other. If a charged object is placed in an electric field, then it will experience a force.

Just like with gravitational fields, electric fields can be represented by field lines. Electric field lines are drawn to show the direction of the force that would act on a positive charge — see p.134 and p.137.

Electric field strength

Electric field strength, ***E***, is defined as the force per unit positive charge. It's the force that a charge of +1 C would experience if it was placed in an electric field.

$$E = \frac{F}{Q}$$

E = electric field strength in NC^{-1}

F = force on the charged object in N

Q = charge of the object in C

E is a vector pointing in the direction that a positive charge would move. The units of ***E*** are newtons per coulomb (NC^{-1}). The field strength may depend on where you are in the field (see radial fields on page 135).

Be careful — in this equation ***F*** is the force acting on a charge *Q* which is in the electric field. *Q* is not causing the electric field. Don't confuse it with the *Q* in the electric field strength equation on page 135.

Example — Maths Skills

Find the force acting on an electron in an electric field with a field strength of 5000 NC⁻¹. The charge on an electron is –1.60 × 10⁻¹⁹ C.

Just rearrange the equation for electric field strength and put in the numbers:

$$E = \frac{F}{Q} \Rightarrow F = E \times Q = 5000 \times (-1.60 \times 10^{-19}) = -8 \times 10^{-16}\ \text{N}$$

Learning Objectives:

- Understand the concept of electric fields as being one of a number of forms of field giving rise to a force.
- Know that electric fields are due to charges.
- Understand how electric field lines are used to map electric fields.
- Know what electric field strength is.
- Know how to use the equation $E = \frac{F}{Q}$.
- Know that a uniformly charged sphere can be modelled as a point charge at its centre.
- Know how to calculate the force between two point charges using Coulomb's law, $F = \frac{Qq}{4\pi\varepsilon_0 r^2}$.
- Know how to calculate electric field strength using $E = \frac{Q}{4\pi\varepsilon_0 r^2}$ for a point charge.

Specification References 6.2.1 and 6.2.2

Tip: Remember, vector means it has a magnitude and a direction.

Tip: Electric field strength can be measured in Vm^{-1} as well as NC^{-1}.

Radial fields

Point charges have a **radial electric field** (see Figure 1). Remember, the field lines show the direction in which a positive charge (shown by $+q$ in Figure 1) would feel a force when placed in the electric field. So for a positive point charge, $+Q$, the field lines point away from the point charge, and for a negative point charge, $-Q$, they point towards it.

Tip: The charge $+q$ in Figure 1 also has an electric field — it just hasn't been drawn here.

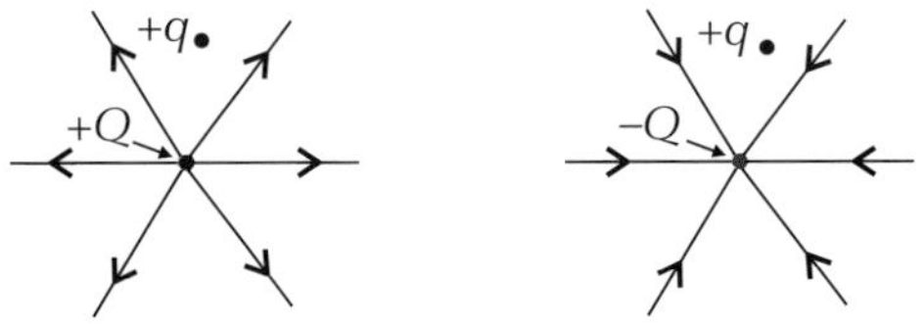

Figure 1: *Electric field lines for a positive point charge and a negative point charge.*

Figure 2: *You can charge a balloon by rubbing it against a material. The charge of the balloon can then be treated as a point charge at the balloon's centre (assuming the balloon is spherical and the charge is evenly distributed over the balloon's surface).*

Notice how the field lines get further apart as you move further from the charge — this shows that the electric field strength is getting weaker.

A uniformly charged sphere has its charge evenly distributed across its surface. You can treat this as if it is a point charge with all of its charge concentrated at the centre of the sphere. Any body that behaves as if all its charge is concentrated at the centre has a radial field.

Coulomb's law

You can calculate the force on a charged object in a radial electric field using Coulomb's law. It gives the force of attraction or repulsion between two point charges, Q and q, in a vacuum:

$$F = \frac{Qq}{4\pi\varepsilon_0 r^2}$$

F = force on the object in N

Q and q = charges of the two objects in C

ε_0 = "epsilon-nought", the permittivity of free space = 8.85×10^{-12} $C^2N^{-1}m^{-2}$

r = distance between Q and q in m

The force on Q is always equal and opposite to the force on q — the direction depends on the charges.

Tip: It's the interaction between the electric fields of Q and q that causes the force between them.

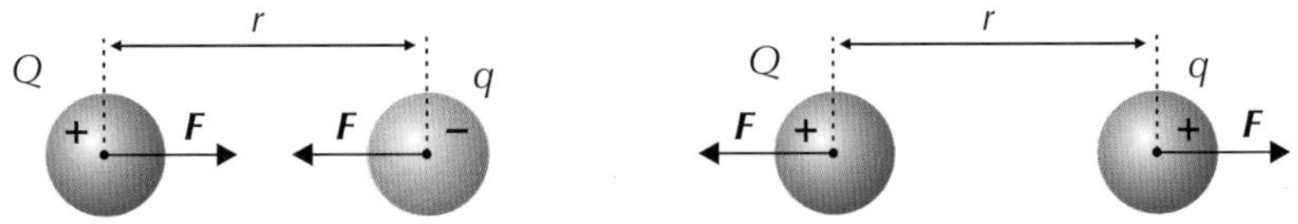

*If the charges are opposite then the force is attractive. **F** will be negative.*

*If the charges are alike then the force is repulsive. **F** will be positive.*

Figure 3: *The direction of the forces on two charged objects.*

Coulomb's law is another case of an inverse square law, so $F \propto \frac{1}{r^2}$ (page 64). The further apart the charges, the weaker the force between them. The size of the force F also depends on the permittivity, ε, of the material between the two charges. For free space (a vacuum), the permittivity is $\varepsilon_0 = 8.85 \times 10^{-12}$ $C^2N^{-1}m^{-2}$. If the two charges weren't in a vacuum, you would replace ε_0 with the ε for the material they're in (see page 137 for more on permittivity).

Tip: You can also give the units of ε_0 as Fm^{-1}.

Example — Maths Skills

Find the acceleration experienced by a free electron 2.83 mm from the centre of a sphere carrying a charge of +0.510 μC. The charge of an electron is –1.60 × 10⁻¹⁹ C and the mass of an electron is 9.11 × 10⁻³¹ kg. (You may assume the sphere has a radius less than 2.83 mm.)

The acceleration of an object is given by $\boldsymbol{F} = m\boldsymbol{a} \Rightarrow \boldsymbol{a} = \frac{\boldsymbol{F}}{m}$.

You know the mass, so find the force using Coulomb's law:

$$\boldsymbol{F} = \frac{Qq}{4\pi\varepsilon_0 r^2} = \frac{(0.510 \times 10^{-6}) \times (-1.60 \times 10^{-19})}{4\pi\varepsilon_0 \times (2.83 \times 10^{-3})^2}$$
$$= -9.161... \times 10^{-11}\ \text{N}$$

Then use this to find the acceleration:

$$\boldsymbol{a} = \frac{\boldsymbol{F}}{m} = \frac{-9.161... \times 10^{-11}}{9.11 \times 10^{-31}} = -1.00564... \times 10^{20}$$
$$= -1.01 \times 10^{20}\ \text{ms}^{-2}\ \text{(to 3 s.f.)}$$

Exam Tip
Always assume the charges are in a vacuum if the question doesn't say otherwise.

Tip: The acceleration is negative because it's towards the sphere.

Electric field strength in radial fields

When the electric field is being generated by a point charge, we call the charge generating the field Q and redefine the charge experiencing the force as q (as in Figure 1 on p.134). In a radial field, $\boldsymbol{E}$ is equal to:

$$\boldsymbol{E} = \frac{Q}{4\pi\varepsilon_0 r^2}$$

$\boldsymbol{E}$ = electric field strength in NC^{-1}

$\boldsymbol{Q}$ = point charge in C

ε_0 = the permittivity of free space in $C^2N^{-1}m^{-2}$

$\boldsymbol{r}$ = distance from the point charge in m

Tip: Q is the point charge creating the radial field, not the charge experiencing a force inside the field.

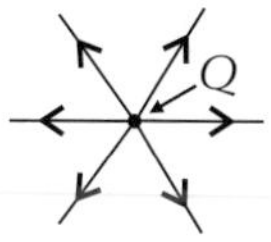

Example — Maths Skills

The electric field strength 0.15 m away from the centre of a charged sphere is 44 000 NC⁻¹. What's the charge on the sphere? (You may assume the sphere has a radius less than 0.15 m.)

$$\boldsymbol{E} = \frac{Q}{4\pi\varepsilon_0 r^2} \Rightarrow Q = (4\pi\varepsilon_0 r^2)\boldsymbol{E} = 4\pi \times (8.85 \times 10^{-12}) \times 0.15^2 \times 44\,000$$
$$= 1.10100... \times 10^{-7}$$
$$= 1.1 \times 10^{-7}\ \text{C (to 2 s.f.)}$$

Tip: Remember — a uniformly charged sphere can be treated as if all its charge acts at its centre (p.134).

This is another case of the inverse square law — $\boldsymbol{E} \propto \frac{1}{r^2}$. Field strength decreases as you go further away from Q. On a diagram, the field lines of a radial field get further apart (see Figure 1 on p.134), and if you plot the electric field strength against r you get the same shape as for gravitational field strength on page 66. If the charge isn't a point charge (e.g. if it's a charged metal sphere), then the electric field strength inside the object doesn't have the same $\boldsymbol{E} \propto \frac{1}{r^2}$ relation (you don't need to worry what the electric field is inside a charged object, it just doesn't follow an inverse square law).

Tip: Graphs of $\boldsymbol{E}$ against r usually only show the magnitude of $\boldsymbol{E}$ (so they're always positive). But if $\boldsymbol{E}$ was negative, the graph would just be reflected in the horizontal axis.

Tip: The graph of $\boldsymbol{g}$ against r is the same shape, but it's always under the r axis as $\boldsymbol{g}$ is always negative (p.66).

Tip: This only works if the point you're interested in lies on the same line as the two charges. Anywhere else and you have to start finding resultant vectors.

Tip: You can also find the total force, $\boldsymbol{F}$, on a charge that's between two charges. It's very similar to this method, since $\boldsymbol{F} = \boldsymbol{E}q$.

Tip: You could also have said that towards Q_2 was positive. You'd get a positive answer, showing that a positive charge at the mid point would move towards Q_2.

Tip: The charge on an electron, $-e$, is -1.60×10^{-19} C.

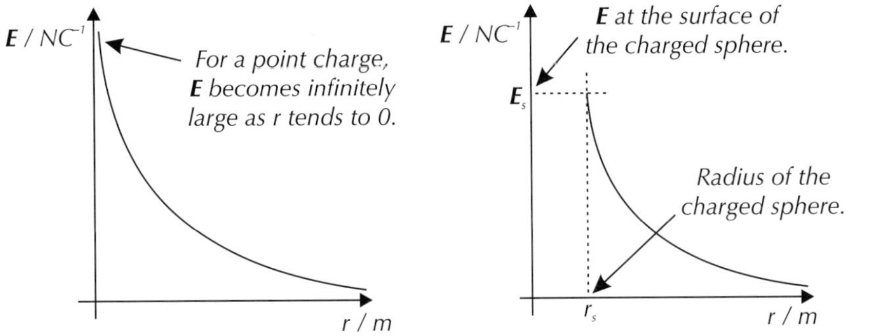

Figure 4: *Graphs of* ***E*** *against r for a point charge (left) and a charged sphere (right).*

You can also work out the combined electric field strength for a point that lies between two different charges, Q_1 and Q_2. Remember, $\boldsymbol{E}$ is a vector, so you need to pick a positive direction.

Example — Maths Skills

Find the electric field strength at a point halfway between point charges of $Q_1 = +1.7\ \mu C$ and $Q_2 = +1.2\ \mu C$, separated by 2.4×10^{-2} m.

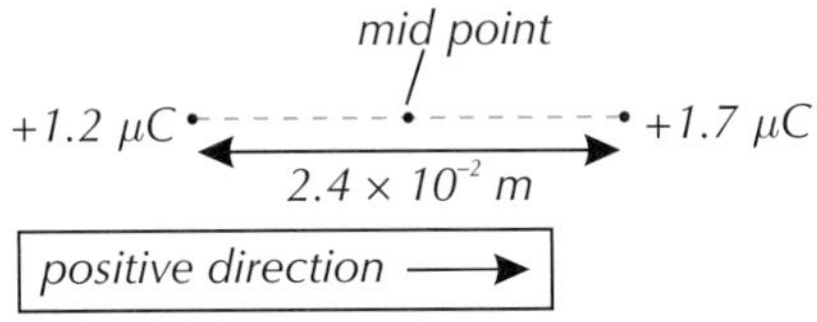

You're working with vectors, so choose a positive direction.

Let's say towards Q_1 is positive.

Both charges would repel a positive charge, so $\boldsymbol{E}$ due to Q_2 is positive and $\boldsymbol{E}$ due to Q_1 is negative. So the electric field strength, $\boldsymbol{E}$, is equal to $\boldsymbol{E}$ due to Q_2, minus $\boldsymbol{E}$ due to Q_1:

$$\boldsymbol{E} = \frac{1.2 \times 10^{-6}}{4\pi \times (8.85 \times 10^{-12}) \times (1.2 \times 10^{-2})^2} - \frac{1.7 \times 10^{-6}}{4\pi \times (8.85 \times 10^{-12}) \times (1.2 \times 10^{-2})^2}$$

$$= \frac{1.2 \times 10^{-6} - 1.7 \times 10^{-6}}{4\pi \times (8.85 \times 10^{-12}) \times (1.2 \times 10^{-2})^2} = \frac{-0.5 \times 10^{-6}}{4\pi \times (8.85 \times 10^{-12}) \times (1.2 \times 10^{-2})^2}$$

$$= -3.122... \times 10^{7} = -3.1 \times 10^{7}\ \text{NC}^{-1}\ \text{(to 2 s.f.)}$$

Practice Questions — Application

Q1 Two electrons are fired towards each other and reach a separation of 5.22×10^{-13} m. What's the force on each electron at this point?

Q2 The electric field generated by a charged sphere with charge 4.15 μC is measured as 15 000 NC^{-1}. How far from the centre of the sphere must the measuring instrument be?

Q3 A particle with charge 5.0×10^{-5} C is in an electric field and experiences a force of 0.080 N. Calculate the magnitude of the electric field strength.

Practice Questions — Fact Recall

Q1 What's $\boldsymbol{E}$ a measure of?

Q2 Draw the electric field generated by a positive point charge.

Q3 Give the equation of Coulomb's law.

Q4 Sketch a graph of $\boldsymbol{E}$ and r for a positive point charge, where r is the distance from the charge.

2. Uniform Electric Fields

You've met radial electric fields, now it's time for uniform electric fields. You get uniform fields in parallel plate capacitors, which you met on p.111.

Learning Objectives:

- Know what a parallel plate capacitor is.
- Know how to calculate uniform electric field strength using $E = \frac{V}{d}$.
- Know what is meant by the permittivity of a material.
- Know and be able to use the equations $C = \frac{\varepsilon_0 A}{d}$ and $C = \frac{\varepsilon A}{d}$ where $\varepsilon = \varepsilon_r \varepsilon_0$.
- Be able to describe the motion of charged particles in a uniform electric field.

Specification Reference 6.2.3

Electric field strength in uniform fields

A **uniform electric field** is one with the same electric field strength everywhere. It can be produced by connecting two parallel plates to the opposite poles of a battery — see Figure 1. This is a parallel plate capacitor. The field lines point from the plate with the more positive potential to the plate with the less positive potential.

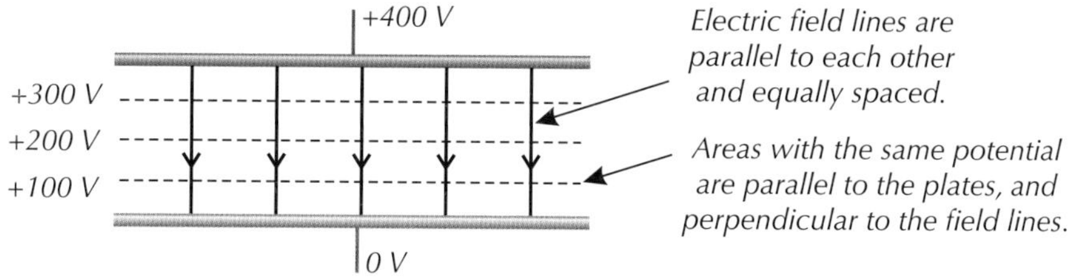

+400 V
+300 V
+200 V
+100 V
0 V
Electric field lines are parallel to each other and equally spaced.
Areas with the same potential are parallel to the plates, and perpendicular to the field lines.

Figure 1: *Electric field lines between parallel plates.*

The field strength $\boldsymbol{E}$ is the same at all points between the two plates and is given by:

$$E = \frac{V}{d}$$

$\boldsymbol{E}$ = electric field strength in Vm^{-1}

$\boldsymbol{V}$ = potential difference between the plates in V

$\boldsymbol{d}$ = distance between the plates in m

Tip: $\boldsymbol{E}$ can be in Vm^{-1} or NC^{-1}.

Note that here the potential difference between the plates is the same as the potential of the top plate. This is because for this example the potential of the bottom plate is 0 V. This won't always be the case, so you might find it easier to think of it as ΔV.

Figure 2: *Electric field lines between two plates shown by the alignment of pepper flakes in oil. The straight electric field lines between the plates show the area with the uniform electric field.*

Example — Maths Skills

What's the electric field strength between two parallel plates 0.15 m apart with a potential of +650 V and +200 V respectively?

$$E = \frac{\Delta V}{d} = \frac{650 - 200}{0.15} = 3000\ Vm^{-1}$$

Capacitance of parallel plate capacitors

The capacitance (p.111) of a capacitor depends on the **permittivity** of the material between the plates — how easy it is to generate an electric field in the material. It also depends on the dimensions of the capacitor. When there's a vacuum between the plates, capacitance can be calculated by using:

$$C = \frac{\varepsilon_0 A}{d}$$

$\boldsymbol{C}$ = capacitance of a capacitor in F

$\boldsymbol{\varepsilon_0}$ = permittivity of free space in Fm^{-1}

$\boldsymbol{A}$ = area of the plates in m^2

$\boldsymbol{d}$ = separation of the plates in m

Tip: Remember, capacitance is measured in farads (F). This is a large unit though, so you'll often see nano- or picofarad capacitors.

If the plates have a material in between them instead of a vacuum, ε_0 is replaced with permittivity, ε:

$$C = \frac{\varepsilon A}{d}$$ where ε is given by: $$\varepsilon = \varepsilon_r \varepsilon_0$$

$\boldsymbol{\varepsilon}$ = permittivity of material in Fm^{-1}

$\boldsymbol{\varepsilon_r}$ = relative permittivity

Tip: ε_r doesn't have any units because it's a ratio of the size of the electric field generated in the material, compared to if it was generated in a vacuum.

Tip: You can assume that the paper fully fills the space between the plates.

Tip: You could also give the answer as 2.9 pF.

Example — Maths Skills

A capacitor is made up of two parallel plates, each with an area of 12 mm². The plates are separated by a 0.10 mm thick piece of paper which has a relative permittivity of 2.7. Calculate the capacitance of this capacitor.

Permittivity of paper $= \varepsilon = \varepsilon_r \varepsilon_0 = 2.7 \times 8.85 \times 10^{-12} = 2.3895 \times 10^{-11}$ Fm^{-1}

Then calculate the capacitance:

$$C = \frac{\varepsilon A}{d} = \frac{2.3895 \times 10^{-11} \times 12 \times 10^{-6}}{0.10 \times 10^{-3}} = 2.8674 \times 10^{-12} = 2.9 \times 10^{-12} \text{ F (to 2 s.f.)}$$

Charged particles in a uniform electric field

A uniform electric field can be used to determine whether a particle is charged or not. The path of a charged particle moving through an electric field (at an angle to the field lines) will bend — the direction depends on whether it's a positive or negative charge.

A charged particle entering an electric field at right angles to the field (as in Figure 3) feels a constant force, equal to $\boldsymbol{F} = \boldsymbol{E}Q$, parallel to the electric field lines. If the particle is positively charged then the force acts on it in the same direction as the field lines. If it's negatively charged, the force is in the opposite direction to the field lines. The work done on the particle by this force ($W = Fd$) increases its kinetic energy and causes it to accelerate at a constant rate in the direction of the force (Newton's second law). If the particle's velocity initially has a component at right angles to the field lines, this component will remain unchanged and the velocity in this direction will be uniform. That's Newton's first law. The combined effect of constant acceleration and constant velocity at right angles to one another is a curved path.

Tip: Remember, the equation for the kinetic energy of an object is $E_k = \frac{1}{2}mv^2$.

Tip: This motion is similar to a projectile in a uniform gravitational field, which you met in year 1 of A-level.

Tip: An electron (which is negatively charged) would accelerate upwards rather than downwards in the electric field in Figure 3.

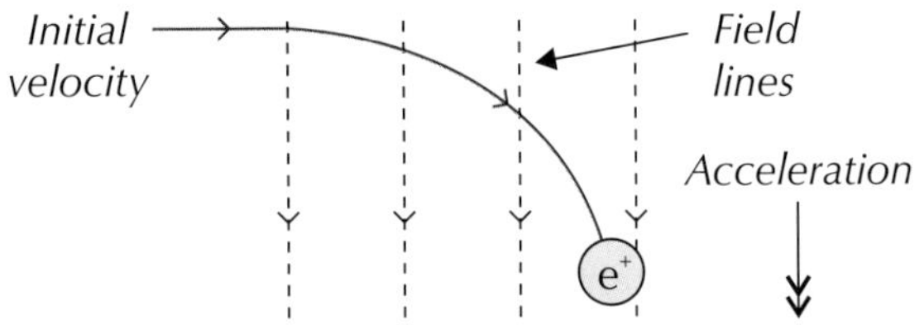

Figure 3: *Motion of a positron (which is positively charged) in a uniform electric field.*

Practice Questions — Application

Q1 Copy the diagram on the right and complete the path of the electron as it travels through the electric field.

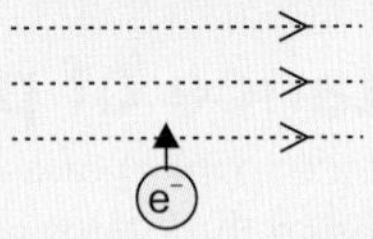

Q2 An alpha particle with a charge of +2e and a mass of 6.64×10^{-27} kg is suspended freely between two parallel plates. The top plate has no charge and the bottom plate is charged to +5.00 nV. How far apart are the plates if the alpha particle isn't moving?

Q3 A capacitor is made up of two parallel plates, each with a 7.0×10^{-4} m² area. The vacuum gap between the plates is 8.5×10^{-6} m wide. Calculate the capacitance of the capacitor.

Tip: For Q2, if the particle isn't moving, the upwards force from the electric field must equal the downwards force from gravity (the weight, $\boldsymbol{W} = mg$).

Tip: 1 nV = 1×10^{-9} V.

Practice Question — Fact Recall

Q1 Give the equation that relates the permittivity of a material to the permittivity of free space. Define all symbols.

3. Electric Potential

Just like the gravitational potential of a mass in a gravitational field, you can find the electric potential of a charge in an electric field. As electric forces can be attractive or repulsive, electric potential can be positive or negative.

What is electric potential?

All points in an electric field have an **electric potential**, V. This is equal to the work done bringing a unit positive charge (+1 C) from a point infinitely far away to that point in the electric field. This means that at infinity, the electric potential will be zero.

So electric potential is the potential energy per unit charge. In a radial field around a point charge, electric potential is given by:

$$V = \frac{Q}{4\pi\varepsilon_0 r}$$

V *= electric potential in V*

Q *= point charge creating the electric field in C*

r *= distance from the point charge in m*

The sign of V depends on the charge Q — V is positive when Q is positive and the force is repulsive, and negative when Q is negative and the force is attractive. The absolute magnitude of V is greatest on the surface of the charge, and decreases as the distance from the charge increases.

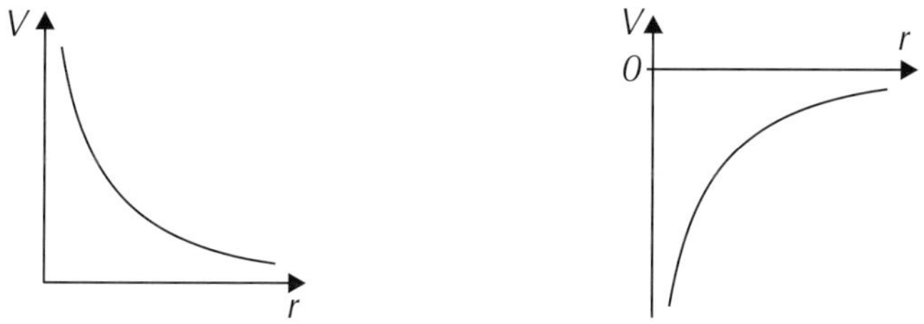

V changing with r for a positive charge (repulsive force). V is initially positive and tends to zero as r increases towards infinity.

V changing with r for a negative charge (attractive force). V is initially negative and tends to zero as r increases towards infinity.

Figure 1: *V changing with r for a positive charge (left) and for a negative charge (right).*

If a charge is moved within an electric field, it can experience a change in electric potential. For example, if a point charge is moved away from or towards a spherical charge, or if a point charge is moved from one plate of a parallel plate capacitor to the other plate.

Example — Maths Skills

A positively charged particle is placed 0.035 m from the centre of a sphere with a charge of +3.1 μC and radius 0.02 m. If the particle is then repelled by 0.19 m, what change in potential does it experience?

Change in potential = final potential – initial potential

$$= \frac{(3.1 \times 10^{-6})}{4\pi\varepsilon_0(0.035 + 0.19)} - \frac{(3.1 \times 10^{-6})}{4\pi\varepsilon_0 \times 0.035}$$

$= -6.72529... \times 10^5 = -6.7 \times 10^5$ V (to 2 s.f.)

Learning Objectives:

- Know that the electric potential at a point in an electric field is the work done in bringing a unit positive charge from infinity to the point.
- Know that electric potential is zero at infinity.
- Be able to use the electric potential equation $V = \frac{Q}{4\pi\varepsilon_0 r}$ at a distance r from a point charge.
- Understand changes in electric potential.
- Be able to use $\frac{Qq}{4\pi\varepsilon r}$ or $\frac{Qq}{4\pi\varepsilon_0 r}$ to calculate electric potential energy of a point charge q at a distance r from a point charge Q.
- Be able to sketch a force-distance graph for a point or spherical charge and know that work done to move a charge is equal to the area under the graph.
- Know and be able to use the equation $C = 4\pi\varepsilon_0 R$ for the capacitance of an isolated sphere, and be able to derive it.

Specification Reference 6.2.4

Tip: If you have more than one charge creating an electric field, you can find the total electric potential at a single point by adding the electric potentials at that point due to each charge.

Tip: You can give electric potential in V or JC^{-1}.

Electric potential energy

You saw on the previous page that electric potential is the electric potential energy that a unit positive charge (+1 C) would have at a certain point. This means you can find the **electric potential energy** for any charge, q, at that point in the electric field by multiplying the electric potential, V, by the value of the charge. So electric potential energy = Vq. By substituting in the equation for V from the previous page, you get:

$$\text{electric potential energy} = \frac{Qq}{4\pi\varepsilon_0 r}$$

Q *= point charge creating the electric field in C*

q *= point charge in the electric field in C*

r *= distance between Q and q in m*

ε_0 *= permittivity of free space in Fm⁻¹*

Tip: You might see electric potential energy labelled as E — <u>don't</u> confuse this E with the $\boldsymbol{E}$ for electric field strength on p.133.

Tip: If the charges weren't in a vacuum, then ε_0 would just be replaced with ε (see p.137).

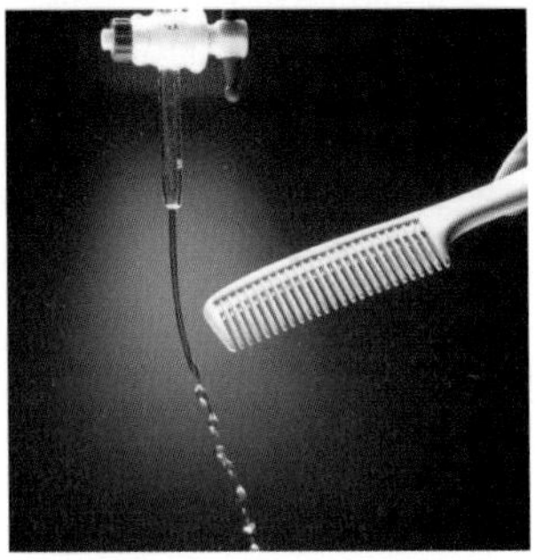

Figure 2: *Static charge on a comb doing work against gravity to lift a trickle of water.*

If you move a unit charge in an electric field and change its electric potential energy, then work has to be done by a force. The work done is equal to the change in the charge's electric potential energy. Both energy and work done are measured in joules (J).

For a radial field (e.g. of a point or spherical charge) you can plot the force applied to the unit charge being moved, $\boldsymbol{F}$, against the distance, r, from the charge producing the electric field. ($\boldsymbol{F} \propto \frac{1}{r^2}$ — see p.134.) The area under the graph between two points gives the work done, W, to move a unit charge between those two points — see Figure 3.

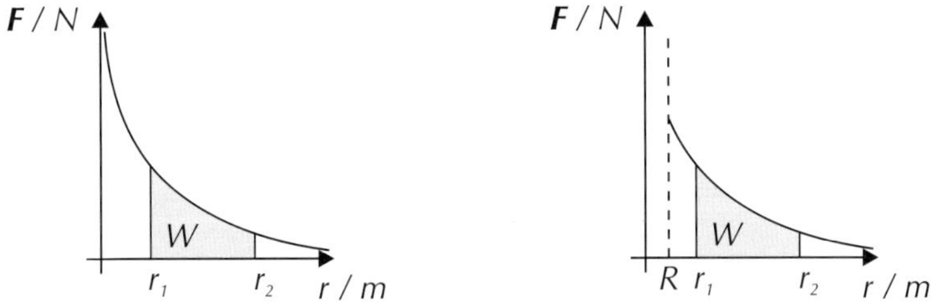

Figure 3: *The area under a graph of* ***F*** *against r is used to find the work done, W, to move a unit charge between r_1 and r_2, in the electric field of a point charge (left) and of a uniformly charged sphere with radius R (right).*

Tip: If you're asked to estimate W from a $\boldsymbol{F}$–r graph, you'll probably have to work out the area under the graph by counting the number of squares on the graph paper or by splitting the area up into trapeziums.

Tip: As with the $\boldsymbol{E}$-r graphs on p.136, you don't need to worry about the force, $\boldsymbol{F}$, <u>inside</u> a charged sphere, just that it doesn't follow an inverse square law.

Example — Maths Skills

A +1.6 μC point charge is in a vacuum.
A –0.80 μC point charge is 0.072 m from the positive charge.

a) Calculate the electric potential energy of the negative charge.

$$\text{Electric potential energy} = \frac{Qq}{4\pi\varepsilon_0 r} = \frac{(1.6\times10^{-6})\times(-0.80\times10^{-6})}{4\pi\times8.85\times10^{-12}\times0.072}$$

$$= -0.15985... = -0.16 \text{ J (to 2 s.f.)}$$

b) Calculate the work done to move the negative charge to be 0.015 m away from the positive charge.

Work done = change in electric potential energy

= final electric potential energy – initial electric potential energy

$$= \frac{(1.6\times10^{-6})\times(-0.80\times10^{-6})}{4\pi\times8.85\times10^{-12}\times0.015} - (-0.15985...)$$

$$= -0.6074... = -0.61 \text{ J (to 2 s.f.)}$$

Tip: There's a force of attraction between the two charges in this example, which is why the work done to move them closer together is negative.

Capacitance of a charged sphere

You can use the formula for the electric potential of a radial field and the fact that $Q = CV$ (p.111) for capacitors to derive an expression for the capacitance of an isolated charged sphere, assuming the charge is evenly distributed.

As it is a charged sphere, you can assume all of its charge is at its centre and treat it like a point charge.

Substitute $V = \frac{Q}{C}$ into $V = \frac{Q}{4\pi\varepsilon_0 R}$, where R is the radius of the sphere:

$$\frac{Q}{C} = \frac{Q}{4\pi\varepsilon_0 R}$$

Cancel out the Q's:

$$\frac{1}{C} = \frac{1}{4\pi\varepsilon_0 R}$$

Rearrange for C, which gives:

$$C = 4\pi\varepsilon_0 R$$

ε_0 = permittivity of free space in Fm^{-1}

C *= capacitance of charged sphere in F*

R *= radius of the charged sphere in m*

Tip: For an 'isolated' charged sphere, you can assume the sphere is infinitely far away from all other charges.

Tip: We put the radius R for the distance, because we want an equation for the electric potential at the surface of the sphere.

Tip: You need to be able to derive this equation from $Q = CV$ and $V = \frac{Q}{4\pi\varepsilon_0 R}$.

Practice Question — Application

Q1 A small metal sphere is being held stationary directly next to a charged metal sphere, as shown on the right.

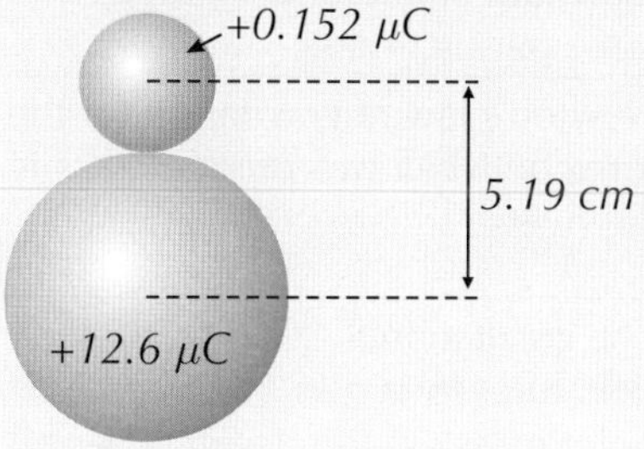

a) What's the electric potential due to the larger sphere at the centre of the smaller metal sphere?

b) The smaller sphere is released so that it can move freely, and is repelled by 12.9 cm before being stopped again. How much work is done in moving the sphere by this distance?

c) The smaller sphere has a radius of 2.11 cm. Calculate its capacitance when it's in isolation.

Tip: The value of ε_0 is 8.85×10^{-12} Fm^{-1}.

Practice Questions — Fact Recall

Q1 What is meant by the electric potential of a point?

Q2 Sketch the graph of electric potential V against distance r in a radial field for:

a) a positive point charge.

b) a negative point charge.

Q3 State the two equations needed to derive an equation for the capacitance of a charged sphere in terms of the radius of the sphere.

Learning Objective:

- Know and understand the similarities and differences between the gravitational field of a point mass and the electric field of a point charge.

Specification Reference 6.2.2

4. Comparing Electric and Gravitational Fields

You might have thought a lot of the formulas from the last topic looked familiar — electric and gravitational fields are more similar than you might think...

The main similarities

A lot of the formulas used for electric fields are the same as those used for gravitational fields but with Q (or q) instead of M (or m) and $\frac{1}{4\pi\varepsilon_0}$ instead of G:

Gravitational field strength: $\boldsymbol{g} = \frac{\boldsymbol{F}}{m}$

Newton's law of gravitation: $\boldsymbol{F} = -\frac{GMm}{r^2}$

Gravitational field strength for a radial field: $\boldsymbol{g} = -\frac{GM}{r^2}$

Gravitational potential: $V_g = -\frac{GM}{r}$

Electric field strength: $\boldsymbol{E} = \frac{\boldsymbol{F}}{Q}$

Coulomb's law: $\boldsymbol{F} = \frac{Qq}{4\pi\varepsilon_0 r^2}$

Electric field strength for a radial field: $\boldsymbol{E} = \frac{Q}{4\pi\varepsilon_0 r^2}$

Electric potential: $V = \frac{Q}{4\pi\varepsilon_0 r}$

Tip: The graphs of $\boldsymbol{g}$ against r (p.66) and V_g against r for gravitational fields (p.68) are also very similar to the graphs of $\boldsymbol{E}$ against r (p.136) and V against r for electric fields (p.139).

Other similarities that are useful to know include:

Gravitational field strength, $\boldsymbol{g}$, is force per unit mass.	Electric field strength, $\boldsymbol{E}$, is force per unit positive charge.
Newton's law of gravitation for the force between two point masses is an inverse square law. $\boldsymbol{F} \propto \frac{1}{r^2}$	Coulomb's law for the electric force between two point charges is also an inverse square law. $\boldsymbol{F} \propto \frac{1}{r^2}$
The gravitational field lines for a point mass... 	The electric field lines for a negative point charge... 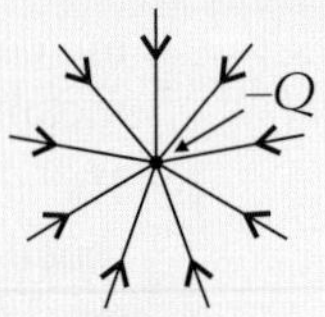
Gravitational potential, V_g, is potential energy per unit mass and is zero at infinity.	Electric potential, V, is potential energy per unit positive charge and is zero at infinity.

Tip: Don't forget — a spherical mass acts as a point mass and a spherical charge acts as a point charge.

The main differences

Although gravitational and electric fields are similar, they're not the same:

- Gravitational forces are always attractive. Electric forces can be either attractive or repulsive.
- Objects can be shielded from electric fields, but not gravitational fields.
- The size of an electric force depends on the medium between the charges, e.g. plastic or air. For gravitational forces, this makes no difference.

Practice Questions — Fact Recall

Q1 List three similarities between electric and gravitational fields.

Q2 Give two differences between electric and gravitational fields.

Section Summary

Make sure you know...

- That an electric field is one type of field that gives rise to a force.
- That you get electric fields around charges.
- That the electric field strength, $\boldsymbol{E}$, is defined as the force per unit positive charge in an electric field.
- How to use the equation $\boldsymbol{E} = \frac{\boldsymbol{F}}{Q}$ to calculate electric field strength.
- That an electric field around a point charge is radial.
- How to represent radial electric fields using field lines, where the direction of the field lines show the direction in which a positive charge would move if it was placed in the field.
- That the further apart electric field lines are, the weaker the electric field at that point.
- That a charged sphere can be treated as if all of its charge is concentrated at the centre of the sphere, and so has a radial electric field.
- How to use Coulomb's law, $\boldsymbol{F} = \frac{Qq}{4\pi\varepsilon_0 r^2}$, to find the force between two point charges in a vacuum.
- That ε_0 is the permittivity of free space.
- How to use the equation $\boldsymbol{E} = \frac{Q}{4\pi\varepsilon_0 r^2}$ to calculate the magnitude of $\boldsymbol{E}$ in a radial field.
- What a graph of $\boldsymbol{E}$ against r for a radial field looks like.
- What is meant by a uniform electric field.
- What a parallel plate capacitor is, and that it contains a uniform electric field when there is a potential difference across the plates.
- How to use the equation $\boldsymbol{E} = \frac{V}{d}$ to calculate the electric field strength of a uniform electric field.
- That the equation used to calculate the capacitance of a parallel plate capacitor is $C = \frac{\varepsilon_0 A}{d}$ when there is a vacuum between the two plates, and that the equation becomes $C = \frac{\varepsilon A}{d}$ when there is another material between the two plates.
- The equation $\varepsilon = \varepsilon_r \varepsilon_0$ and how to use it to calculate the permittivity of a material.
- That a charged particle follows a curved path when entering an electric field at right angles.
- What is meant by the electric potential at a point, and that electric potential becomes zero at an infinite distance from a point charge.
- How to use the equation $V = \frac{Q}{4\pi\varepsilon_0 r}$ to calculate the electric potential at a given point in a radial electric field.
- That a charge can experience a change in electric potential when moving through an electric field.
- How to use the formula $\frac{Qq}{4\pi\varepsilon_0 r}$ to calculate the electric potential energy at any distance from a point charge.
- How to sketch a force-distance graph for both a point and a spherical charge, and that the area under this graph between two distances will give the work done to move a unit charge between those two distances.
- How to derive the equation $C = 4\pi\varepsilon_0 R$ and how to use it to calculate the capacitance of an isolated charged sphere.
- The similarities and differences between gravitational and electric fields.

Exam-style Questions

1 The diagram shows a (positively charged) alpha particle being fired in a straight line between charged parallel plates, travelling parallel to the plates. Which of the following statements is correct?

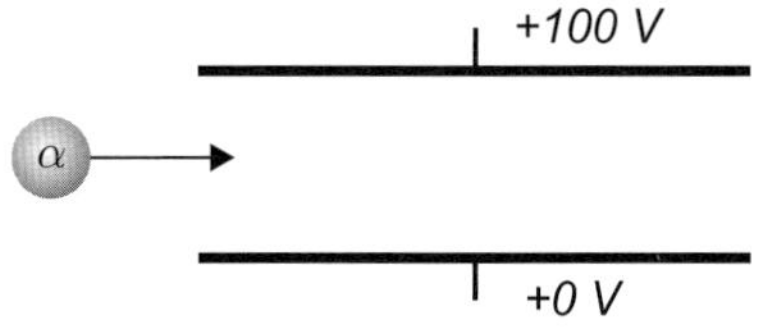

A The particle's path will be deflected downwards.

B The particle's path will be deflected upwards.

C The particle's path will not be deflected.

D The particle will be forced back in the direction it came.

(1 mark)

2 Which of the following is true for an isolated charged sphere in a vacuum?

A The capacitance of the sphere is only dependent on its mass.

B The capacitance of the sphere is only dependent on its radius.

C The capacitance of the sphere is only dependent on its charge.

D The capacitance of the sphere is only dependent on its charge and radius.

(1 mark)

3 Which of the following graphs shows how the magnitude of the force, F, needed to move a point charge away from a charged sphere with radius R, varies with distance, r.

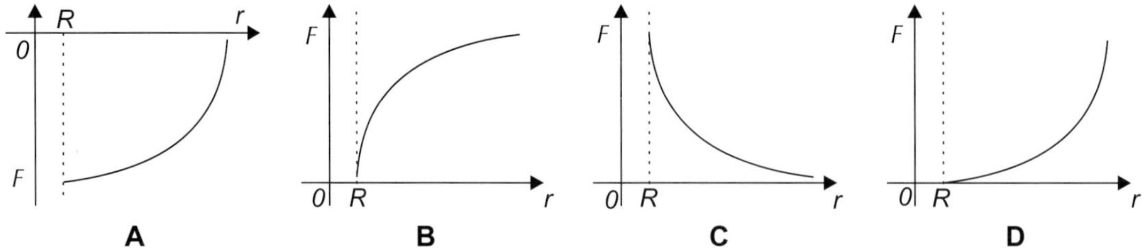

(1 mark)

4 Point A is exactly halfway between a lithium nucleus (containing 3 protons and 2 neutrons) and an electron. The electric field strength at point A is $\boldsymbol{E}$.

One of the protons is then removed from the lithium nucleus, making it a helium nucleus. What is the new electric field strength at point A?

A $\frac{3E}{4}$ **B** $\frac{E}{2}$ **C** $\frac{2E}{3}$ **D** $2E$

(1 mark)

5 A charged sphere carrying a charge of –34.7 μC is fixed to the bottom of a plastic tube. Another sphere, with a charge of –92.5 μC, is dropped into the tube and allowed to move freely until it comes to a rest above the fixed sphere. The force due to gravity acting on the top sphere is 1.99 N. This is shown in **Fig 5.1**.

Fig 5.1

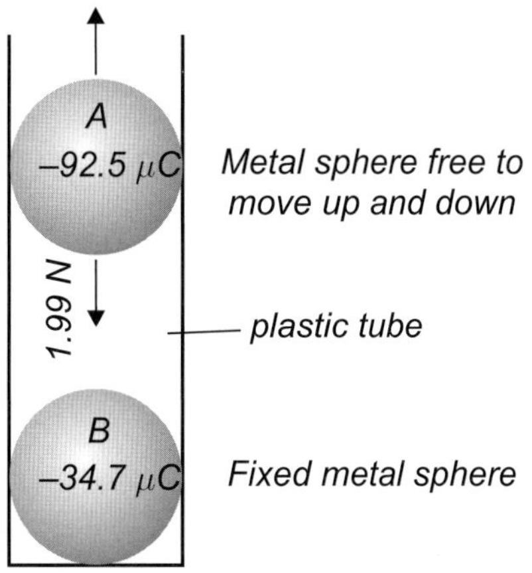

(a) Calculate the distance between the centres of the spheres when sphere *A* comes to rest. You can assume that they do not touch.

(2 marks)

The top sphere is now placed on a smooth, horizontal plane between two parallel plates, as shown (from above) in **Fig 5.2**.

Fig 5.2

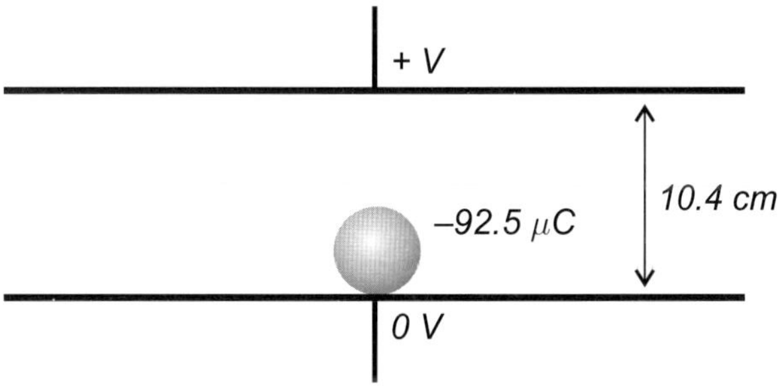

(b) Describe the electric field produced between the two parallel plates.

(2 marks)

(c) The electric field causes a force of 0.18 N to act on the sphere. Calculate the electric potential of the top plate, *V*.

(3 marks)

(d) The capacitance of the two parallel plates is 34.5 pF, and the area of each plate is 0.31 m^2. State whether the space between the plates is a vacuum or not.

(2 marks)

1. Magnetic Fields

Learning Objectives:
- Know that magnetic fields are due to moving charges or permanent magnets.
- Be able to use magnetic field lines to map magnetic fields.
- Be able to sketch the magnetic field patterns for a long straight current-carrying conductor, a flat coil and a long solenoid.
- Be able to use Fleming's left-hand rule.

Specification Reference 6.3.1

You've met gravitational fields, you've met electric fields, and now it's time for the final sort — magnetic fields.

What is a magnetic field?

A **magnetic field** is similar to a gravitational field (page 62) and an electric field (page 133) — it's a region in which a force acts. In a magnetic field, a force is exerted on magnetic materials (e.g. iron).

Magnetic fields exist around permanent magnets and moving charges. Magnetic fields can be represented (mapped) by **field lines**. Field lines go from north to south. The closer the lines, the stronger the field. If the field lines are equally spaced and in the same direction the field is uniform (i.e. the same everywhere).

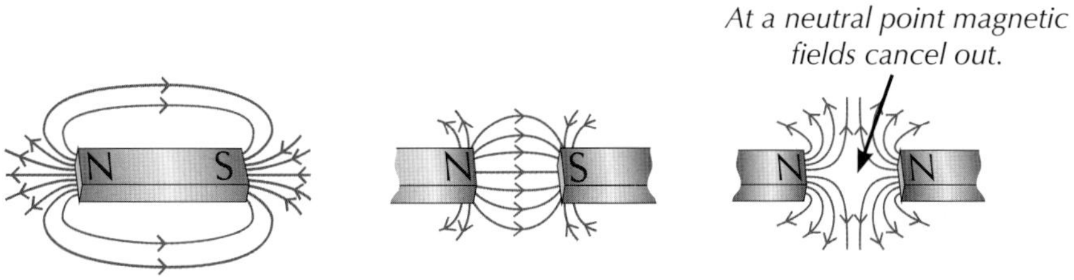

Figure 1: *The magnetic fields created by bar magnets.*

Magnetic fields around a wire

When current flows in a wire or any other long straight conductor, a magnetic field is induced around the wire. The field lines are concentric circles centred on the wire. The direction of a magnetic field around a current-carrying wire can be worked out with the right-hand rule:

- Curl your right hand into a fist and stick your thumb up.
- Point your thumb in the direction of the current through the wire.
- Your curled fingers will then show the direction of the field.

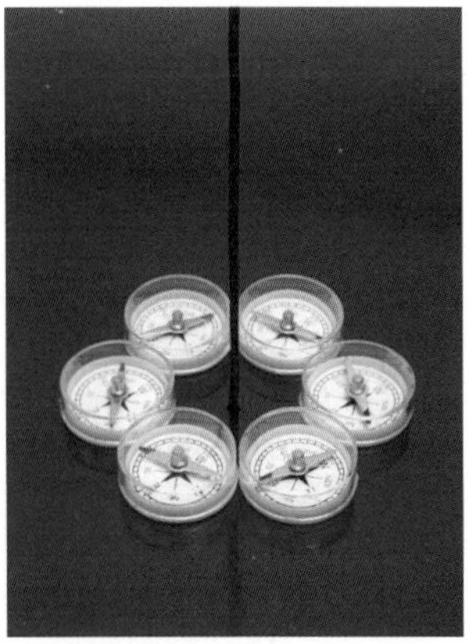

Figure 2: *A current-carrying wire induces a circular magnetic field around it — the needles of the small compasses follow a circle around the wire.*

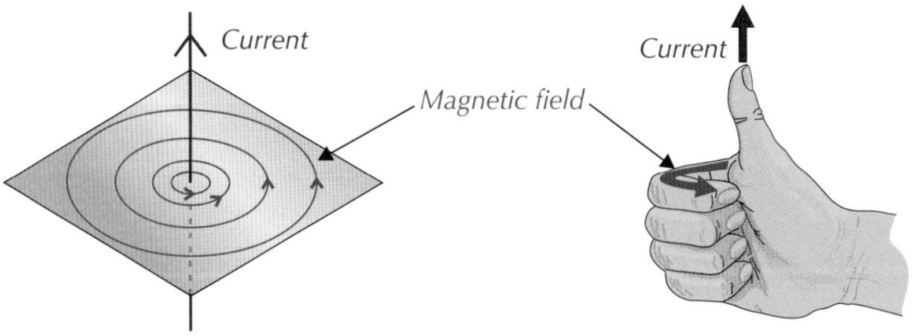

Figure 3: *Diagram to show how your right hand can be used to show the direction of magnetic field lines around a current-carrying wire.*

Flat coils and solenoids

You can loop a current-carrying wire into a coil to get different magnetic field shapes. There are three types of coil you need to know: the single turn coil, the flat coil and the long **solenoid** (see Figure 4).

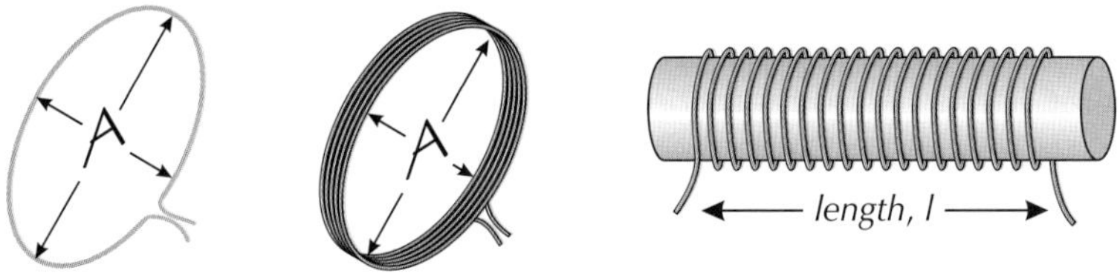

Figure 4: *A single turn coil of area A (left), a flat coil of area A and N turns where N >1 (middle), and a long solenoid of N turns and length l* (right).

The magnetic field surrounding a single turn coil and a flat coil is doughnut shaped, while a solenoid forms a field like a bar magnet (see Figure 5).

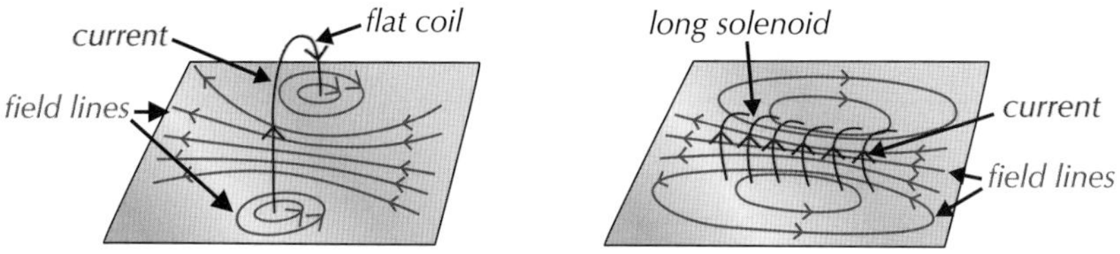

Figure 5: *The magnetic fields created by a current-carrying wire in a flat coil (left) and a solenoid (right).*

Force on a current-carrying wire

If you put a current-carrying wire into an external magnetic field (e.g. between two magnets), the field around the wire and the field from the magnets are added together. This causes a resultant field — lines closer together show where the magnetic field is stronger. These bunched lines cause a 'pushing' force on the wire. Figure 7 shows the resultant field and force when a current-carrying wire is perpendicular to an external magnetic field.

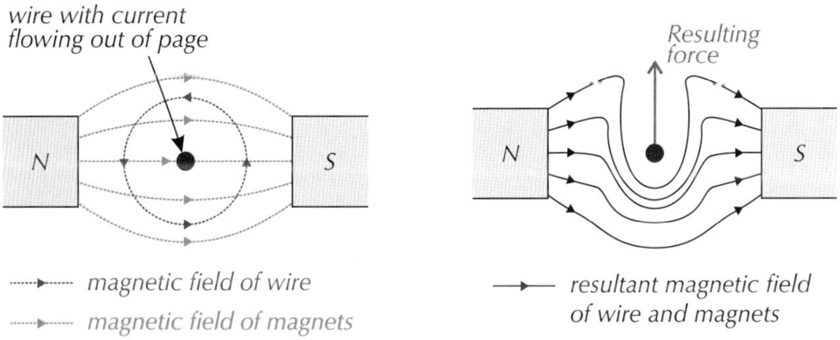

Figure 7: *Diagrams showing the magnetic field of a wire in an external magnetic field (left) and the resultant magnetic field and force produced (right).*

The size of the force depends on the component of the magnetic field that is perpendicular to the current. The direction of the force is always perpendicular to both the current direction and the magnetic field — it's given by Fleming's left-hand rule (see the next page). If the current is parallel to the field lines the size of the force is 0 N — there is no component of the magnetic field perpendicular to the current.

Tip: A solenoid is just a coil with length.

Tip: The right-hand rule also works for a coil. Point your thumb in the direction of the current and you'll see that the induced magnetic field curls around the coil like a doughnut.

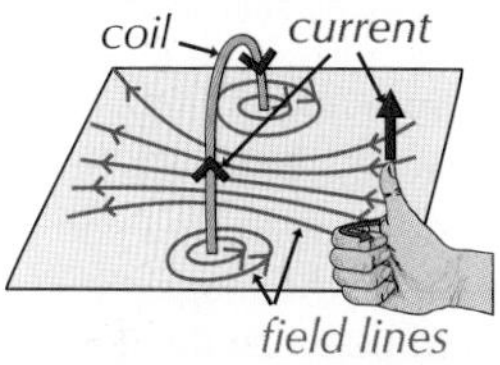

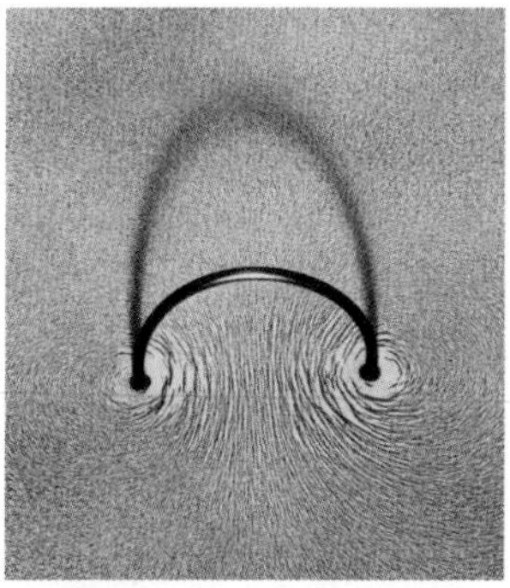

Figure 6: *Steel filings show the magnetic field around a single-turn coil.*

Tip: A solid dot like in Figure 7 (●) shows current flowing out of the page (or towards the reader). A circle with a cross (⊗) shows current going into the page (or away from the reader). This notation can also be used to show field lines going into or coming out of the page.

Tip: The external magnetic field in Fleming's left-hand rule doesn't take into account the field around the wire as in Figure 7. It's just the field before the current was there.

Tip: Conventional current flows in the direction that a positive charge would flow, i.e. from positive to negative.

Tip: As long as the magnetic poles are wide enough, you can assume the magnetic field between them is uniform.

Figure 9: *British physicist John Ambrose Fleming.*

Fleming's left-hand rule

You can use your left hand to find the direction of the current, the direction of the external magnetic field or the direction of the force on the wire (as long as you know the other two). Stretch your thumb, forefinger and middle finger out, as shown in Figure 8, and use the following rules:

- The **F**irst finger points in the direction of the uniform magnetic **F**ield.
- The se**C**ond finger points in the direction of the conventional **C**urrent.
- The thu**M**b points in the direction of the force (the direction of **M**otion).

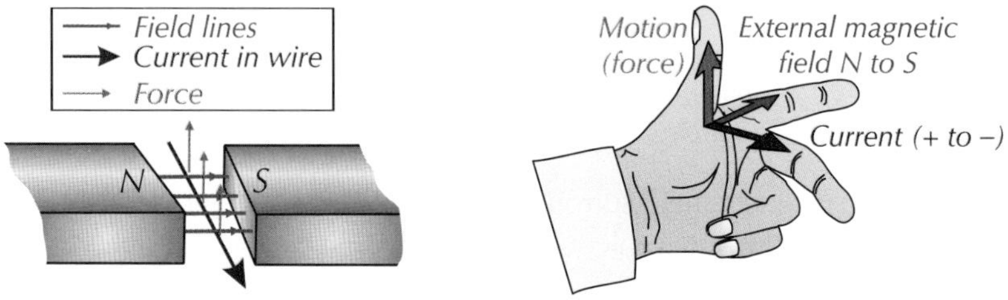

Figure 8: *Fleming's left-hand rule for a current-carrying wire in a magnetic field.*

Example

A current-carrying wire runs between two magnets, as shown on the right. What direction will the force on the wire be?

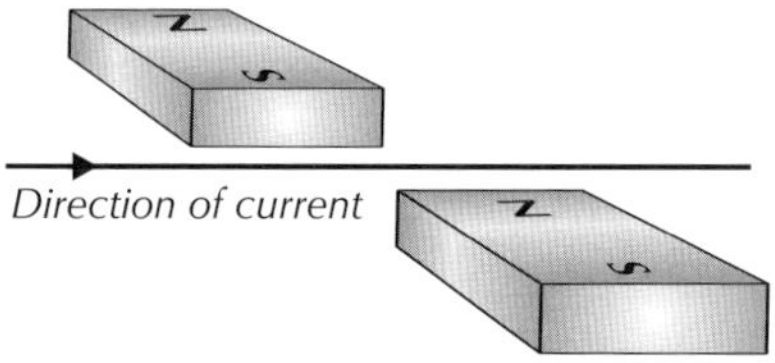

Just use Fleming's left-hand rule to find the direction of the force. Remember, the magnetic field goes from the north pole to the south pole, so the force on the wire acts upwards.

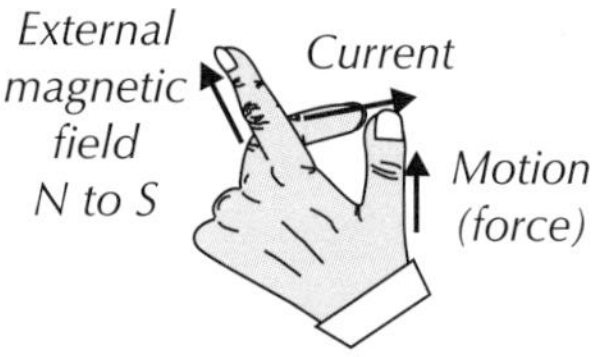

Practice Questions — Application

Q1 A current-carrying wire is shown on the right with the current travelling into the page. Copy the diagram and sketch the magnetic field generated by the current.

Q2 A section of wire with a direct current running through it is fixed at two points and placed in a uniform magnetic field as shown. In what direction will the force on the wire act?

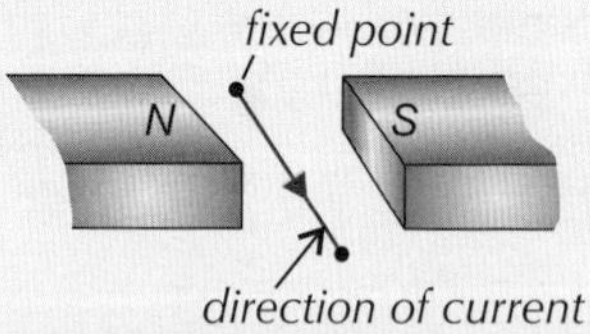

Practice Questions — Fact Recall

Q1 Where are magnetic fields found?

Q2 In which direction do magnetic field lines point?

Q3 State which aspect of a magnetic field is represented by the closeness of magnetic field lines.

Q4 Which hand can you use to find the direction of the force acting on a current-carrying wire at a right angle to a magnetic field? State what each finger represents on this hand.

2. Magnetic Flux Density

Magnetic fields, as you might expect, can have different field strengths. The strength of a magnetic field is known as the magnetic flux density.

Learning Objectives:

- Know what magnetic flux density is, and that its unit is the tesla.
- Know how to calculate the force acting on a current-carrying conductor using $F = BIL\sin\theta$.
- Know techniques and procedures used to determine the uniform magnetic flux density between the poles of a magnet using a current-carrying wire and a digital balance.

Specification Reference 6.3.1

What is magnetic flux density?

The force on a current-carrying wire at a right angle to an external magnetic field is proportional to the **magnetic flux density**, B. Magnetic flux density is sometimes called the strength of the magnetic field. It is defined as:

> The force on one metre of wire carrying a current of one amp at right angles to the magnetic field.

Magnetic flux density is a vector quantity with both a direction and a magnitude. It is measured in teslas, T. One tesla is equal to one newton per amp per metre:

$$1 \text{ tesla} = 1 \frac{\text{N}}{\text{Am}}$$

Tip: It can help to think of flux density as the number of magnetic field lines per unit area.

Tip: One tesla is also equivalent to 1 weber per square metre. Webers (Wb) are the unit of magnetic flux (page 159).

When a current-carrying wire is at 90° to a magnetic field, the size of the force on the wire, F, is proportional to the current, I, the length of wire in the field, L, and the flux density, B, of the external magnetic field. This gives the equation:

$$F = BIL$$

F = force on a current-carrying wire in N

I = current through the wire in A

B = magnetic flux density in T

L = length of the wire in m

Example — Maths Skills

A section of wire carrying a current of 5.2 A is placed at right angles to a uniform magnetic field with a flux density of 19 mT. If the wire experiences a force of 1.2 × 10⁻² N, what length of wire is inside the magnetic field?

First rearrange the formula $F = BIL$ to make L the subject:

$$F = BIL \Rightarrow L = \frac{F}{BI}$$

Put in the numbers:

$$L = \frac{1.2 \times 10^{-2}}{(19 \times 10^{-3}) \times 5.2} = 0.121... = 0.12 \text{ m (to 2 s.f.)}$$

Tip: 1 mT (millitesla) is equal to 1×10^{-3} T.

When a current-carrying wire is at any angle to the magnetic field, the force acting on the wire is caused by the component of the magnetic field which is perpendicular to the wire — $B\sin\theta$, see Figure 1.

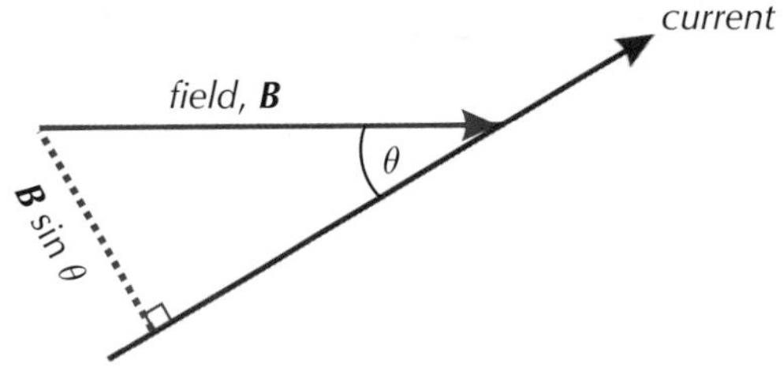

Figure 1: *Diagram showing a magnetic field,* ***B****, at an angle, θ, to a current-carrying wire. The component of the magnetic field which is perpendicular to the current-carrying wire is also shown (=* ***B****sinθ).*

So, for a wire at an angle θ to the field, the force acting on the wire is given by:

$$F = BIL \sin\theta$$

- ***F*** *= force on a current-carrying wire in N*
- ***I*** *= current through the wire in A*
- ***θ*** *= angle between wire and field*
- ***B*** *= magnetic flux density in T*
- ***L*** *= length of the wire in m*

The $\sin\theta$ in the equation means that when the wire and magnetic field are perpendicular to one another ($\theta = 90°$), the force acting on the wire is at a maximum ($\sin\theta = 1$). When the current and field are parallel ($\theta = 0°$) there's no force acting on the wire ($\sin\theta = 0$) — see Figure 3.

Figure 2: *Serbian-American physicist Nikola Tesla, after whom the SI unit of magnetic flux density is named.*

Tip: Remember, $\sin 90° = 1$ and $\sin 0° = 0$.

Exam Tip
You'll be given this equation in the exam data and formulae booklet, but not $F = BIL$. Remember, $F = BIL$ is just a special case where $\sin\theta = \sin 90° = 1$ (see Figure 3).

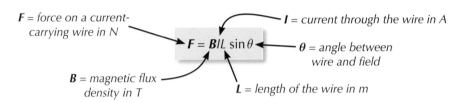

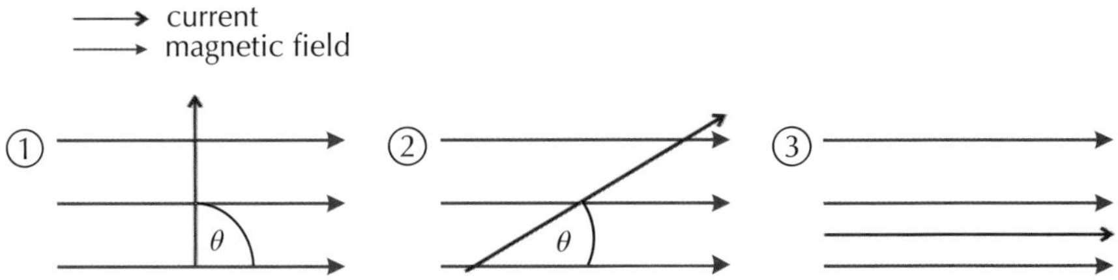

If $\theta = 90°$, $F = BIL$ — If $\theta = 30°$, $F = BIL \times 0.5$ — If $\theta = 0°$, $F = 0$

Figure 3: *The force acting on a current-carrying wire in a uniform magnetic field at 90°, 30° and 0°.*

Example — Maths Skills

A 0.71 m wire carrying a current of 4.1 A lies at 32° to a uniform magnetic field. The magnetic field exerts a force of 0.17 N on the wire. Calculate the magnetic flux density of the magnetic field.

Rearrange $F = BIL\sin\theta$ to make B the subject, then plug in the numbers:

$$F = BIL\sin\theta \Rightarrow B = \frac{F}{IL\sin\theta}$$

$$B = \frac{0.17}{4.1 \times 0.71 \times \sin 32°} = 0.110...$$

$$= 0.11\ \text{T (to 2 s.f.)}$$

Measuring magnetic flux density

You can use a digital balance and the set-up shown in Figure 4 to investigate the uniform magnetic field between the poles of a magnet and obtain a value for flux density, $\boldsymbol{B}$, using $\boldsymbol{F} = \boldsymbol{B}IL$.

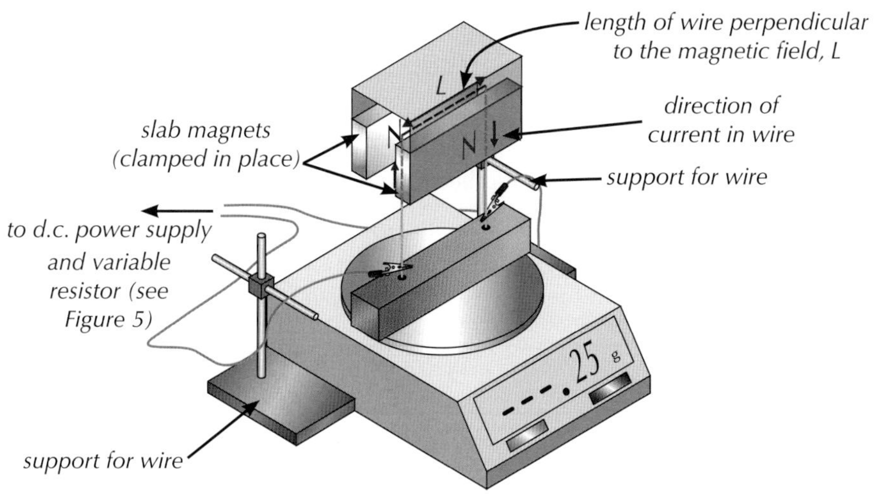

Figure 4: *Set-up for an experiment to measure magnetic flux density.*

Set up the experiment shown in Figure 4. A square hoop of metal wire is positioned so that the top of the hoop, length L, passes through the magnetic field, and is perpendicular to it. When a current flows, the length of wire in the magnetic field will experience a downwards force (Fleming's left-hand rule), which will cause a reading on the mass balance.

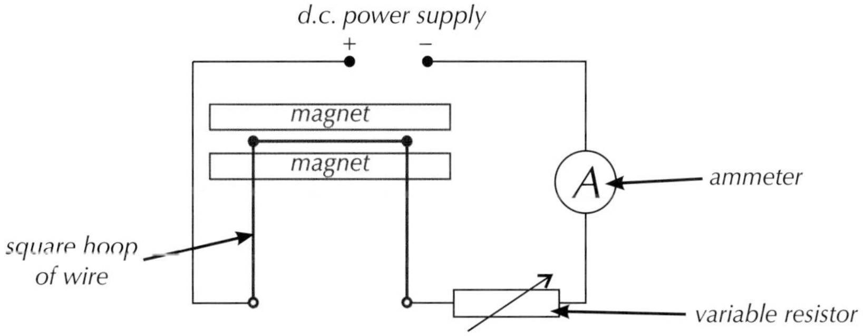

Figure 5: *Circuit diagram for experiment to measure magnetic flux density. The section in red is shown in Figure 4.*

The d.c. power supply should be connected to a variable resistor so that you can alter the current. Zero the digital balance when there is no current through the wire. Turn on the d.c. power supply — if the mass reading is negative, turn off the d.c. power supply and swap the crocodile clips over so that the mass is positive.

Note the mass showing on the digital balance and the current. Then use the variable resistor to change the current and record the new mass reading. Repeat this until you have tested a large range of currents, then do the whole thing twice more so that you have 3 mass readings for each current. Calculate the mean for each mass reading to improve the precision of the results.

Tip: Remember to carry out a full risk assessment before you do this experiment.

Tip: The 'slab' magnets used in this experiment have poles on their largest faces.

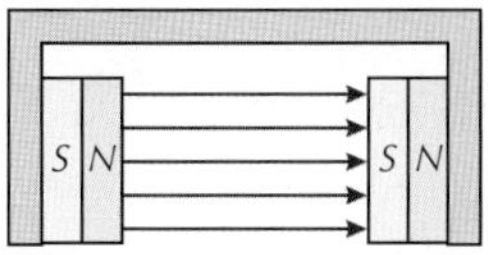

This sort of magnet is sometimes called a 'magnadur magnet'.

Tip: You only need to measure the length of wire shown on the diagram. Even though the vertical parts of the hoop are perpendicular to the field, the forces they feel act horizontally, and so don't affect the mass reading on the balance.

Tip: It's important to zero the balance when no current is flowing so that the mass reading is only due to the force caused by the current in the magnetic field (and not due to the mass of the equipment).

Tip: A d.c. power supply is used so that the direction of the force is constant — if an a.c. supply was used, the direction of the force would keep changing.

Convert your mass readings into force using $\boldsymbol{F} = m\boldsymbol{g}$, then plot your data on a graph of force $\boldsymbol{F}$ against current I. Draw a line of best fit. You should find that you get a graph through the origin showing that force is proportional to current. Because $\boldsymbol{F} = \boldsymbol{B}IL$, the gradient of the line of best fit is equal to $\boldsymbol{B}L$. Because the length is constant, you can divide the gradient by your value for L to get the magnetic flux density, $\boldsymbol{B}$.

Tip: Remember $\boldsymbol{g} = 9.81\ \text{Nkg}^{-1}$.

Tip: Give the forces to the least number of s.f. that the data is given to.

Example — Maths Skills

A student carries out the experiment described above to find the magnetic flux density, *B*. Complete the data table below and plot the results on a graph. Describe the relationship between force and current, and estimate a value for *B*.

Convert the masses in g to kg by dividing them by 1000.
Then use $\boldsymbol{F} = m\boldsymbol{g}$ to turn the masses into forces and draw the graph:

current / A	*mean mass recorded on mass balance / g*	*force acting on current-carrying wire / N*
1.0	*0.20*	*0.0020*
2.0	*0.41*	*0.0040*
3.0	*0.61*	*0.0060*
4.0	*0.82*	*0.0080*

Length L = 0.050 m throughout.

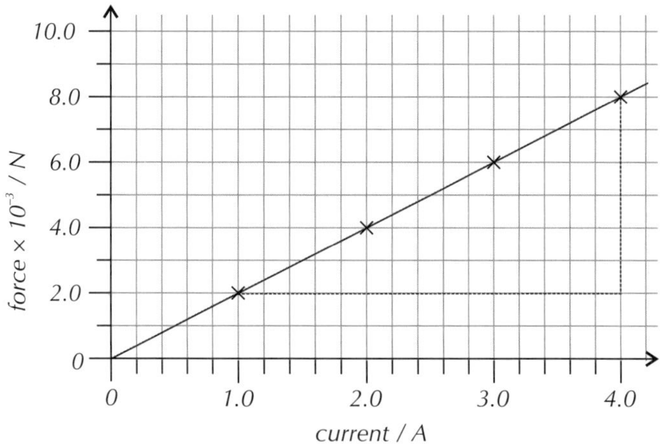

The graph is a straight line through the origin, so force is directly proportional to current.

$$\boldsymbol{B}L = \text{gradient} = \frac{\Delta y}{\Delta x} = \frac{(8.0 \times 10^{-3}) - (2.0 \times 10^{-3})}{4.0 - 1.0} = 2.0 \times 10^{-3}$$

$$\boldsymbol{B} = \text{gradient} \div L = 2.0 \times 10^{-3} \div 0.050 = 0.040\ \text{T}$$

Practice Questions — Application

Q1 a) A 25 cm straight wire carrying a current of 4.0 A is placed into a magnetic field of magnetic flux density 0.20 T. The angle between the magnetic field lines and the wire is 53°. Calculate the force acting on the wire due to the field.

b) The wire is rotated so that it's parallel to the magnetic field lines. What is the force acting on the wire now?

Q2 The diagram below shows an experiment set-up to determine the magnetic flux density of a uniform magnetic field between two slab magnets. A square wire hoop is placed such that the horizontal length *L* is perpendicular to the field between the magnets. The wire carries a current of 3.0 A. The hoop is connected to a d.c. power supply, a variable resistor and ammeter.

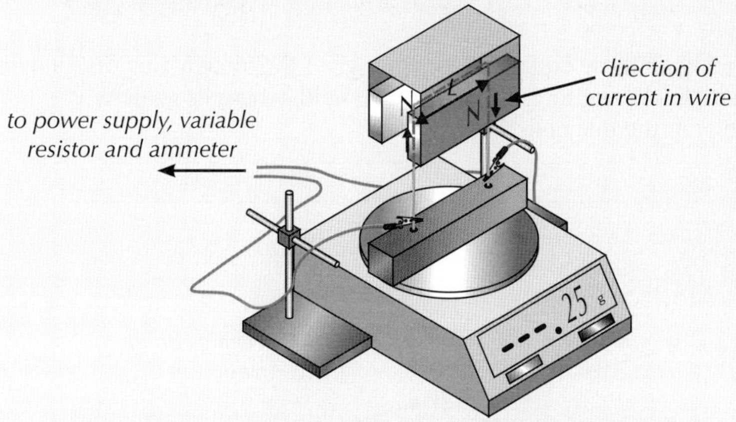

a) Draw a circuit diagram of the set-up described above.

b) Explain why a variable resistor is used in the circuit.

c) Explain why the vertical lengths of wire are unimportant and only length *L* is measured.

d) The mass shown on the balance is recorded for a range of currents. Explain how this information can be used to determine the magnetic flux density.

Practice Questions — Fact Recall

Q1 Give the condition needed to use the equation $\boldsymbol{F} = \boldsymbol{B}I L$ to calculate the force on a current-carrying wire in a magnetic field.

Q2 Explain why the force acting on a wire due to an external magnetic field is at a maximum when the wire and the magnetic field are perpendicular to one another.

Learning Objectives:

- Be able to use $F = BQv$ to calculate the force on a charged particle travelling at right angles to a uniform magnetic field.
- Understand the effect on a charged particle moving in a magnetic field, and how this can lead to circular orbits of charged particles.
- Know what happens to charged particles moving in a region occupied by both electric and magnetic fields, and how this is used in a velocity selector.

Specification Reference 6.3.2

3. Forces on Charged Particles

Any charged particle in a magnetic field feels a force as long as its moving — you need to know how to calculate the size and direction of the force.

Charged particles in a magnetic field

A force acts on a charged particle moving in a magnetic field. This is why a current-carrying wire experiences a force in a magnetic field (page 147) — electric current in a wire is the flow of negatively charged electrons.

- The force on a current-carrying wire in a magnetic field perpendicular to the current is given by $F = BIL$ (page 149).
- Electric current, I, is the flow of charge, Q, per unit time, t. So $I = \frac{Q}{t}$.
- A charged particle which moves a distance L in time t has a velocity, $v = \frac{L}{t}$. So $L = vt$.

Putting all these equations together gives the force acting on a single charged particle moving through a magnetic field, where its velocity is perpendicular to the magnetic field:

$$F = BIL = B\frac{Q}{\cancel{t}}v\cancel{t}$$

$$F = BQv$$

F = force in N

B = magnetic flux density in T

Q = charge on the particle in C

v = velocity of the particle in ms^{-1}

Exam Tip

In many exam questions, Q is the size of the charge on the electron, which is 1.60×10^{-19} C.

Example — Maths Skills

An electron travels at a velocity of 2.00×10^4 ms^{-1} perpendicular to a uniform magnetic field of strength 2.00 T. What is the magnitude of the force acting on the electron? (The magnitude of the charge on an electron is 1.60×10^{-19} C.)

Just use the equation $F = BQv$ and put the correct numbers in:

$$F = BQv$$
$$= 2.00 \times (1.60 \times 10^{-19}) \times (2.00 \times 10^4)$$
$$= 6.40 \times 10^{-15} \text{ N}$$

Figure 1: Circular tracks made by charged particles in a cloud chamber with an applied magnetic field.

The circular path of particles

Charged particles travelling perpendicular to a magnetic field travel in a circular path — see Figure 2. By Fleming's left-hand rule the force on a moving charge travelling perpendicular to a magnetic field is always perpendicular to its direction of travel. Mathematically, that is the condition for circular motion (page 42).

To use Fleming's left-hand rule (page 148) for charged particles, use your second finger (normally current) as the direction of motion for a positive charge. If the particle carries a negative charge (i.e. an electron), point your second finger in the opposite direction to its motion.

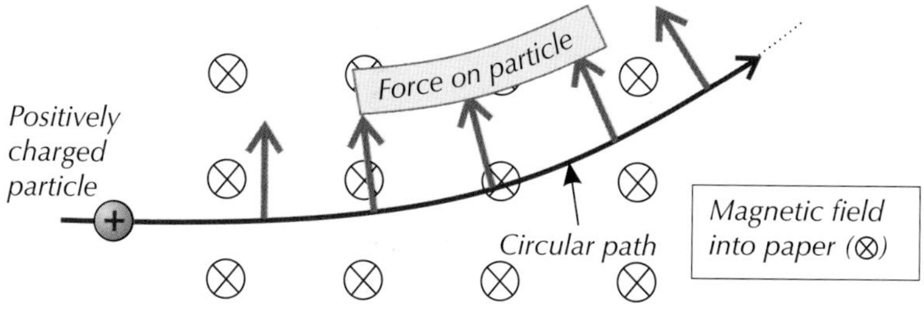

Figure 2: *A charged particle moving perpendicular to a magnetic field follows a circular path.*

> **Tip:** You'll often need to draw 3D situations in 2D like this (or interpret 2D diagrams as a 3D situation). Just remember that a cross always means "into the page" and a dot means "out of the page".

This effect is used in particle accelerators such as cyclotrons and synchrotrons, which use magnetic fields to accelerate particles to very high energies along circular paths.

HOW SCIENCE WORKS

It's also used in mass spectrometers to analyse chemical samples. Ions (charged particles) with the same velocity are made to enter a magnetic field which deflects them in a curved path towards a detector. The radius of curvature depends on the charge and mass of the particles (see equation below). The identity of the ions reaching the detector can be deduced from their mass to charge ratio.

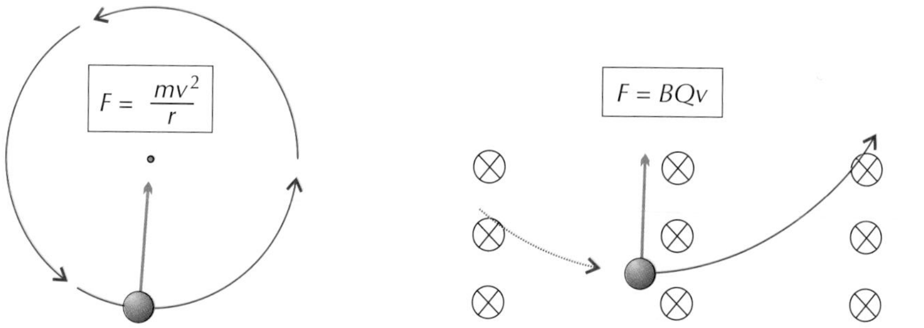

Figure 3: *Diagrams showing the force acting on a particle in a circular orbit (left) and the force acting on a particle in a magnetic field (right).*

The radius of the circular path followed by charged particles in a magnetic field can be found by combining the equations for the force on a charged particle in a magnetic field and for the force on a particle in a circular orbit.

> **Tip:** For more on circular motion, see pages 38-44.

$$F = \frac{mv^2}{r} \quad \text{and} \quad F = BQv \quad \Rightarrow \quad \frac{mv^2}{r} = BQv$$

This rearranges to:

$$r = \frac{mv}{BQ}$$

r *= radius of circular path in m*

m *= mass of the particle in kg*

v *= magnitude of the velocity of the particle in ms^{-1}*

B *= magnitude of the magnetic flux density in T*

Q *= charge on the particle in C*

> **Tip:** A 'v' is cancelled from the top and bottom in the final rearrangement here.

This means that:

- The radius of curvature increases (i.e. the particle is deflected less) if the mass or velocity of the particle increase.
- The radius of curvature decreases (i.e. the particle is deflected more) if the strength of the magnetic field or the charge on the particle increase.

Tip: Protons have a mass of 1.673×10^{-27} kg and a charge equal to the elementary charge $e = 1.60 \times 10^{-19}$ C. You'll be given these in your data and formulae booklet in the exam.

Example — Maths Skills

A proton travels through a magnetic field of flux density 6.0 mT at a velocity of 110 ms⁻¹. The direction of travel of the proton is perpendicular to the magnetic field. Calculate the radius of the circular path of the proton.

$$r = \frac{mv}{BQ} = \frac{(1.673\times10^{-27})\times110}{(6.0\times10^{-3})\times(1.60\times10^{-19})} = 1.916...\times10^{-4} = 0.19\,\text{mm (to 2 s.f.)}$$

Examples — Maths Skills

For each of the following, say which of the two particles would follow a circular path with the smaller radius in a magnetic field of flux density *B*.

a) A carbon-12 nucleus with velocity *v*, relative mass of 12 and relative charge of +6, and a carbon-14 nucleus with velocity *v*, relative mass of 14 and relative charge of +6.

The radius of the circular path followed by the particles is given by:

$$r = \frac{mv}{BQ}$$

In this case v, B and Q are identical for both particles but m is larger for the carbon-14 nucleus. As r is directly proportional to m, the carbon-12 nucleus will follow the circular path with the smaller radius.

b) A carbon-14 nucleus with velocity *v*, relative mass of 14 and relative charge of +6, and a nitrogen-14 nucleus with velocity *v*, relative mass of 14 and relative charge of +7.

m, v and B are identical for both particles but Q is larger for the nitrogen nucleus. As r is inversely proportional to Q, the nitrogen nucleus will follow the circular path with the smaller radius.

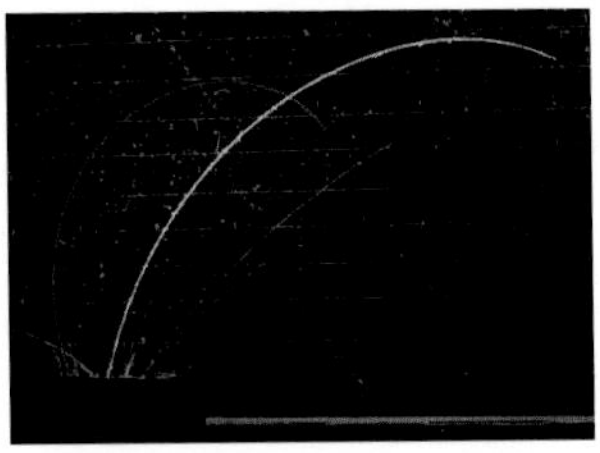

Figure 4: False-colour cloud chamber image showing the different paths of a proton (red) and an alpha particle (yellow) in the chamber's magnetic field.

Velocity selectors

Velocity selectors are used to separate out charged particles of a certain velocity from a stream of accelerated charged particles moving at a range of speeds. They do this by applying both a magnetic and an electric field at the same time perpendicular to each other, while a stream of particles is fired perpendicularly to both fields at a device with a narrow gap called a collimator.

Particles fired into the velocity selector experience opposing forces from the electric and magnetic fields (see Figure 5). So, for positively-charged particles travelling through an electric field that goes from top to bottom and a magnetic field that goes straight into the page:

- The magnetic field tries to deflect the particles upwards — check this with Fleming's left-hand rule. The force on each particle is $F = BQv$ (see page 154).
- The electric field tries to deflect particles downwards (opposite charges attract, like charges repel). The force on the particle is $F = EQ$ (see p.138).

Tip: Negatively-charged particles would experience these forces in the opposite directions. Here, the magnetic field would try to deflect the particles downwards and the electric field would try to deflect the particles upwards.

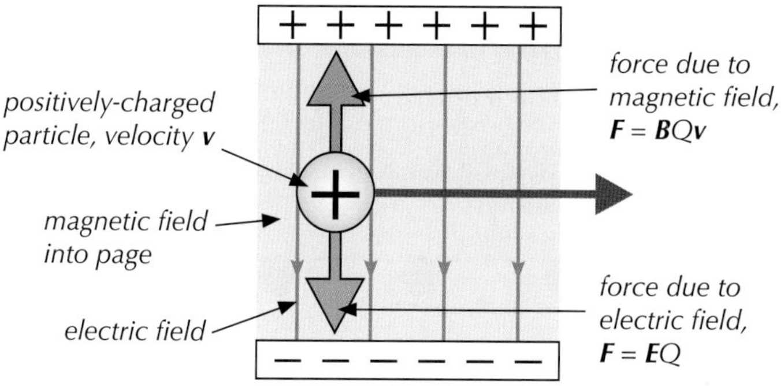

***Figure 5:** A charged particle travelling through an electric field that goes from top to bottom and a magnetic field that goes straight into the page. The particle experiences opposing forces from the two fields.*

> **Tip:** For the examples on this page, the electric field, ***E***, goes top to bottom (i.e. down the page), and the magnetic field, ***B***, goes straight into the page. The particles are positively-charged.

Particles will be deflected unless the forces balance (i.e. $BQv = EQ$). Cancelling Qs and rearranging gives:

$$v = \frac{E}{B}$$

- v = velocity of the particle in ms^{-1}
- E = electric field strength in NC^{-1}
- B = magnetic flux density in T

So only particles with velocity $v = \frac{E}{B}$ as given above will travel in a straight line to pass through the gap in the collimator.

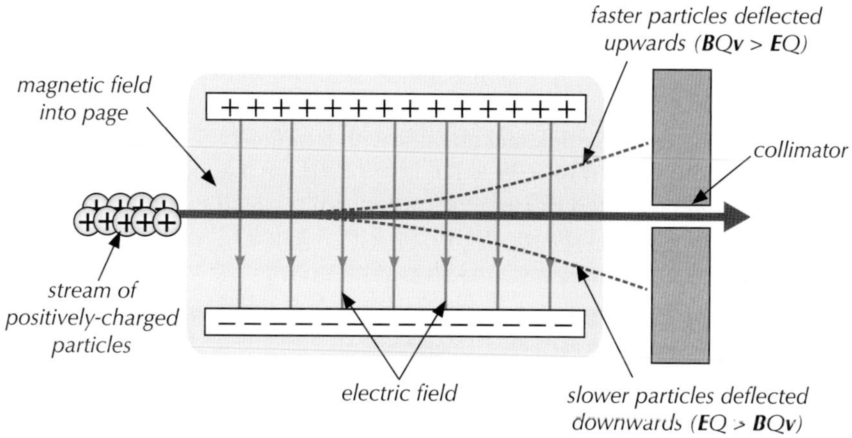

***Figure 6:** Diagram showing the possible paths of positively-charged particles through a velocity selector.*

> **Tip:** The directions of deflection are reversed for negatively-charged particles.

You can select and vary the velocity of the particles that get through the collimator by changing the strength of the magnetic or electric fields. Velocity selectors are often used in mass spectrometers to ensure that the accelerated particles entering the magnetic field have the same velocity.

> **Tip:** A mass spectrometer is a device that identifies particles by their mass to charge ratio (see p.155).

Example — Maths Skills

Calculate the magnetic flux density required in a velocity selector to isolate charged particles travelling through a 3.4×10^4 NC^{-1} electric field at 2.1×10^5 ms^{-1}.

Rearrange $v = \frac{E}{B}$ and put the correct numbers in:

$$v = \frac{E}{B} \Rightarrow B = \frac{E}{v} = \frac{3.4 \times 10^4}{2.1 \times 10^5} = 0.1619... = 0.16 \text{ T (to 2 s.f.)}$$

Tip: Neutrons, along with protons, are one of the nucleons found in atoms. They have a mass of 1.675×10^{-27} kg and carry no charge.

Practice Questions — Application

Q1 Why would a neutron moving through a magnetic field perpendicular to its direction of motion not experience a force?

Q2 In which direction will the force due to the magnetic field act on the electron in the diagram below?

Q3 Find the force that acts on a particle with a charge of 3.2×10^{-19} C travelling at a velocity of 5.5×10^{3} ms^{-1} perpendicular to a magnetic field with a flux density of 640 mT.

Q4 A charged particle is moving at right angles to both an electric field and a magnetic field, which are also perpendicular to each other. The flux density of the magnetic field is 20.0 mT. Calculate the magnitude of the electric field if the particle is moving at 5.2×10^{6} ms^{-1} in a straight line.

Q5 a) The size of the force needed to keep an object in circular motion is given by $F = \frac{mv^2}{r}$. Combine this with $F = BQv$ to find an expression for the magnetic flux density B needed to keep a charged particle in circular motion for a given radius r and orbital speed v.

b) In a circular particle accelerator with a radius of 5.49 m, protons are accelerated to 1.99×10^{7} ms^{-1}. Find the magnetic flux density B that's required to keep the protons following the circular path of the accelerator. $m_p = 1.673 \times 10^{-27}$ kg and $Q_p = 1.60 \times 10^{-19}$ J.

Practice Question — Fact Recall

Q1 If an electron is travelling through a uniform magnetic field perpendicular to its velocity, what shape will its path take? (Assume it has infinite space to move into.)

4. Magnetic Flux and Flux Linkage

Magnetic flux and flux density are used to describe the magnetic field strength (magnetic flux density) in a defined area. If the magnetic flux of a field is cut by a conductor, an e.m.f. is induced in the conductor.

Learning Objectives:

- Know what magnetic flux, ϕ, is and that it's measured in webers.
- Know what magnetic flux linkage is.
- Be able to calculate magnetic flux using $\phi = \boldsymbol{B}A\cos\theta$.

Specification Reference 6.3.3

Magnetic flux

You met magnetic flux density, $\boldsymbol{B}$, on p.149 — it's a measure of the strength of a magnetic field (or you can think of it as the number of field lines per unit area). The total **magnetic flux**, ϕ, passing through an area, A, perpendicular to a magnetic field, $\boldsymbol{B}$, is defined as:

$$\phi = \boldsymbol{B}A$$

ϕ = magnetic flux in Wb (webers)
$\boldsymbol{B}$ = magnetic flux density in T
A = area in m^2

You can think of flux as the number of field lines. But remember that flux is continuous — field lines are just a way of drawing it. Figure 1 shows magnetic flux inside a single loop coil. The flux inside the coil is $\phi = \boldsymbol{B}A$, where A = area of coil.

Tip: You can only use the equation $\phi = \boldsymbol{B}A$ if $\boldsymbol{B}$ is normal to A (see Figure 1) — otherwise there's an extra term in the equation, which you'll see on page 161.

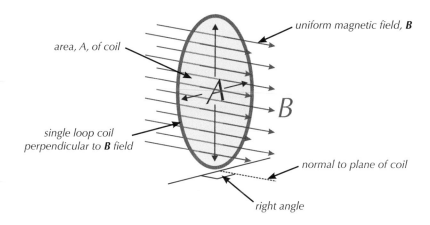

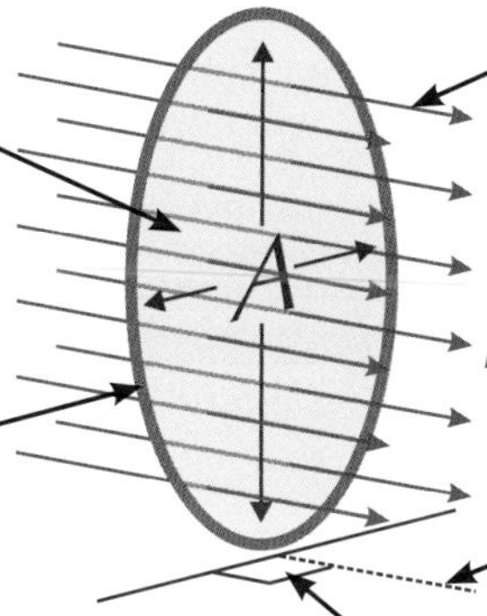

***Figure 1:** Diagram showing a single loop coil which has an area A perpendicular to a magnetic field of magnetic flux density **B**.*

***Figure 2:** German physicist Wilhelm Eduard Weber made major advances in the field of electromagnetism. The unit of magnetic flux is named after him.*

Example — Maths Skills

A square with sides of length 4.5 cm is placed in a magnetic field, normal to the field's direction. Find the magnetic flux passing through the square if the magnetic flux density is 0.92 T.

Start by finding the area of the square, remembering to convert the units:

$$A = (4.5 \times 10^{-2}) \times (4.5 \times 10^{-2}) = 2.025 \times 10^{-3}\ \text{m}^2$$

Then just put the numbers into the equation above:

$$\phi = \boldsymbol{B}A = 0.92 \times (2.025 \times 10^{-3}) = 1.863 \times 10^{-3}$$
$$= 1.9 \times 10^{-3}\ \text{Wb (to 2 s.f.)}$$

Tip: E.m.f. is another way of saying voltage — you saw it in year 1 of A-Level physics if you need a reminder. See pages 163-172 for more on induced e.m.f.

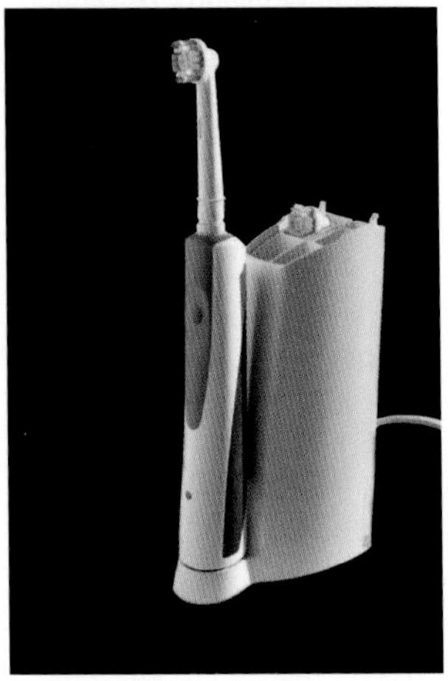

Figure 4: Electromagnetic induction is used to charge electric toothbrushes, avoiding the need for exposed wires.

Tip: The e.m.f. can either be positive or negative — it depends on the direction of movement and the way you connect the wires.

If the wires are left connected in the same way, the e.m.f. will alternate as the magnet moves back and forth.

Electromagnetic induction

If there is relative motion between a conducting rod and a magnetic field, the electrons in the rod will experience a force (see p.154), which causes them to accumulate at one end of the rod.

This induces an **electromotive force (e.m.f.)** across the ends of the rod exactly as connecting a battery to it would — this is called **electromagnetic induction**. If the rod is part of a complete circuit, then an induced current will flow through it.

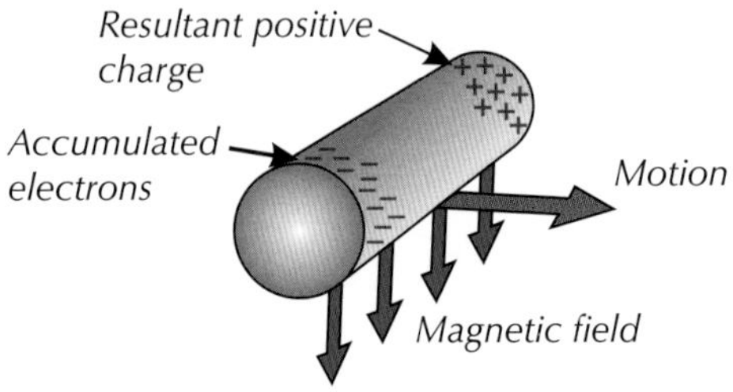

Figure 3: *A conducting rod moving through a magnetic field.*

An e.m.f. is induced when the conductor cuts the magnetic flux. The conductor can move and the magnetic field stay still or the other way round — you get an e.m.f. either way.

You can induce an e.m.f. in a flat coil or solenoid in the same way. In either case, the e.m.f. is caused by the magnetic field (or 'magnetic flux') that passes through the coil changing (see Figure 5). If the coil is part of a complete circuit, an induced current will flow through it.

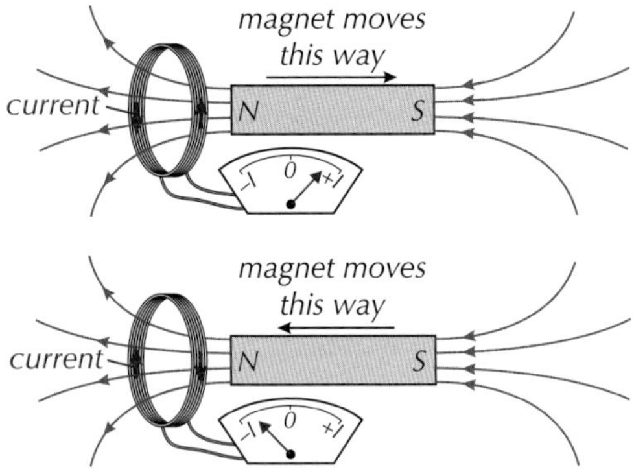

Figure 5: *Current and e.m.f. induced when a magnet is moved towards and away from a coil.*

Magnetic flux linkage

When a wire coil is moved in a magnetic field, the size of the e.m.f. induced depends on the magnetic flux passing through the coil, ϕ, and the number of turns on the coil cutting the flux (p.163). The product of these is called the **magnetic flux linkage** (or just flux linkage), $N\phi$. For a coil of N turns normal to $\boldsymbol{B}$, the flux linkage is given by:

$$\text{Flux linkage} = N\phi$$

Flux linkage in Wb

ϕ = magnetic flux in Wb

N = number of turns on the coil cutting the flux

You saw on the last page that magnetic flux, ϕ, is equal to $\boldsymbol{B}A$, so it follows that magnetic flux linkage, $N\phi$, is equal to $\boldsymbol{B}AN$.

Example — Maths Skills

The flux linkage of a coil with a cross-sectional area of 0.33 m² normal to a magnetic field of flux density 0.15 T is 4.0 Wb. How many turns are in the coil?

Just rearrange the equation for flux linkage to make N the subject and then put the numbers in:

$$[N\phi] = \mathbf{B}AN \Rightarrow N = \frac{[N\phi]}{\mathbf{B}A}$$

$$= \frac{4.0}{0.15 \times 0.33} = 80.808... = 81 \text{ turns (to 2 s.f.)}$$

Tip: Don't be tempted to cancel down the 'N's here — consider the flux linkage, $N\phi$, as a stand-alone term.

Flux linkage at an angle

When the magnetic flux isn't perpendicular to the area you're interested in (e.g. Figure 6), you need to use trigonometry to resolve the magnetic field vector into components that are parallel and perpendicular to the area.

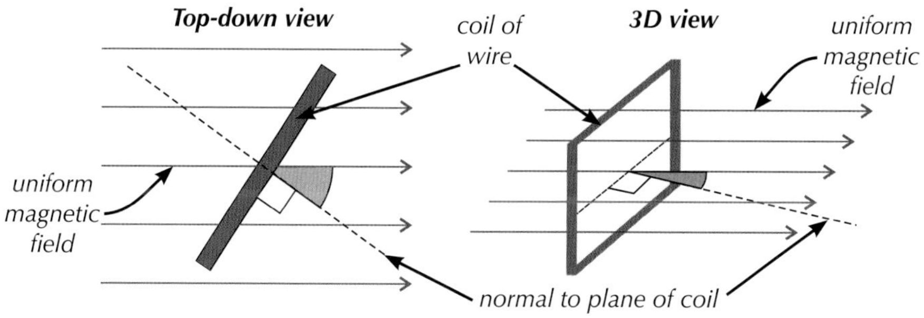

Figure 6: *A coil at an angle to a uniform magnetic field.*

Tip: Pulling the coil in Figure 6 out of the magnetic field would induce an e.m.f.

To find the magnetic flux passing through a coil, you're interested in the component of the magnetic field perpendicular to the area of the coil. By trigonometry, this is equal to $\mathbf{B}\cos\theta$ — see Figure 7.

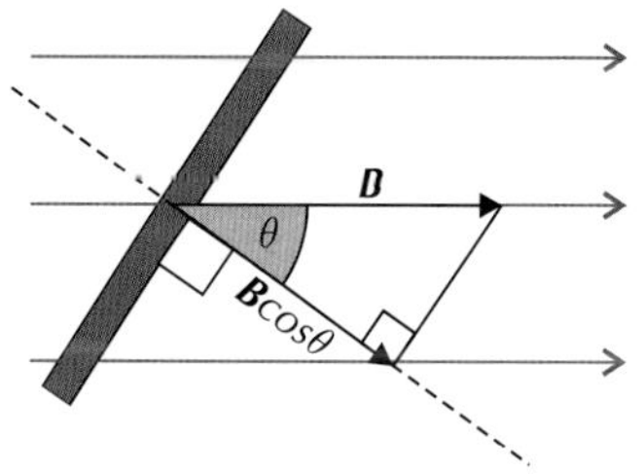

Figure 7: *For a coil at an angle to a magnetic field, the component of the field perpendicular to the area of the coil is* $\mathbf{B}\cos\theta$*, where* θ *is the angle between the field and the normal to the plane of the coil.*

Tip: Remember SOC CAH TOA: $\cos\theta$ = adjacent ÷ hypotenuse

See page 246 for more.

So for a single loop of wire when $\mathbf{B}$ is not perpendicular to area, you can find the magnetic flux using this equation:

$$\phi = \mathbf{B}A\cos\theta$$

- ϕ = magnetic flux in Wb
- $\mathbf{B}$ = magnetic flux density in T
- θ = angle between the normal to the plane of the coil and the magnetic field in °
- A = area of the coil in m²

Exam Tip
You won't be given this in the data and formulae booklet, but you will be given $\phi = \boldsymbol{B}A\cos\theta$, so you just need to remember that flux linkage = $N\phi$.

And for a coil with N turns, you can find the flux linkage with the equation:

$$\text{Flux linkage} = N\phi = \boldsymbol{B}AN\cos\theta$$

$\boldsymbol{B}$ = *magnetic flux density in T*
$\boldsymbol{A}$ = *area of the coil in m²*
$\boldsymbol{\theta}$ = *angle between the normal to the plane of the coil and the magnetic field in °*
$\boldsymbol{N\phi}$ = *Flux linkage in Wb*
$\boldsymbol{N}$ = *number of turns on the coil cutting the flux*

Example — Maths Skills

A rectangular coil of wire with exactly 200 turns and sides of length 5.00 cm and 6.51 cm is rotating in a magnetic field with $\boldsymbol{B} = 8.56 \times 10^{-3}$ T. Find the flux linkage of the coil when the normal to the area of the coil is at 12.6° to the magnetic field, as shown below.

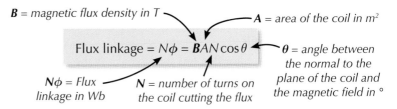

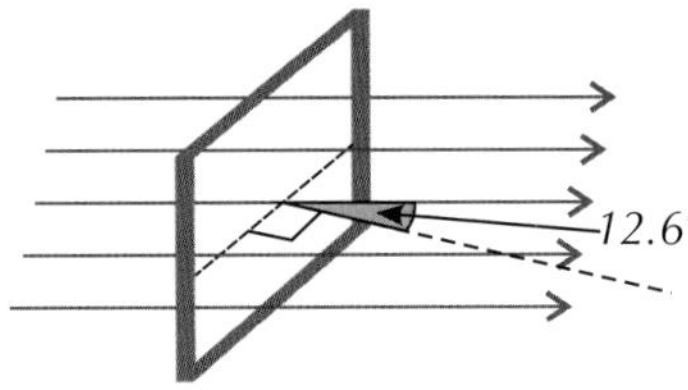

First find the area of the coil:

Area = $(5.00 \times 10^{-2}) \times (6.51 \times 10^{-2}) = 3.255 \times 10^{-3}$ m²

Then just put the numbers into the equation:

$N\phi = \boldsymbol{B}AN\cos\theta$
$= (8.56 \times 10^{-3}) \times (3.255 \times 10^{-3}) \times 200 \times \cos 12.6° = 5.438... \times 10^{-3}$
$= 5.44 \times 10^{-3}$ Wb (to 3 s.f.)

Practice Questions — Application

Q1 A student sets up an experiment to measure the strength of the Earth's magnetic field, $\boldsymbol{B}$. He does this by measuring the flux linkage of a rectangular wire coil.

a) Explain why the flux linkage of the coil changes when he rotates the coil.

b) The student finds the highest value of flux linkage to be 1.3×10^{-6} Wb. If he used a coil with an area of 25 cm² and exactly 10 turns, find the local value of $\boldsymbol{B}$.

Q2 A coil of wire with 355 turns and an area of 145 cm² is placed in a 2.06 mT magnetic field. Calculate what angle the normal to the plane of the coil would need to be at to the field in order for the coil to have a magnetic flux linkage of 10.0×10^{-3} Wb.

Practice Questions — Fact Recall

Q1 What are the units of magnetic flux?

Q2 What is induced when a magnetic field through a conductor changes?

Q3 What must ϕ be multiplied by to calculate magnetic flux linkage?

5. Faraday's Law and Lenz's Law

There's more electromagnetic induction coming up for you to sink your teeth into. In this topic, you'll see how Faraday and Lenz tried to explain the phenomenon better with the inventively named Faraday's law and Lenz's law.

Learning Objectives:

- Know Faraday's law of electromagnetic induction.
- Know that e.m.f. = −rate of change of magnetic flux linkage, and be able to calculate induced e.m.f. using $\varepsilon = -\frac{\Delta(N\phi)}{\Delta t}$.
- Know techniques and procedures used to investigate magnetic flux using search coils.
- Know Lenz's law.

Specification Reference 6.3.3

Faraday's law

Faraday's law links the rate of change of flux linkage with e.m.f.:

> Induced e.m.f. is directly proportional to the rate of change of flux linkage.

It can be written as:

$$\varepsilon = -\frac{\Delta(N\phi)}{\Delta t} = -\frac{N\Delta\phi}{\Delta t}$$

$\Delta(N\phi)$ = change in flux linkage in Wb

N = number of turns on the coil cutting the flux

ε = induced e.m.f. in V

$\Delta\phi$ = change in magnetic flux in Wb

Δt = time taken for flux linkage to change in s

For a coil, induced e.m.f. depends on the number of turns and how fast flux through the coil is changing. The unit of flux, the weber (Wb), is defined in terms of the e.m.f. induced:

> A change in flux linkage of one weber per second will induce an electromotive force of 1 volt in a loop of wire.

Tip: The minus sign is Lenz's law – see page 166.

Example — Maths Skills

A coil with 75 turns and an area of 12 cm² is placed between two electromagnets. The flux density of the magnetic field between the electromagnets is then uniformly increased from 0.15 T to 1.5 T over 0.20 seconds. If the plane of the coil is perpendicular to the magnetic field lines of the field, calculate the e.m.f. induced across the coil.

Convert area to m²: 12 cm² = 12×10^{-4} = 0.0012 m²

Then start with Faraday's law: $\varepsilon = -\frac{\Delta(N\phi)}{\Delta t}$

You saw on page 160 that $N\phi = \boldsymbol{B}AN$, so:

$$\varepsilon = -\frac{\Delta(\boldsymbol{B}AN)}{\Delta t} = -\frac{(1.5 - 0.15) \times 0.0012 \times 75}{0.20} = -0.6075$$

$$= -0.61 \text{ V (to 2 s.f.)}$$

Tip: An electromagnet is an electrical magnet that can be turned on and off, and its field strength can also be varied.

When using the formula above for a conducting rod, think of flux change as field lines being 'cut' as the rod moves. Remember that magnetic flux $\phi = \boldsymbol{B}A$ (see p.159) — here, think of A as the area of flux cut in a certain time.

Tip: In this first example, it's the changing magnetic flux density ($\boldsymbol{B}$) that causes the change in flux linkage. In this second example, it's the changing area (A) that causes the change in flux linkage.

Example — Maths Skills

A copper bar in a uniform magnetic field of 0.11 T is moved perpendicular to the field for 0.63 seconds. The total area that is covered by the bar in this time is 42 cm². Calculate the e.m.f. induced across the bar.

$\varepsilon = -\frac{\Delta(N\phi)}{\Delta t}$ and $\phi = \boldsymbol{B}A$, so:

$$\varepsilon = -\frac{\Delta(\boldsymbol{B}AN)}{\Delta t} = -\frac{0.11 \times (42 \times 10^{-4})}{0.63} = -7.33\ldots \times 10^{-4} = -0.73 \text{ mV (to 2 s.f.)}$$

Tip: There's only 1 'turn' in a conducting rod, so you can ignore the N in Faraday's law.

Exam Tip
You could be asked to find the e.m.f. induced on something other than a rod, e.g. the Earth's magnetic field across the wingspan of a plane. Just think of it as a moving rod and use the equation as usual.

Figure 1: British physicist Michael Faraday.

Faraday's equation can be used to find the e.m.f. in terms of a rod's velocity through a magnetic field.

Example — Maths Skills

The diagram below shows a conducting rod of length L moving through a perpendicular uniform magnetic field, B, at a constant velocity, v. Show that e.m.f. induced in the rod is equal to $-BLv$.

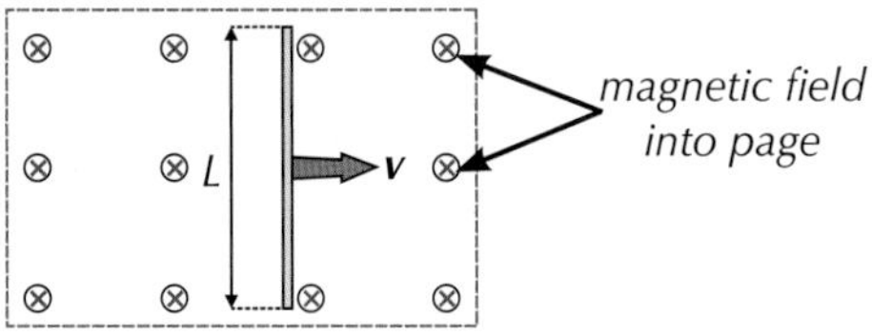

Displacement of rod is $\mathbf{s} = \mathbf{v}\Delta t$ (displacement = velocity × time). Area of flux it cuts, $A = L \times \mathbf{s} = L\mathbf{v}\Delta t$. Total magnetic flux cut through, $\Delta\phi = \mathbf{B}A = \mathbf{B}L\mathbf{v}\Delta t$

Faraday's law gives $\varepsilon = -\frac{\Delta(N\phi)}{\Delta t} = -\frac{\Delta\phi}{\Delta t}$ (since $N = 1$)

So the induced e.m.f., $\varepsilon = -\frac{\Delta\phi}{\Delta t} = -\frac{\mathbf{B}L\mathbf{v}\Delta t}{\Delta t} = -\mathbf{B}L\mathbf{v}$

Electromagnetic induction graphs

The e.m.f. induced in a conductor by a changing external magnetic field is equal to the negative of the gradient of a graph of flux linkage ($N\phi$) against time. The area under a graph of e.m.f. against time gives the negative of the change in flux linkage — see Figure 2.

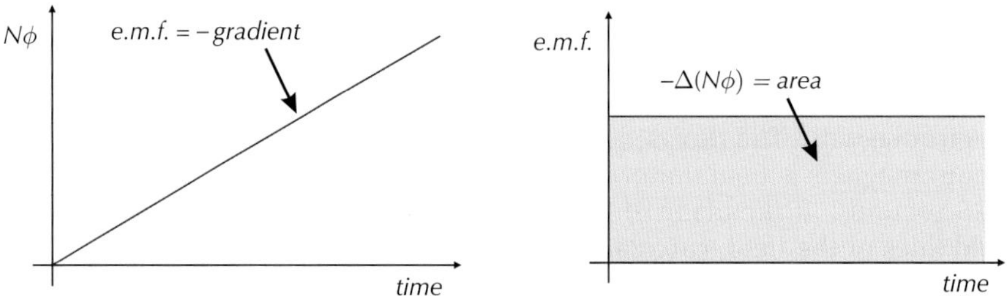

Figure 2: Graphs of flux linkage against time (left) and e.m.f. against time (right) for a conductor in a changing external magnetic field.

Example — Maths Skills

A coil is placed into a uniform magnetic field. The magnetic field density of the magnetic field is then steadily increased over time. The graph below shows the flux linkage of the coil against time. Calculate the magnitude of the e.m.f. induced in the coil.

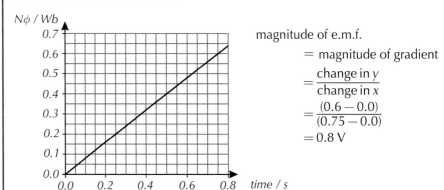

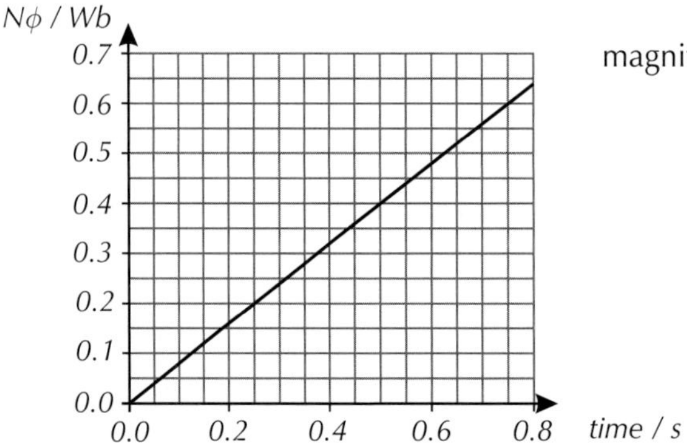

magnitude of e.m.f.

$= $ magnitude of gradient

$= \frac{\text{change in } y}{\text{change in } x}$

$= \frac{(0.6 - 0.0)}{(0.75 - 0.0)}$

$= 0.8$ V

Investigating magnetic flux

You can find the magnetic flux, ϕ, passing through an area between two bar magnets using the set-up in Figure 3.

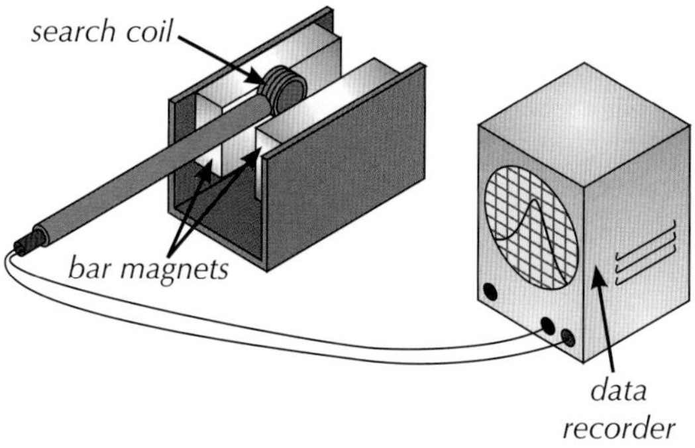

Figure 3: *Set-up for experiment to investigate magnetic flux.*

Place two bar magnets a small distance apart with opposite poles facing each other — they should be far enough apart not to snap together, but otherwise as close as possible to give a uniform field.

Get a search coil — this is a small coil of wire with a known number of turns (N) and a known area (A). Connect it to a data recorder and set the recorder to measure the induced e.m.f. with a very small time interval between readings.

Place the search coil in the middle of the magnetic field so that the area of the coil is parallel to the surface of the magnets. Start the data recorder. Keeping the coil in the same orientation, immediately move the coil out of the field. An e.m.f. will be induced due to the magnetic flux linkage through the coil changing from maximum, $N\phi$, to zero as you remove the coil from the field.

Use your data or the data recorder to plot a graph of induced e.m.f. against time. Using Faraday's Law, estimating the area under the graph of e.m.f. against time and taking the negative gives you an estimate for the total flux linkage change (p.164). The final flux linkage is zero since the magnetic flux density, $\boldsymbol{B}$, is zero away from the magnets, so the change in flux linkage is equal to the flux linkage in the uniform magnetic field. Flux linkage = $N\phi$, so to find ϕ in area A, divide the total flux linkage change by the number of turns on the search coil (N).

Repeat this experiment several times and find the mean of your values for ϕ.

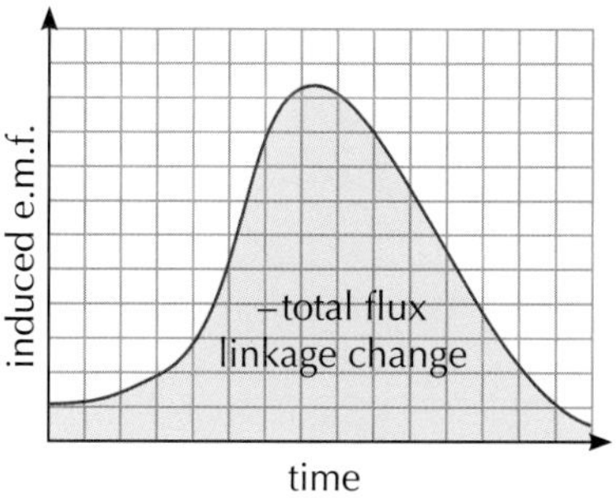

Figure 4: *An example of a graph of e.m.f. against time obtained by moving a coil out of a magnetic field.*

Tip: Remember to carry out a risk assessment before starting this experiment.

Tip: This set-up uses magnets with poles on their largest faces again — as on p.151.

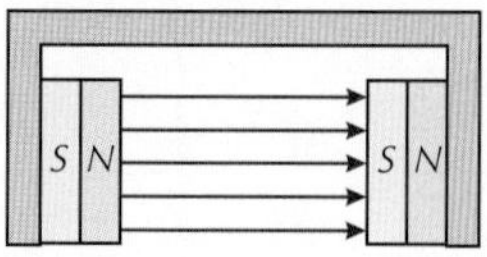

Tip: You can also use this method to find the magnetic flux density, $\boldsymbol{B}$, between the magnets. Flux linkage = $\boldsymbol{B}AN$ (p.160), so just divide the flux linkage by AN, where A is the area of the search coil.

Tip: Remember, you can estimate the area under a graph by counting the number of squares under the graph then multiplying by the value of one square (see p.239).

Tip: This is why there's a minus sign in Faraday's Law.

Tip: Remember, a current is only induced if the conductor is part of a complete circuit — if it isn't an e.m.f. will be induced but no current will flow.

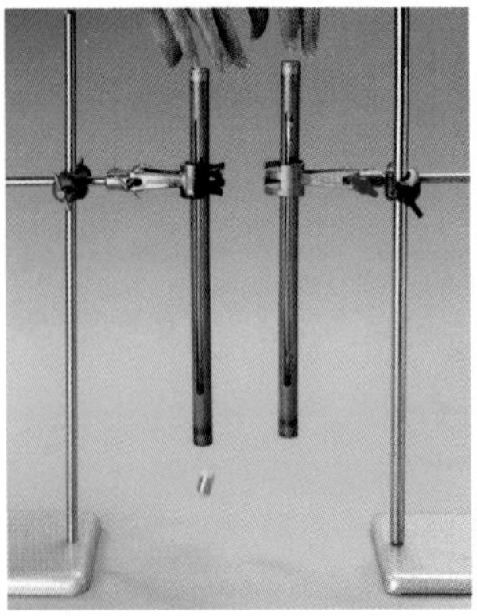

Figure 5: A magnetic weight (right) takes longer to fall down a copper tube than a non-magnetic weight (left) due to Faraday's and Lenz's laws. The falling magnet induces a current in the tube that generates a magnetic field which opposes the magnetic field of the magnet, slowing it down.

Lenz's law

The direction of an induced e.m.f. (and current) are given by **Lenz's law**:

> The induced e.m.f. is always in such a direction as to oppose the change that caused it.

A changing magnetic field induces an e.m.f. in a coil (see page 160). If the coil is part of a complete circuit, a current is induced in the same direction as the induced e.m.f. The induced current then produces its own magnetic field (p.146). Lenz's law says:

- If the original magnetic field is getting stronger, the current will be induced in the direction that generates a magnetic field in the opposite direction to the external field, to try to weaken it.
- If the original magnetic field is getting weaker (collapsing), the current will be induced in the direction that generates a magnetic field in the same direction as the external field, to try to maintain it.

Example

The area of a flat coil is perpendicular to a magnetic field as it collapses by 50% as shown below. What will be the direction of the current induced in the loop?

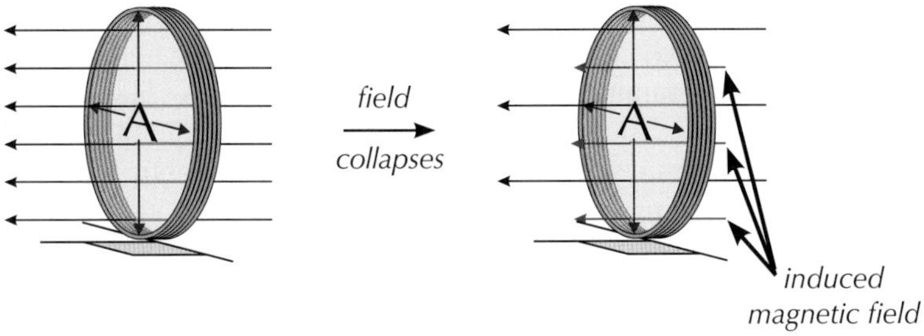

A collapsing field means the field is getting weaker and the field lines are getting further apart (p.146). So by Lenz's law the current induced in the coil will induce a magnetic field in the same direction as the collapsing field to try to maintain the original field. Use the right-hand rule (p.146) to find the direction of the induced current.

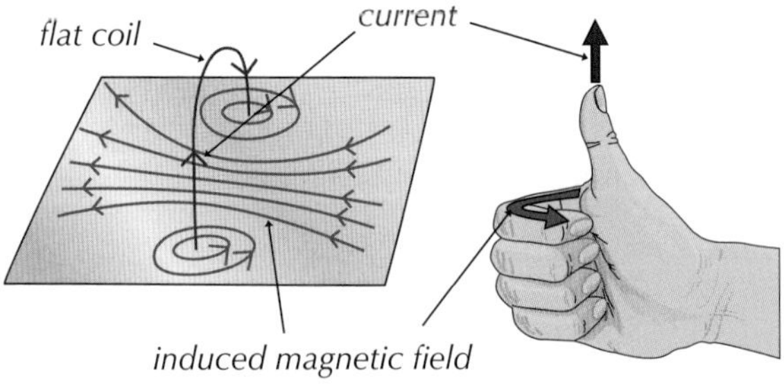

The induced field is to the left, so the induced current is clockwise when viewed from the right.

Example

Lenz's law says that the induced e.m.f. will produce a force that opposes the motion of the conductor — in other words a resistance. Picture a straight conductor being moved down through a perpendicular magnetic field.

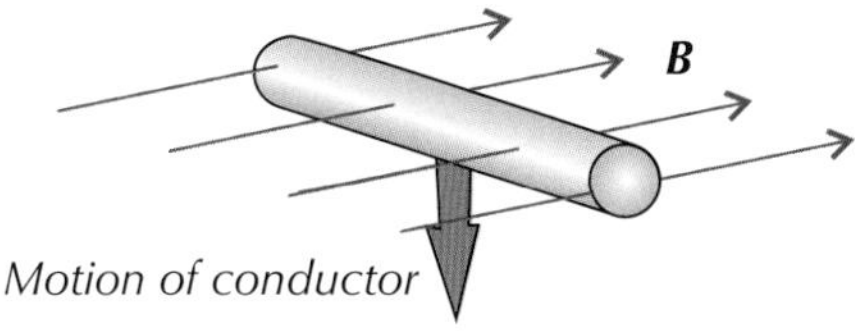

Using Fleming's left-hand rule (see p.148), point your thumb in the direction of the force of resistance — which is in the opposite direction to the motion of the conductor. Point your first finger in the direction of the field. Your second finger will now give you the direction of the induced e.m.f.

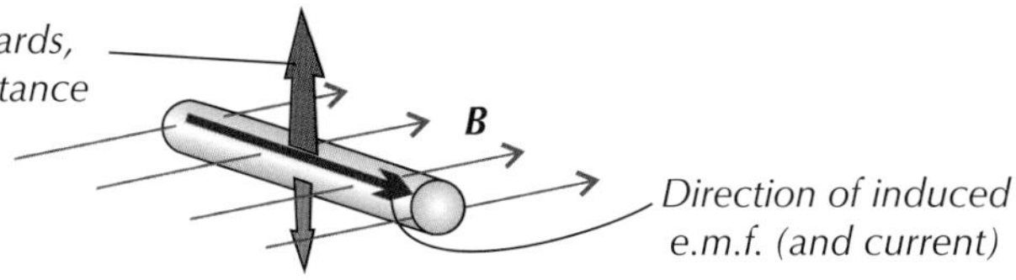

If the conductor is connected as part of a circuit, a current will be induced in the same direction as the induced e.m.f.

Tip: If you've forgotten which finger is which, here's a reminder:

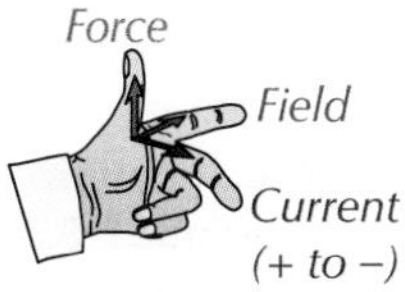

Practice Questions — Application

Q1 Explain how an e.m.f. can be induced across the wings of a plane as it is flown.

Q2 A wire coil with exactly 50 turns and an area of 0.24 m^2 is placed perpendicular to a magnetic field with flux density 1.5 T. A few minutes later the magnetic field is switched off. Calculate the magnitude of the e.m.f. induced in the coil if the magnetic flux density decreased to 0 T in 0.32 s.

Q3 The diagram on the right shows a 1.2 m long copper wire moving upwards at a velocity of 0.50 ms^{-1} through a perpendicular magnetic field of flux density 5.4 mT.

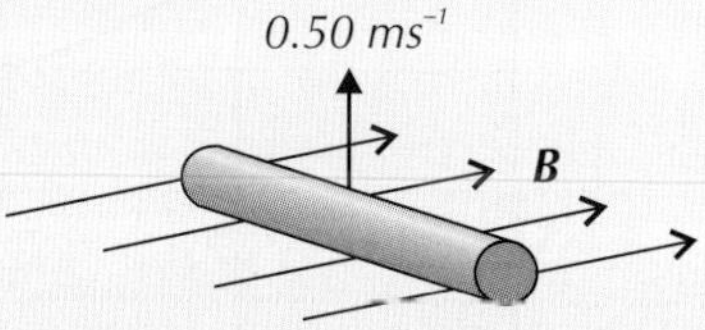

a) The wire is part of a complete circuit. State the direction of the current induced in the wire.

b) Calculate the magnitude of the e.m.f. induced in the wire.

Practice Questions — Fact Recall

Q1 What's Faraday's law?

Q2 What is the gradient of a graph of flux linkage against time equal to?

Q3 A student places a search coil, with area A, connected to a data recorder between two bar magnets. Keeping the area of the coil parallel to the surfaces of the magnets, he pulls the coil out of the field. The data recorder plots a graph of the e.m.f. induced in the coil against time. Explain how the student can use this graph to work out the magnetic flux passing through area A between the two magnets.

Q4 What's Lenz's law?

6. Uses of Electromagnetic Induction

Learning Objectives:
- Know how a simple a.c. generator works.
- Know how a simple laminated iron-cored transformer works.
- Be able to use $\frac{n_s}{n_p} = \frac{V_s}{V_p} = \frac{I_p}{I_s}$ for an ideal transformer.
- Know techniques and procedures used to investigate transformers.

Specification Reference 6.3.3

Electromagnetic induction allows electricity to be produced on a large scale. You'll need to know how it's is used in generators and transformers.

Induced e.m.f. in a rotating coil

The movement of a conductor in a magnetic field is used to generate electricity. When a coil such as that in Figure 1 rotates uniformly (at a steady speed) in a magnetic field, the coil cuts the flux and an alternating e.m.f. is induced.

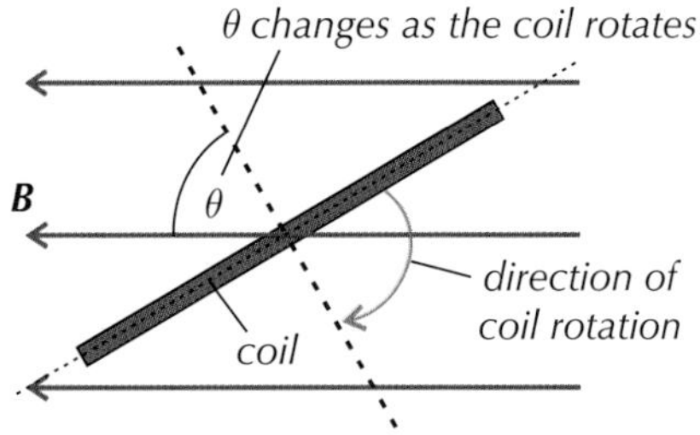

Figure 1: *The value of θ changes as the coil rotates.*

The amount of flux cut by the coil (flux linkage) is given by $N\phi = \boldsymbol{B}AN\cos\theta$ (see page 162).

Tip: Note the axis of rotation is perpendicular to the magnetic field.

Tip: Figure 1 is a top-down view of the coil, as in Figure 6 on p.161.

As the coil rotates, θ changes so the flux linkage varies sinusoidally between $+\boldsymbol{B}AN$ and $-\boldsymbol{B}AN$. Because the flux linkage is varying sinusoidally (see Figure 2), the rate of change of flux linkage is always changing, so the induced e.m.f., ε, is also changing (due to Faraday's law, p.163). You saw on p.164 that the induced e.m.f. is equal to the negative of the gradient of a graph of $N\phi$ against time. So the e.m.f. also varies sinusoidally — it is at a maximum when the plane of the coil is parallel to the lines of flux, and is zero when the coil is perpendicular to the lines of flux — see Figure 2.

Tip: 'Sinusoidally' means it follows the same pattern as a sin (or cos) curve.

Tip: You don't need to know these graphs, they're just here to help you understand generators (see next page).

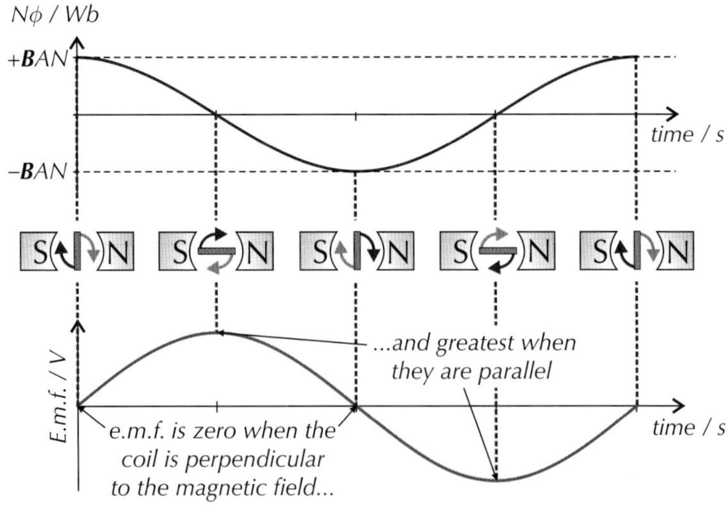

Figure 2: *The sinusoidal graphs produced by plotting flux linkage and induced e.m.f. against time for a rotating coil in a uniform magnetic field.*

The e.m.f. induced can be altered by changing the speed of rotation or the size of the magnetic field:

- Increasing the speed of rotation will increase the frequency of rotation. This in turn causes an increase in the rate of change of flux linkage, and therefore increases the maximum e.m.f..
- Increasing the magnetic flux density $\boldsymbol{B}$ will increase the maximum e.m.f., but will have no effect on the frequency of rotation.

Tip: The frequency of the rotation of the coil, f, is related to the period of the coil, T, by the equation $f = \frac{1}{T}$. You saw this on p.40.

Generators

Generators, or dynamos, convert kinetic energy into electrical energy — they induce an electric current by rotating a coil in a magnetic field.

Figure 4 shows a simple alternator — a generator of alternating current (a.c.). It has slip rings and brushes to connect the coil to an external circuit. The output voltage and current change direction with every half rotation of the coil, producing an alternating current.

Figure 3: *The Gramme dynamo was one of the first electric generators used to deliver power for industry.*

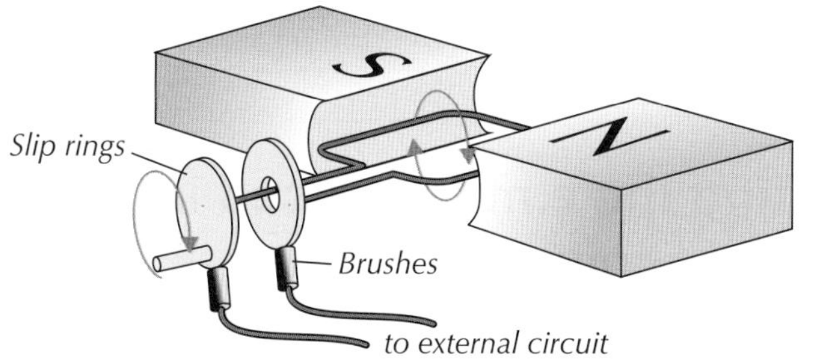

Figure 4: *A simple alternator.*

What's a transformer?

Transformers are devices that make use of electromagnetic induction to change the size of the voltage for an alternating current. They consist of two coils of wire wrapped around an iron core — see Figure 5.

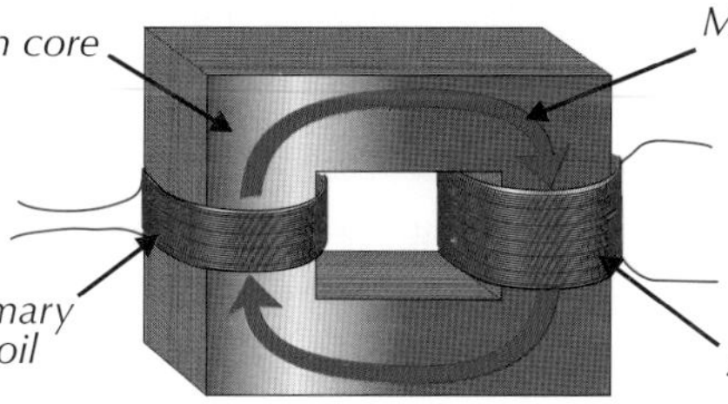

Figure 5: *The basic structure of a (step-up) transformer.*

An alternating current flowing in the primary (or input) coil produces an alternating magnetic field, causing the core to magnetise, demagnetise and remagnetise continuously in opposite directions. The changing magnetic field is passed through the iron core to the secondary (or output) coil, where the rapidly changing flux linkage through the coil induces an alternating voltage (e.m.f.) due to Faraday's law — see page 163. The e.m.f. produced is of the same frequency as the input voltage.

The magnitude of the induced e.m.f. is proportional to the change in flux linkage, $\Delta(N\phi)$, so the size of the e.m.f. induced in the secondary coil is determined by the ratio of the number of turns on each coil along with the voltage across the primary coil. Step-up transformers increase the voltage by having more turns on the secondary coil than the primary. Step-down transformers reduce the voltage by having fewer turns on the secondary coil.

Figure 6: *An early version of a transformer.*

Power loss in transformers

Real-life transformers aren't 100% efficient — some power is always lost. The metallic core is being cut by the continuously changing flux, which induces an e.m.f. in the core. In a continuous core this causes currents called eddy currents, which cause it to heat up and energy to be lost.

Eddy currents are looping currents induced by the changing magnetic flux in the core. They create a magnetic field that acts against the field that induced them, reducing the field strength. They also dissipate energy by generating heat. The effect of eddy currents can be reduced by laminating the core — this involves having layers of the core separated out by thin layers of insulator, so a current can't flow.

Tip: There's more on power loss in transformers on page 172.

Calculating the e.m.f. induced in transformer coils

From Faraday's law (page 163), the induced e.m.f.s in both the primary (*p*) and secondary (*s*) coils can be calculated:

Primary coil: $V_p = -\frac{n_p \Delta\phi}{\Delta t}$

Secondary coil: $V_s = -\frac{n_s \Delta\phi}{\Delta t}$

V_p = voltage across primary coil in V

$\frac{\Delta\phi}{\Delta t}$ = rate of change of flux in Wb s^{-1}

n_p = number of turns on primary coil

V_s = voltage across secondary coil in V

n_s = number of turns on secondary coil

Ideal transformers are 100% efficient, so power in equals the power out. Power is current × voltage, so for an ideal transformer $I_pV_p = I_sV_s$, or:

$$\frac{I_p}{I_s} = \frac{V_s}{V_p}$$

Tip: In an ideal transformer, the flux through the secondary coil is the same as the flux through the primary coil and no energy is lost in the transfer.

This can be combined with the equations for induced e.m.f. in each coil:

$$\frac{I_p}{I_s} = \frac{V_s}{V_p} = \frac{-n_s\frac{\Delta\phi}{\Delta t}}{-n_p\frac{\Delta\phi}{\Delta t}}$$

Which gives the transformer equation:

$$\frac{n_s}{n_p} = \frac{V_s}{V_p} = \frac{I_p}{I_s}$$

n_s = number of turns on secondary coil

n_p = number of turns on primary coil

V_p = voltage across primary coil in V

V_s = voltage across secondary coil in V

I_p = current in primary coil in A

I_s = current in secondary coil in A

Exam Tip
This is how the equation will appear in your data and formulae booklet.

Example — Maths Skills

What is the output voltage for a transformer with a primary coil of 120 turns, a secondary coil of 350 turns and an input voltage of 230 V?

Just use the equation for an ideal transformer and rearrange it to make the output voltage (V_s) the subject:

$$\frac{n_s}{n_p} = \frac{V_s}{V_p} \Rightarrow V_s = \frac{V_p n_s}{n_p} = \frac{230 \times 350}{120} = 670.83... = 670 \text{ V (to 2 s.f.)}$$

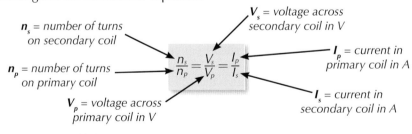

Example — **Maths Skills**

The secondary coil of a step-up transformer has 95 turns. Calculate the number of turns on the primary coil if the current across the transformer decreases from 22 A to 4.0 A.

Rearrange the ideal transformer equation to make n_p the subject:

$$\frac{n_s}{n_p} = \frac{I_p}{I_s} \Rightarrow n_p = \frac{n_s I_s}{I_p} = \frac{95 \times 4.0}{22} = 17.27... = 17 \text{ turns (to 2 s.f.)}$$

Investigating transformers

You can investigate the relationship between the number of turns and the voltages across the coils of a transformer by setting up the equipment as shown in Figure 7.

Put two C-cores together and wrap wire around each to make the coils. Begin with 5 turns in the primary coil and 10 in the secondary coil (a ratio of 1:2). Turn on the a.c. supply to the primary coil. Use a low voltage — remember transformers increase voltage, so make sure you keep it at a safe level. Record the voltage across each coil.

Keeping V_p the same so it's a fair test, repeat the experiment with different ratios of turns. Try 1:1 and 2:1. Divide n_s by n_p and V_s by V_p.

You should find that for each ratio of turns, $\frac{n_s}{n_p} = \frac{V_s}{V_p}$.

Tip: Make sure you carry out a risk assessment before carrying out any experiment.

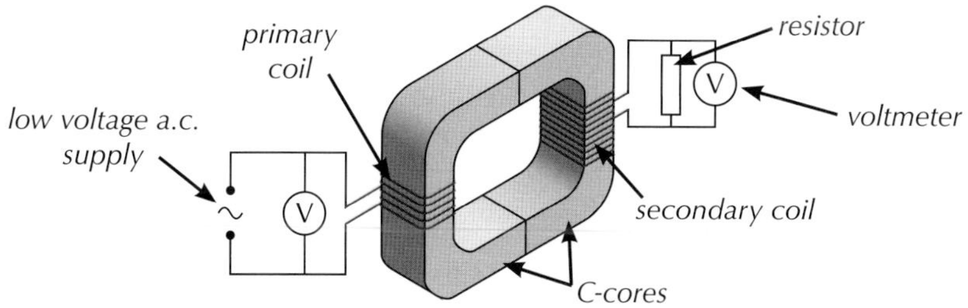

Figure 7: *Experimental set-up for investigating the relationship between the number of turns and voltages in transformers.*

Tip: The resistor is connected to the secondary coil to stop large currents from being generated in the circuit.

To investigate the relationship between current and voltage of the transformer coils for a given number of turns in the coil you can use the same equipment as before, but add a variable resistor to the primary coil circuit and an ammeter to both circuits as shown in Figure 8.

Turn on the power supply and record the current through and voltage across each coil. Leaving the number of turns constant, adjust the variable resistor to change the input current. Record the current and voltage for each coil, then repeat this process for a range of input currents.

You should find that for each current, $\frac{n_s}{n_p} = \frac{V_s}{V_p} = \frac{I_p}{I_s}$.

Tip: These formulas won't quite work in your investigation because real transformers aren't 100% efficient.

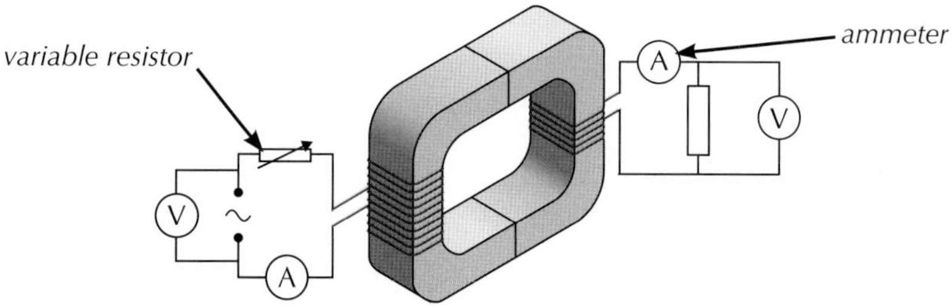

Figure 8: *Set-up for investigating the relationship between current and voltage in a transformer.*

Tip: You saw $P = I^2R$ in year 1 of A-Level physics.

Tip: Although some energy is lost inside the transformers at each end, it's nowhere near as much as the energy that would be lost if the electricity were transmitted at 230 V.

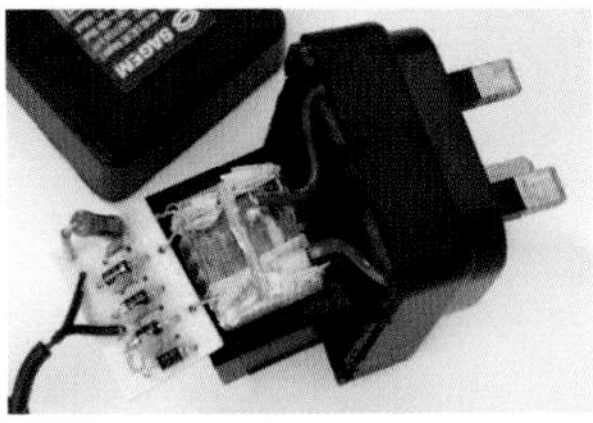

Figure 9: Plugs can contain transformers to reduce the voltage taken from the mains.

Transformers in the national grid

Transformers are an important part of the national grid. Electricity from power stations is sent round the country in the national grid at the lowest possible current. This is because a high current causes greater energy losses due to heating in the cables. The power losses due to the resistance of the cables is equal to $P = I^2R$ — so if you double the transmitted current, you quadruple the power lost. Using cables with the lowest possible resistance can also reduce energy loss.

Since power = current × voltage, a low current means a high voltage for the same amount of power transmitted. Transformers allow us to step up the voltage to around 400 000 V for transmission through the national grid.

High voltages raise safety and insulation issues, so the voltage has to be stepped back down to a safer 230 V at local substations before it can be used in homes.

Other uses of transformers

Transformers are used in many household devices. Lots of electronic devices like laptops, mobiles, monitors and speakers can't function using a standard 230 V mains supply — they need a much lower voltage (and usually a d.c. supply too).

The chargers for these devices contain transformers to adjust the voltage — they're contained in the plug or a box in the cable.

Practice Questions — Application

Q1 a) The primary coil of a transformer has 250 turns. The secondary coil has 420 turns. If the voltage across the primary coil is 190 V, calculate the voltage across the secondary coil.

b) There is a current of 13 A in the secondary coil of a different transformer. The voltage across the primary coil is 120 V and the voltage across the secondary coil is 75 V. Calculate the current in the primary coil.

Q2 Laptop chargers contain a transformer that changes the voltage from the mains supply. Mains electricity in the UK supplies 230 V.

a) Low resistance wires can be used to increase the efficiency of a transformer. State one other way in which the efficiency of a transformer can be improved.

b) The laptop requires a voltage of 19 V.

(i) Does the charger use a step-up or step-down transformer?

(ii) The primary coil in the transformer has 110 turns. Calculate the number of turns in the secondary coil to the nearest whole number.

Practice Questions — Fact Recall

Q1 Describe how a simple generator produces an alternating current.

Q2 How does a transformer change the voltage of an electricity supply?

Q3 Wire can be wrapped around two C-cores to make a transformer. This transformer can then be used to investigate the relationship between the ratio of turns on the primary and secondary coils and the output voltage. This is done by varying the number of secondary coils whilst keeping the number of primary coils the same. State one other factor which should be kept constant in this experiment.

Section Summary

Make sure you know...

- That magnetic fields exist around permanent magnets and moving charges.
- That magnetic field lines point from north to south, and the closer the field lines the stronger the field.
- How to find the direction of a magnetic field around a current-carrying wire with the right-hand rule.
- The shape and direction of the magnetic fields around a long straight current-carrying wire, a flat coil and a long solenoid.
- How to use Fleming's left-hand rule to find the direction of the current, the direction of the external magnetic field or the direction of the force on a wire, provided the other two are given.
- That magnetic flux density is the force on one metre of wire carrying a current of one amp at right angles to a magnetic field.
- That the unit for magnetic flux density is the tesla, T, and that one tesla is equal to one newton per amp per metre.
- How to calculate the force acting on a current-carrying conductor at an angle θ to a magnetic field using $\boldsymbol{F} = \boldsymbol{B}IL\sin\theta$.
- How the uniform magnetic flux density between the poles of a magnet can be measured using a current-carrying wire and a digital balance.
- How to calculate the force acting on a single charged particle moving through a magnetic field, where its velocity is perpendicular to the magnetic field, using $\boldsymbol{F} = \boldsymbol{B}Q\boldsymbol{v}$.
- That the force on a moving charge travelling perpendicular to a magnetic field is always perpendicular to its direction of travel, so charged particles in a uniform magnetic field can move in circular orbits.
- How electric and magnetic fields are used in velocity selectors to separate out particles of a certain velocity from a stream of accelerated charged particles moving at a range of speeds.
- That the total magnetic flux, ϕ, passing through an area that is perpendicular to a magnetic field is equal to the area multiplied by the magnetic flux density.
- That magnetic flux is measured in webers.
- That the magnetic flux linkage of a coil in a magnetic field is the product of the magnetic flux passing through the coil multiplied by the number of turns in the coil.
- How to calculate the magnetic flux through an area that is not perpendicular to a magnetic field using $\phi = \boldsymbol{B}A\cos\theta$.
- That Faraday's law states that induced e.m.f. is directly proportional to the rate of change of flux linkage, and be able to calculate induced e.m.f. using $\varepsilon = -\frac{\Delta(N\phi)}{\Delta t}$.
- That the gradient of a graph of flux linkage against time is equal to the negative of the induced e.m.f..
- That the area under a graph of induced e.m.f. against time is equal to $-\Delta(N\phi)$.
- How a search coil connected to a data recorder can be used to investigate magnetic flux between two bar magnets by moving the coil out of the field and plotting a graph of induced e.m.f. against time.
- That Lenz's law states that the induced e.m.f. is always in such a direction as to oppose the change that caused it.
- How a simple a.c. generator converts kinetic energy into electrical energy by rotating a coil in a magnetic field.
- How simple laminated iron-cored transformers work, and how the size of the voltage induced in the secondary coil is determined by the ratio of the number of turns on each coil and the voltage across the primary coil.
- How to use the equation $\frac{n_s}{n_p} = \frac{V_s}{V_p} = \frac{I_p}{I_s}$ for an ideal transformer.
- How to investigate the relationship between voltage, current and the number of turns in a transformer.

Exam-style Questions

1 A 33 cm long conducting rod moves perpendicular to a magnetic field of magnetic flux density 21 mT. An e.m.f. of 4.5 mV is generated across its length. At what velocity is the bar moving?

A 0.071 ms^{-1}

B 14.1 ms^{-1}

C 0.65 ms^{-1}

D 1.5 ms^{-1}

(1 mark)

2 In a velocity selector, electrons are travelling perpendicularly through an electric field and a magnetic field, which are at right angles to each other. The velocity selector is used to select electrons moving with a velocity of 42 kms^{-1} by varying the magnetic flux density, so only electrons travelling at this speed will travel in a straight line and pass through a hole in a collimator. The field strength of the electric field is 37×10^3 NC^{-1}. What is the magnetic flux density of the magnetic field?

A 0.88 T

B 1.1 T

C 8.8×10^2 T

D 1.1×10^{-3} T

(1 mark)

3 An electron is fired into a uniform magnetic field with a flux density of 0.93 T at a speed of 8.1×10^7 m, as **Fig 3.1** shows.

(a) What shape will the electron's path take? You can assume that the electron's velocity is perpendicular to the magnetic field.

(1 mark)

Fig 3.1

(b) Calculate the magnitude of the force the electron will experience, and state its direction.

(2 marks)

(c) An alpha particle of charge +2*e* is fired into the same magnetic field as the electron at the same speed. State the magnitude and direction of the force experienced by the alpha particle.

(2 marks)

4 An experiment is set up as shown in **Fig 4.1** to measure the flux density between two magnets. A stiff metal wire is clamped in place so that it passes through a metal cradle resting on a top pan balance. A magnet is attached to either side of the metal cradle, with opposite sides facing each other so that there is a uniform magnetic field between them. Crocodile clips connect the stiff metal wire to a circuit containing an ammeter and a variable d.c. power supply. The balance is zeroed before the power supply is switched on.

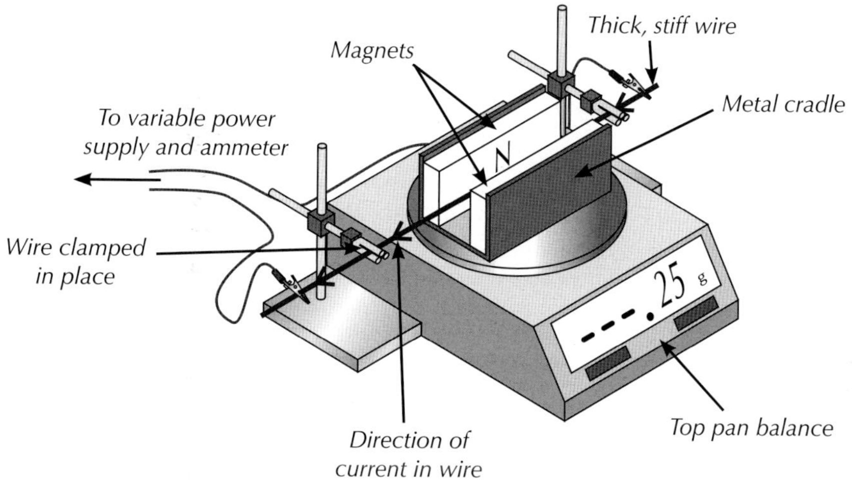

Fig 4.1

(a) State the direction of the force acting on the metal cradle due to the current in the wire.

(1 mark)

(b) Explain how the set-up shown in **Fig 4.1** could be used to determine the magnetic flux density of the magnetic field between the magnets, while keeping the length of wire constant.

(4 marks)

(c) A 52 cm long current-carrying wire is in a uniform magnetic field with a flux density of 44 mT. The wire is perpendicular to the lines of magnetic flux. If the wire experiences a force of 68 mN, calculate the current passing through the wire.

(2 marks)

5 The voltage across the primary coil of a step-down transformer is 230 V and the voltage across the secondary coil is 7 V.

(a) Suggest why the transformer won't work if the current supplied to the primary coil is a direct current.

(2 marks)

(b) The sum of the turns in the primary and secondary coils around the iron core equals 250. Calculate the number of turns of the secondary coil around the core.

(2 marks)

(c)* Describe an experiment that can be used to investigate the relationship between the current and voltage of transformer coils. Include a labelled diagram of the apparatus used and suggest how the accuracy of the experiment could be improved.

(6 marks)

* The quality of your response will be assessed in this question.

Module 6 Section 4: Nuclear and Particle Physics

Learning Objectives:

- Understand the alpha-particle scattering experiment and how this provided evidence of the atom containing a small, charged nucleus.
- Understand the simple nuclear model of the atom, including protons, neutrons and electrons.
- Understand what is meant by the proton number and nucleon number of a nucleus.
- Understand the ${}^{A}_{Z}X$ notation for the representation of nuclei.
- Know what is meant by an isotope.

Specification Reference 6.4.1

Tip: The circular fluorescent screen extends almost all the way around the other equipment.

Figure 2: *Hans Geiger and Ernest Rutherford, pictured in their lab with some equipment they used to detect alpha particles.*

1. Atomic Structure

Our ideas about the structure of the atom have changed a lot over the years. But we began to settle on our current model at the start of the last century. It's thanks to some clever folks called Rutherford, Geiger and Marsden...

The development of the atomic model

Following his discovery of the electron in the late 19th century, J.J. Thomson proposed the Thomson model of the atom, also known as the 'plum pudding' model. This model said that atoms were made up of a globule of positive charge, with negatively charged electrons sprinkled in it, like fruit in a plum pudding. It was widely accepted at the time, until the Rutherford alpha-scattering experiment of 1909.

The Rutherford alpha-scattering experiment

In Rutherford's laboratory, Hans Geiger and Ernest Marsden studied the scattering of alpha particles by thin metal foils — see Figure 1.

A stream of alpha particles from a radioactive source was fired at very thin gold foil, which was surrounded by a circular fluorescent screen. When alpha particles from a radioactive source strike a fluorescent screen, a tiny visible flash of light is produced. Geiger and Marsden recorded these flashes, and counted the number of alpha particles scattered at different angles.

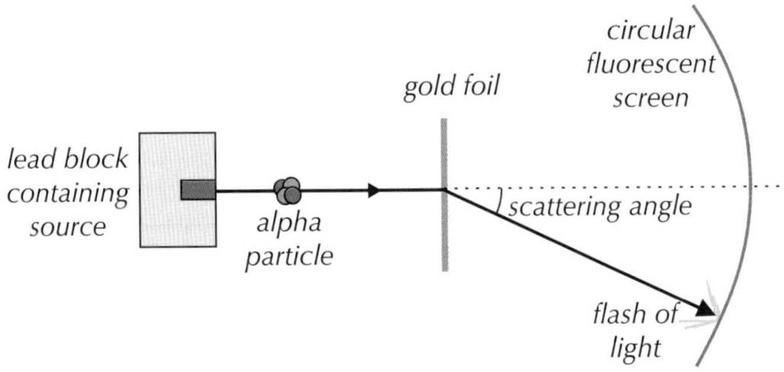

Figure 1: *The experimental set-up for detecting the deflection of alpha particles by thin gold foil.*

If the Thomson model was right, all the flashes should have been seen within a small angle of the beam, as shown in Figure 3. However, this wasn't what they saw.

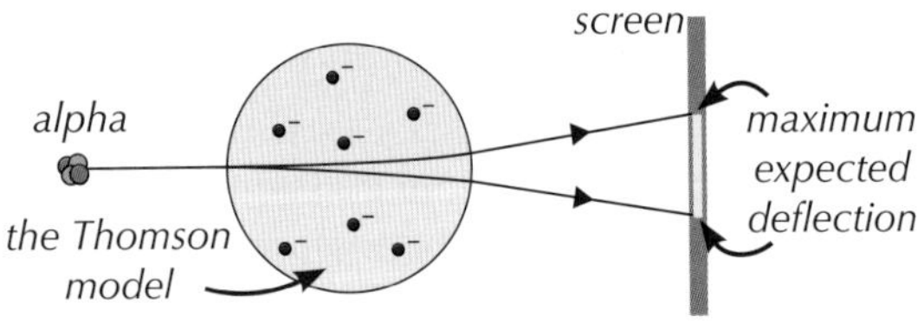

Figure 3: *The expected scattering of alpha particles from the Thomson model of the atom.*

Geiger and Marsden observed that most alpha particles went straight through the foil, but a few scattered at angles greater than 90°, sending them back the way they came. These results led Rutherford to some important conclusions:

- Most of the alpha particles went straight through the foil. So the atom is mainly empty space.
- Some of the alpha particles were deflected through large angles, so the centre of the atom must have a highly positive charge to repel them. Rutherford named this the nucleus.
- Very few particles were deflected by angles greater than 90°, so the nucleus must be tiny.
- Most of the mass must be in the nucleus, since the fast alpha particles (with high momentum) were deflected backwards by the nucleus.

So most of the mass and the positive charge in an atom must be contained within a tiny, central nucleus. This led to the **nuclear model of the atom**.

Tip: The idea of most of the mass being in the nucleus might be easier to visualise with larger objects. For example, imagine firing a golf ball at a bowling ball — the golf ball would just bounce straight backwards, and the bowling ball would barely move at all, because the mass of the bowling ball is so much bigger than the mass of the golf ball. It's the same with firing an alpha particle at a nucleus — the alpha particle is deflected straight back because the mass of the nucleus is so large.

The simple nuclear model of the atom

Inside every atom, there's a positive nucleus containing **neutrons** (which have no charge) and positively charged **protons**. Protons and neutrons are both known as **nucleons**. Orbiting this core are the negatively charged electrons.

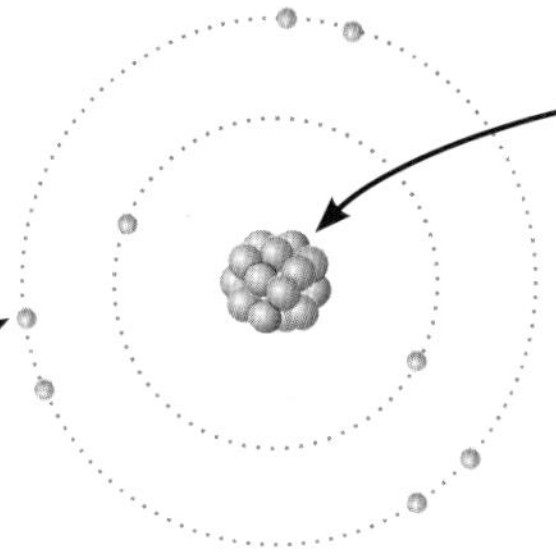

Figure 4: *An oxygen atom, with eight protons, eight neutrons and eight electrons.*

The particles in an atom have different charges and masses, shown in Figure 5. Charge is measured in coulombs (C) and mass is measured in kilograms (kg).

Particle	Charge (C)	Mass (kg)
Proton	$+1.60 \times 10^{-19}$	1.673×10^{-27}
Neutron	0	1.675×10^{-27}
Electron	-1.60×10^{-19}	9.11×10^{-31}

Figure 5: *Masses and charges of atomic particles.*

Tip: The proton and neutron masses given in Figure 5 are for when they're in isolation i.e. when they're not part of a nucleus. Once part of a nucleus, their masses change (see p.207).

Exam Tip
You'll be given the masses (in kg) of all three particles and the elementary charge, e, in the data and formulae booklet in the exam.

1.60×10^{-19} C is the elementary charge, e. The charge on an electron, –e, is equal and opposite to the charge on a proton, +e.

The nucleus only makes up a tiny proportion of an atom — it's only about one 10 000th of the diameter of the whole atom. The electrons orbit at relatively vast distances from the nucleus, so most of the atom is empty space.

The proton and neutron are roughly 2000 times more massive than the electron, so the nucleus makes up nearly all of the mass of the atom.

Proton and nucleon number

The **proton number** of an atom is the number of protons in its nucleus. It is sometimes called the atomic number, and has the symbol Z. It's the proton number that defines the element — no two different elements will have the same number of protons.

In a neutral atom, the number of electrons equals the number of protons. The element's reactions and chemical behaviour depend on the number of electrons. So the proton number tells you a lot about its chemical properties.

Tip: A neutral atom must have the same number of protons and electrons so that the positive and negative charges cancel each other out.

The **nucleon number** is also called the mass number, and has the symbol A. It tells you how many protons and neutrons are in the nucleus.

Each proton or neutron in a nucleus has a mass of approximately 1 unified atomic mass unit (= 1.661×10^{-27} kg, see p.192). The mass of an electron compared with a nucleon is virtually nothing, so the number of nucleons is about the same as the atom's mass (in atomic mass units).

Tip: Protons and neutrons have slightly different masses when they are in a nucleus compared to when they are isolated (see page 207). The atomic mass unit is the average nucleon mass, calculated from the mass of a carbon-12 nucleus, divided by 12.

Standard notation

Standard notation summarises all the information about an element's nucleus. Figure 6 shows the standard notation for an element X, with nucleon number A and proton number Z:

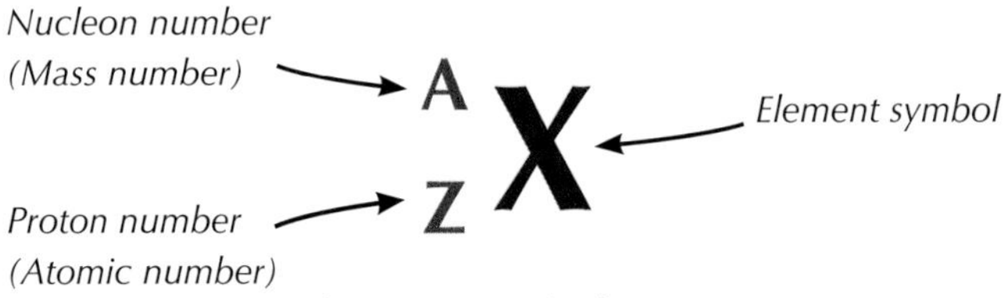
Nucleon number
(Mass number)
A
X
Element symbol
Proton number
(Atomic number)
Z

Figure 6: *Standard notation.*

Example

A carbon-12 atom has 6 protons and 6 neutrons.

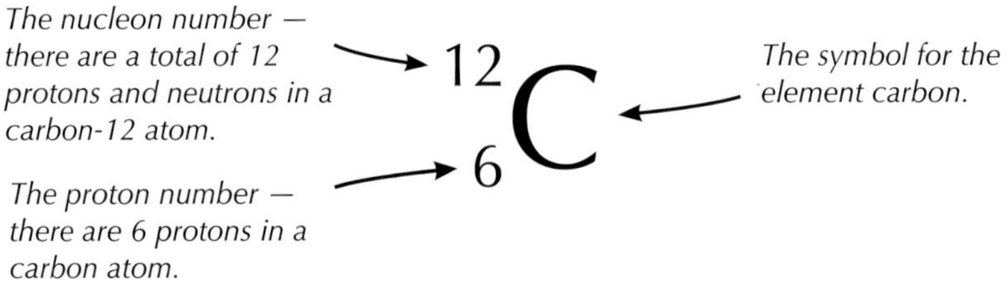
The nucleon number — there are a total of 12 protons and neutrons in a carbon-12 atom.
12
C
6
The proton number — there are 6 protons in a carbon atom.
The symbol for the element carbon.

Isotopes

Atoms with the same number of protons but different numbers of neutrons are called **isotopes**. Changing the number of neutrons doesn't affect the atom's chemical properties. The number of neutrons affects the stability of the nucleus though. Too few or too many neutrons can make a nucleus unstable. Unstable nuclei may be radioactive and decay to make themselves more stable (see page 192).

Isotopes are often named using their nucleon number, e.g. carbon's isotopes include carbon-12 and carbon-13. Some isotopes of elements have special names e.g. hydrogen's isotopes include deuterium and tritium.

Example

Hydrogen has three isotopes — protium, deuterium and tritium.

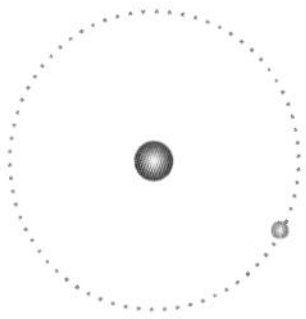

Protium has 1 proton and 0 neutrons.

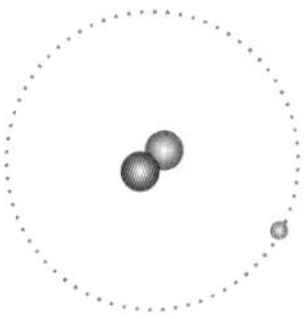

Deuterium has 1 proton and 1 neutron.

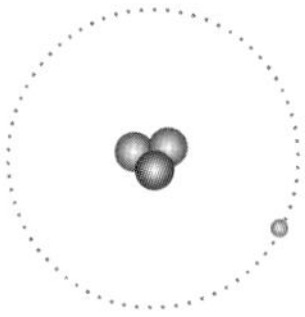

Tritium has 1 proton and 2 neutrons.

Figure 7: *Tritium can be used to illuminate fire exit signs and watch faces without the need for electricity.*

Practice Questions — Application

Q1 An atom of oxygen, O, has 8 protons and 8 neutrons.

a) What is the proton number of this atom?

b) What is the nucleon number of this atom?

c) Write this information in standard notation.

Q2 Element X has 21 protons and 24 neutrons.

a) Write this in standard notation.

b) Suggest the standard notation for a different isotope of element X.

Q3 Helium is written in standard notation as: ${}^{4}_{2}\text{He}$

a) How many protons does an atom of helium have?

An isotope of helium has a nucleon number of 3.

b) How many protons does this isotope have?

Practice Questions — Fact Recall

Q1 Explain the conclusions about the structure of the atom that could be drawn from the results of the alpha-particle scattering experiment.

Q2 Describe the simple nuclear model of the atom.

Q3 What is the proton number of an atom?

Q4 What is the nucleon number of an atom?

Q5 What are isotopes?

2. The Nucleus

The nucleus may be tiny, but it contains most of the mass of the atom. It takes a lot to hold it all together in such a small space...

Learning Objectives:

- Know the relative sizes of the atom and the nucleus.
- Know that the radius of the nucleus can be calculated using the equation $R = r_0A^{1/3}$, where r_0 is a constant and A is the nucleon number.
- Know how to calculate the mean densities of atoms and nuclei.
- Know what is meant by the strong nuclear force.
- Understand the short-range nature of the strong nuclear force.
- Know that the strong nuclear force is repulsive at distances below about 0.5 fm, and attractive up to a distance of about 3 fm.

Specification Reference 6.4.1

The size of the nucleus and atom

By probing atoms using scattering and diffraction methods, we know that the diameter of an atom is about 0.1 nm (1×10^{-10} m) and the diameter of the smallest nucleus is a few fm (1 fm = 1×10^{-15} m). So nuclei are really, really tiny compared with the size of the whole atom.

To make this easier to visualise, try imagining a large Ferris wheel as the size of an atom. If you then put a grain of rice in the centre, this would be approximately the size of the atom's nucleus.

Molecules are just a number of atoms joined together. As a rough guide, the size of a molecule equals the number of atoms in it multiplied by the size of one atom.

Nuclear radius

As you know from p.177, the particles that make up the nucleus (i.e. protons and neutrons) are called nucleons. The number of nucleons in an atom is called the nucleon (or mass) number, A. Unsurprisingly, as more nucleons are added to the nucleus, it gets bigger.

You can measure the size of a nucleus by firing particles at it. If you plot the radius of the nucleus, R, against the nucleon number, you get a graph like the one shown in Figure 1.

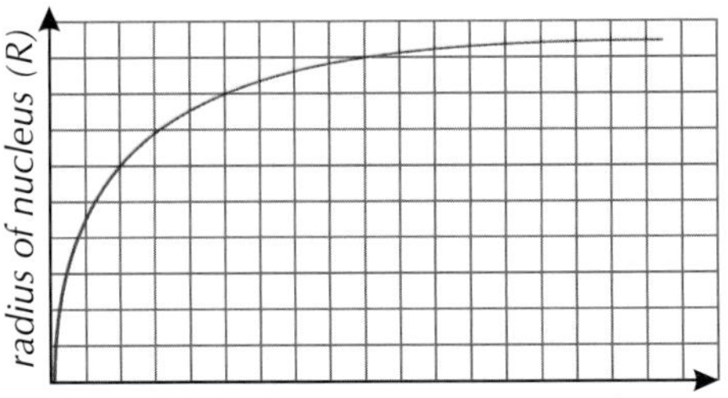

Figure 1: *A graph of nuclear radius, R, against nucleon number, A.*

In fact, nuclear radius increases roughly as the cube root of the nucleon number. You can see this by plotting nuclear radius against the cube root of the nucleon number — as in Figure 2.

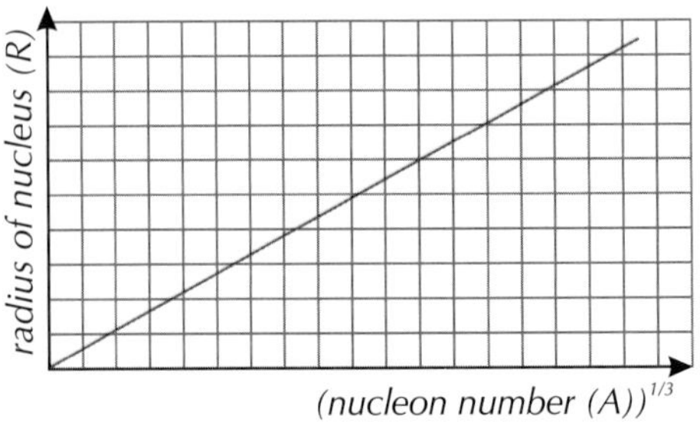

Figure 2: *A graph of nuclear radius, R, against the cube root of nucleon number, A.*

The fact that this graph is a straight line through the origin shows that nuclear radius is directly proportional to the cube root of the nucleon number. This relationship can be written as: $R \propto A^{1/3}$.

By introducing a constant, r_0, we can make this into an equation:

R = *nuclear radius* → $R = r_0 A^{1/3}$ ← **A** = *nucleon number*

Some current experimental results suggest that r_0 is about 1.4 fm, whilst others can give slightly different values.

Tip: You don't need to learn the value of r_0. If you're asked to do calculations like this in the exam, you'll be given all the information you need to answer the question.

Example — Maths Skills

An atom of rubidium has a nucleon number of 85. Calculate the radius of its nucleus. r_0 = 1.4 fm.

$R = r_0 A^{1/3}$

$= 1.4 \times 10^{-15} \times 85^{1/3}$

$= 6.1555... \times 10^{-15}$ m = 6.2 fm (to 2 s.f.)

Tip: Since A is just a number, with no units, the units of R will depend on what units you use for r_0. If you plug in r_0 in fm, your answer for R will be in fm too.

Density of the nucleus

To calculate the mean density of a nucleus you need to know the mass and volume of the nucleus — and the equation for density:

$$\rho = \frac{m}{V}$$

If you're asked to estimate mean nuclear density, you might have to work out the volume of the nucleus from its radius. Assume the nucleus is a sphere and substitute the radius into the equation:

$$V = \frac{4}{3}\pi r^3$$

Nuclear density is pretty much the same, regardless of the element — roughly 10^{17} kg m^{-3}. Nuclear density is much higher than atomic density. This suggests that:

- Most of an atom's mass is in its nucleus.
- The nucleus is small compared to the atom.
- An atom must contain a lot of empty space.

Tip: The symbol for density, ρ, is the Greek letter 'rho'. Its unit is kgm^{-3}.

Tip: Be careful of units if you're calculating densities — lengths need to be in metres and masses need to be in kilograms.

Tip: You can find the atomic density in the same way as you find the density of the nucleus. Just use the radius of the atom instead of the radius of the nucleus. You can assume the electron mass is negligible for density calculations, so the mass will be the same in both calculations.

Examples — Maths Skills

A carbon atom has a mass of 2.00×10^{-26} kg, an atomic radius of 7.0×10^{-11} m, and a nuclear radius of 3.2×10^{-15} m. Calculate its mean atomic density.

$V_{atom} = \frac{4}{3}\pi r^3 = \frac{4}{3} \times \pi \times (7.0 \times 10^{-11})^3 = 1.436... \times 10^{-30}$ m^3

$\rho_{atom} = \frac{m}{V}$ so $\rho = \frac{2.00 \times 10^{-26}}{1.436... \times 10^{-30}} = 1.392... \times 10^4$

$= 1.4 \times 10^4$ kg m^{-3} (to 2 s.f.)

Calculate its mean nuclear density.

$V_{nucleus} = \frac{4}{3}\pi r^3 = \frac{4}{3} \times \pi \times (3.2 \times 10^{-15})^3 = 1.372... \times 10^{-43}$ m^3

$\rho_{nucleus} = \frac{m}{V}$ so $\rho = \frac{2.00 \times 10^{-26}}{1.372... \times 10^{-43}} = 1.457... \times 10^{17}$

$= 1.5 \times 10^{17}$ kg m^{-3} (to 2 s.f.)

Forces in the nucleus

There are several different forces acting on the nucleons in a nucleus. The electrostatic force causes the positively charged protons in the nucleus to repel each other (p.134). The gravitational force causes all the nucleons in the nucleus to attract each other due to their mass (p.62).

However, the repulsion from the electrostatic force is much, much bigger than the gravitational attraction. If these were the only forces acting in the nucleus, the nucleons would fly apart. So there must be another attractive force that holds the nucleus together — called the **strong nuclear force**.

The strong nuclear force

To hold the nucleus together, the strong nuclear force must be an attractive force that overcomes the electrostatic force. Experiments have shown that the strong nuclear force between nucleons has a short range. It can only hold nucleons together when they are separated by up to a few femtometres — the size of a nucleus. The strength of the strong nuclear force between nucleons quickly falls beyond this distance.

Experiments also show that the strong nuclear force works equally between all nucleons. This means that the size of the force is the same whether for proton-proton, neutron-neutron or proton-neutron interactions.

At very small separations, the strong nuclear force must be repulsive — otherwise there would be nothing to stop it crushing the nucleus to a point.

Tip: The gravitational attraction in the nucleus is so small compared to the other forces, you can just ignore it.

Comparing strong nuclear and electrostatic forces

Figure 3 shows how the strong nuclear force changes with the distance between nucleons. It also shows how the electrostatic force changes so that you can see the relationship between these two forces (although only protons feel the electrostatic force).

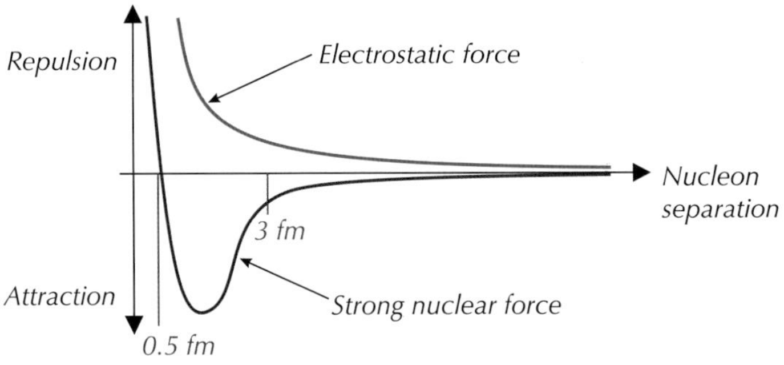

Figure 3: *A graph to show how the strong nuclear and electrostatic forces vary with nucleon separation.*

Figure 4: *The Large Hadron Collider at CERN collides lead ions to try to find out more about the strong nuclear force.*

The graph shows that:

- The strong nuclear force is repulsive for very small separations of nucleons (below about 0.5 fm).
- As nucleon separation increases past about 0.5 fm, the strong nuclear force becomes attractive. It reaches a maximum attractive value and then falls rapidly to zero. After about 3 fm it can no longer hold nucleons together.
- The electrostatic repulsive force extends over a much larger range (indefinitely, actually).

Practice Questions — Application

Q1 The nucleus of an atom of an element has a radius of 7.0 fm and contains 73 neutrons. Find the number of protons in the nucleus.

Q2 Explain how the strong nuclear force acts between nucleons that are:

a) 0.4 fm apart.

b) 1.5 fm apart.

c) 4.2 fm apart.

Q3 A vanadium atom has a mass of approximately 8.5×10^{-26} kg, and an atomic density of 8.4×10^{4} $kg\,m^{-3}$.

a) Calculate the radius of the vanadium atom.

b) The vanadium nucleus has a radius of approximately 5.2×10^{-15} m. Calculate the nuclear density of the vanadium atom.

Tip: You can assume $r_0 = 1.4$ fm for Q1.

Practice Questions — Fact Recall

Q1 Give the approximate diameters of an atom and a small nucleus.

Q2 State how the nuclear radius changes with nucleon number.

Q3 State the shape that a nucleus is assumed to be, and give the equation that can be used to calculate its volume.

Q4 a) Sketch a graph to show how the strong nuclear force changes with nucleon separation, marking on any key separation distances.

b) State the range in which the strong nuclear force is repulsive.

c) State the range in which the strong nuclear force is attractive and can hold nucleons together.

3. Particles and Antiparticles

Particles come in two types — matter and antimatter. Antimatter can be produced in pairs with its corresponding matter, but if the two come into contact, they destroy each other.

Learning Objectives:

- Know that all particles classified as hadrons are subject to the strong nuclear force.
- Know that protons and neutrons are examples of hadrons.
- Know that all particles classified as leptons are subject to the weak nuclear force.
- Know that electrons and neutrinos are examples of leptons.
- Know what particles and antiparticles are, including the pairs electron-positron, proton-antiproton, neutron-antineutron and neutrino-antineutrino.
- Know that particles and their corresponding antiparticles have the same mass.
- Know that an electron and a positron have opposite charges and that protons and antiprotons have opposite charges.
- Know Einstein's mass-energy equation $\Delta E = \Delta mc^2$.
- Understand the creation and annihilation of particle-antiparticle pairs.

Specification References 6.4.2 and 6.4.4

Hadrons

Not all particles can feel the strong nuclear force — the ones that can are called **hadrons**. Hadrons aren't **fundamental particles**. They're made up of smaller particles called quarks (see page 189).

Protons and neutrons are hadrons. This is why they can make atomic nuclei — the nucleus of an atom is made up from protons and neutrons held together by the strong nuclear force (p.182). As well as protons and neutrons, there are other hadrons that you don't get in normal matter, like sigmas (Σ) and mesons — luckily you don't need to know a lot about them.

The decay of hadrons

Most hadrons will eventually decay into other particles.
The exception is protons — most physicists think that protons don't decay.

The neutron is an unstable particle that decays into a proton (along with an electron and an antineutrino). The decay of a neutron is really just an example of β^- decay (see p.192), which is caused by the **weak nuclear force**.

$$n \rightarrow p + e^- + \bar{\nu}$$

neutron → n; *proton* → p; *electron* → e^-; *antineutrino (see page 185)* → $\bar{\nu}$

Free neutrons (i.e. ones not in a nucleus) have a half-life of about 15 minutes. (The neutron is much more stable when it is part of a nucleus.)

Leptons

Leptons are fundamental particles and they don't feel the strong nuclear force. They interact with other particles via the weak nuclear force and gravity (and the electrostatic force if they're charged).

There are two types of lepton you need to know about — electrons (e^-), which should be familiar, and **neutrinos** (ν).

Neutrinos have zero (or almost zero) mass and zero electric charge — so they don't do much. Neutrinos only take part in weak interactions (see p.190). In fact, a neutrino can pass right through the Earth without anything happening to it.

Name	Symbol	Relative charge
electron	e^-	–1
neutrino	ν	0

Figure 1: *The symbols and relative charges of two leptons you need to know — the electron and the neutrino.*

Tip: Relative charge is the charge of the particle expressed a multiple of the elementary charge, *e*. A particle with relative charge –1 has a charge of -1.60×10^{-19} C.

Antimatter

Each particle type has a corresponding **antiparticle** with the same mass but with opposite charge. Antiparticles were predicted before they were discovered. When Paul Dirac wrote down an equation obeyed by electrons, he found a kind of mirror image solution. It predicted the existence of a particle like the electron but with opposite electric charge — the **positron**. The positron turned up later in a cosmic ray experiment. Positrons have identical mass to electrons but they carry a positive charge. It is the antimatter version of the electron.

Figure 3 shows the relative charges of the proton, neutron, electron, neutrino and their antiparticles.

Particle/Antiparticle	Symbol	Relative Charge	Rest Mass (kg)
proton	p	+1	1.673×10^{-27}
antiproton	$\bar{p}$	–1	
neutron	n	0	1.675×10^{-27}
antineutron	$\bar{n}$		
electron	e^-	–1	9.11×10^{-31}
positron	e^+	+1	
neutrino	ν	0	0
antineutrino	$\bar{\nu}$		

Figure 3: *Relative charges and rest masses of particles and their corresponding antiparticles.*

The masses in Figure 3 are all rest masses — the mass of the particle when it's not moving. This is because the masses of objects change when they're moving at very high speeds, but you don't need to know about that.

In the exam, you'll be given the masses of protons, neutrons and electrons. Just remember that the mass of an antiparticle is the same as the mass of its corresponding particle. You need to learn the relative charges on each type of particle though (these are all relative to $e = 1.60 \times 10^{-19}$ C). Neutrinos have zero charge and (almost) zero mass.

Figure 2: *Paul Dirac, the English physicist who predicted the existence of antimatter.*

Tip: All particles are known as matter, and antiparticles are known as antimatter.

Tip: The opposite of a zero charge is just zero. So the antiparticle of any neutral particle will also be neutral.

Tip: Most antiparticles have the same symbol as their corresponding particle, but with a line over the top, i.e. $\bar{x}$. The exceptions to this are charged leptons, like the electron. These tend to have their particles marked with a minus sign, e.g. e^-, to indicate their negative charge, while their antiparticles have a plus sign, e.g. e^+, since they have a positive charge.

Einstein's mass-energy equation

You've probably heard about the equivalence of energy and mass. It all comes out of Einstein's special theory of relativity. Energy can turn into mass and mass can turn into energy, as described by the fantastic and rather famous formula, Einstein's mass-energy equation:

$$\Delta E = \Delta m c^2$$

ΔE = change in energy in J

c = speed of light in ms^{-1} = 3.00×10^8 ms^{-1}

Δm = change in mass in kg

You can use this equation to calculate the energy of a particle at rest with a given mass. This is just how much energy would be released if all the mass was converted to energy.

Tip: In this example, Δm (the change in mass) is just the total mass of the proton, as it's all being converted to energy.

Example — Maths Skills

The mass of a proton at rest is 1.673 × 10^{-27} kg. Calculate the energy that would be released if this proton were completely converted into energy.

$$\Delta E = \Delta mc^2$$
$$= 1.673 \times 10^{-27} \times (3.00 \times 10^8)^2$$
$$= 1.5057 \times 10^{-10}$$
$$= 1.51 \times 10^{-10} \text{ J (to 3 s.f.)}$$

Pair production

If you fire two protons at each other at high speed then you'll end up with a lot of energy at the point of impact. This energy might be converted into more particles.

When energy is converted into mass you get equal amounts of matter and antimatter. So if an extra proton is formed then there will always be an antiproton to go with it. It's called **pair production**.

Tip: Protons repel each other, so it takes a lot of energy to make them collide. The energy supplied is released when they collide, so proton-proton collisions release a lot of energy.

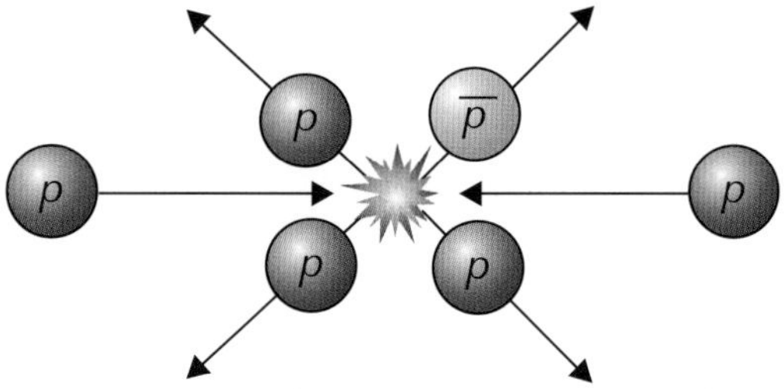

Figure 4: Pair production — two protons colliding and producing a proton-antiproton pair.

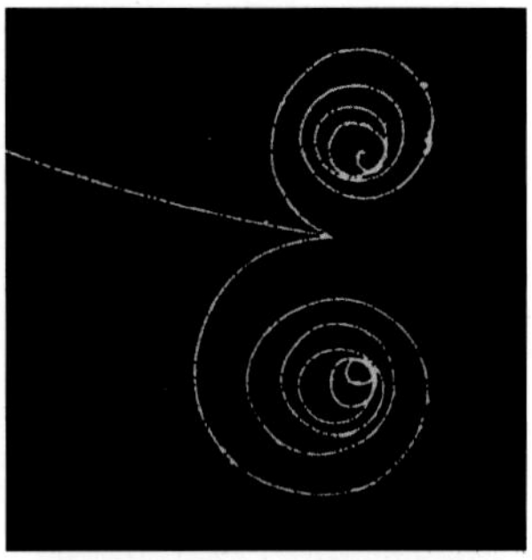

Figure 6: A pair-production event in a bubble tank. A gamma ray photon has come from the right (but is invisible here). The spiral paths show the motion of an electron-positron pair after their production from the photon. The straighter path is another particle produced as a by-product — but you don't need to know what that is.

You can also get pair production from a photon. This only happens if one photon has enough energy to produce that much mass (e.g. a gamma ray photon, γ). It also tends to happen near a nucleus, which helps conserve momentum. You usually get electron-positron pairs produced (rather than any other pair) — because they have a relatively low mass.

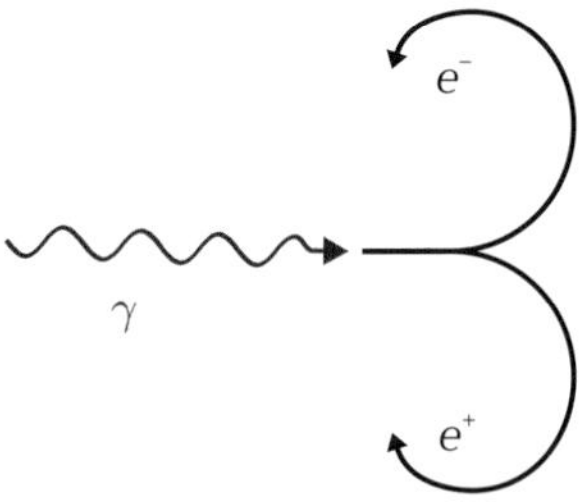

Figure 5: Pair production — an electron-positron pair produced from a gamma ray photon.

Tip: The tracks are curved as there's usually a magnetic field present in experiments. They curve in opposite directions because of their opposite charges — see p.154.

The minimum amount of energy the photon must have is the combined energy of the two particles at rest (i.e. assuming that the particles have negligible kinetic energy). Remember, you can find the energy equal to a particle at rest using $\Delta E = \Delta mc^2$. So you can calculate the minimum energy, E_γ, of the photon.

A particle and its antiparticle have the same rest mass (m), which means that:

$$E_\gamma = 2mc^2$$

E_γ = *minimum energy of photon in J*

c = *speed of light in ms^{-1}*

m = *rest mass of one of the pair-produced particles in kg*

You can go further and find the maximum wavelength, λ_{max}, or minimum frequency, f_{min}, of the photon using the equation for the energy of a photon (which you saw in Year 1):

$$E_\gamma = \frac{hc}{\lambda} = hf$$

E_γ = *energy of photon in J*

λ = *wavelength of photon in m*

f = *frequency of photon in Hz*

h = *Planck's constant in Js*

Just put these two equations for E_γ together and rearrange to find λ_{max} or f_{min}:

$$\lambda_{max} = \frac{h}{2mc} \qquad f_{min} = \frac{2mc^2}{h}$$

Example — Maths Skills

Calculate the minimum energy a photon must have to produce an electron-positron pair.

The mass of an electron (and so positron) is 9.11×10^{-31} kg.

So, $E_\gamma = 2mc^2$

$= 2 \times 9.11 \times 10^{-31} \times (3 \times 10^8)^2$

$= 1.6398 \times 10^{-13} = 1.64 \times 10^{-13}$ J (to 3 s.f.) (= 1.02 MeV)

Calculate the minimum frequency of this photon.

$E_\gamma = hf$, so $f = \frac{E_\gamma}{h}$

$f = \frac{1.6398 \times 10^{-13}}{6.63 \times 10^{-34}} = 2.4733... \times 10^{20} = 2.47 \times 10^{20}$ Hz (to 3 s.f.)

Tip: Energies in particle physics are often given in MeV. To convert J to MeV, divide the value by 1.6×10^{-19} (to change J into eV) and then divide it by 10^6 (to change the eV to MeV).

Annihilation

When a particle meets its antiparticle the result is **annihilation**.
All the mass of the particle and antiparticle gets converted to energy, in the form of a pair of photons. Antiparticles can usually only exist for a fraction of a second before this happens, so you don't get them in ordinary matter.

Tip: The two photons produced in annihilation are always emitted in opposite directions to conserve momentum.

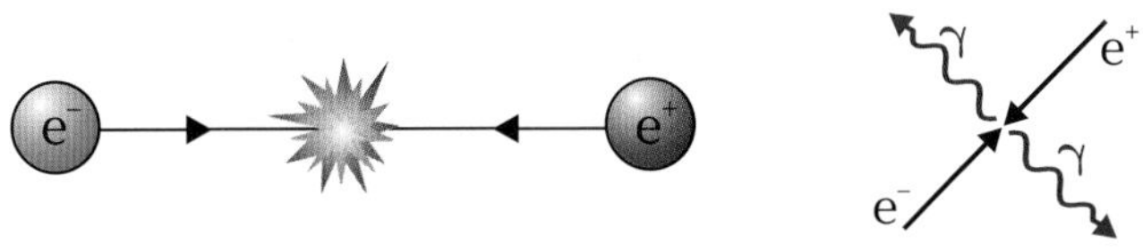

Figure 7: *Electron-positron annihilation.*

Tip: Just like with pair production, you can use the equation $E_\gamma = hf = \frac{hc}{\lambda}$ to find the minimum frequency, or maximum wavelength, of the photons emitted by annihilation events.

Just like with pair production, you can calculate the minimum energy of each photon produced (i.e. assuming that the particles have negligible kinetic energy). The combined energy of the photons will be equal to the combined energy of the particles, so $2E_\gamma = 2mc^2$ and so:

$$E_\gamma = mc^2$$

E_γ = *minimum energy of one photon in J*

c = *speed of light in ms⁻¹*

m = *rest mass of one of the annihilated particles in kg*

Example — Maths Skills

A neutron and antineutron collide and annihilate.
Calculate the minimum energy of one of the photons produced.

The mass of a neutron is 1.675×10^{-27} kg.

So, $E_\gamma = mc^2$
$= 1.675 \times 10^{-27} \times (3.00 \times 10^8)^2$
$= 1.5075 \times 10^{-10}$ J
$= 1.51 \times 10^{-10}$ J (to 3 s.f.) (= 942 MeV)

Exam Tip
Don't worry if you're given a question about a particle you're not familiar with in the exam. You'll be given all the information you need in order to answer the question.

Practice Questions — Application

Q1 A collision between protons can release enough energy to create a proton-antiproton pair. What is the minimum energy in J required for this pair production to take place?

Q2 A Λ^0 particle is a hadron with a mass of 1.983×10^{-27} kg.

a) Calculate the minimum energy of a photon produced when a Λ^0 particle annihilates with its antiparticle.

b) Calculate the minimum frequency and maximum wavelength of this photon.

Q3 A muon is a type of lepton, with a mass of 1.88×10^{-28} kg. Calculate the minimum frequency a photon must have to pair-produce a muon-antimuon pair.

Practice Questions — Fact Recall

Q1 State the type of particle that is affected by the strong nuclear force.

Q2 Name one force through which an electron can interact with other particles.

Q3 Give the relative charge and relative mass of a neutrino, and state whether it is a hadron or lepton.

Q4 a) How does an antiparticle differ from its corresponding particle?

b) How is it the same?

Q5 Name the electron's antiparticle.

Q6 Write down the relative charge of an:

a) antiproton b) antineutron c) antineutrino

Q7 State Einstein's mass-energy equation. Define all symbols used.

Q8 Describe the process of pair production.

Q9 What is produced in the annihilation of a particle and its antiparticle?

4. Quarks and Anti-quarks

Quarks are fundamental particles — they're the smallest amount of matter that you can get that interacts with the strong nuclear force.

Learning Objectives:

- Understand the simple quark model of hadrons in terms of up (u), down (d) and strange (s) quarks and their respective anti-quarks.
- Know the charges of the up (u), down (d), strange (s), anti-up ($\bar{u}$), anti-down ($\bar{d}$) and anti-strange ($\bar{s}$) quarks, as fractions of the elementary charge e.
- Know the quark model of the proton (uud) and of the neutron (udd).
- Understand the decay of particles in terms of the quark model.
- Understand beta-minus (β^-) decay in terms of the quark model.

$$d \rightarrow u + {}^{0}_{-1}e + \bar{\nu}$$

- Understand beta-plus (β^+) decay in terms of the quark model.

$$u \rightarrow d + {}^{0}_{+1}e + \nu$$

- Be able to balance quark transformation equations in terms of charge.

Specification Reference 6.4.2

The simple quark model

Quarks are the building blocks for hadrons like protons and neutrons. In total, there are 6 types of quark. But at A-level, we can limit ourselves to looking at only three types of quark — the up quark (u), the down quark (d), the strange quark (s) and their antiparticles, known as anti-quarks (see Figure 1). The anti-quarks have opposite charges to the quarks. Antiparticles of hadrons (like antiprotons and antineutrons) can be made with anti-quarks.

Name	Symbol	Charge
up	u	$+2/3$ e
down	d	$-1/3$ e
strange	s	$-1/3$ e

Name	Symbol	Charge
anti-up	$\bar{u}$	$-2/3$ e
anti-down	$\bar{d}$	$+1/3$ e
anti-strange	$\bar{s}$	$+1/3$ e

***Figure 1**: Properties of up, down and strange quarks (left), and their anti-quarks (right).*

You'll be given the charges of the quarks in the data and formulae booklet in the exam. You won't be given the charges of the anti-quarks, but remember — the charge on an anti-quark is the same size as the charge on its quark, but with the opposite sign.

Strange quarks and down quarks may look pretty similar, but there are other properties of quarks not given here that distinguish them from each other. Luckily, you don't need to know about these other properties.

Quark composition of hadrons

Evidence for quarks came from hitting protons with high energy electrons. The way the electrons scattered showed that there were three concentrations of charge (quarks) inside the proton. The properties of a particle depend on the properties of the quarks that make it up. A proton has a charge of +1 because the quarks that make it up have an overall charge of +1.

Protons are made of two up quarks and one down quark (uud).

- The total charge of a proton is 2/3 + 2/3 + (– 1/3) = +1.

Neutrons are made up of one up quark and two down quarks (udd).

- The total charge of a neutron is 2/3 + (–1/3) + (–1/3) = 0.

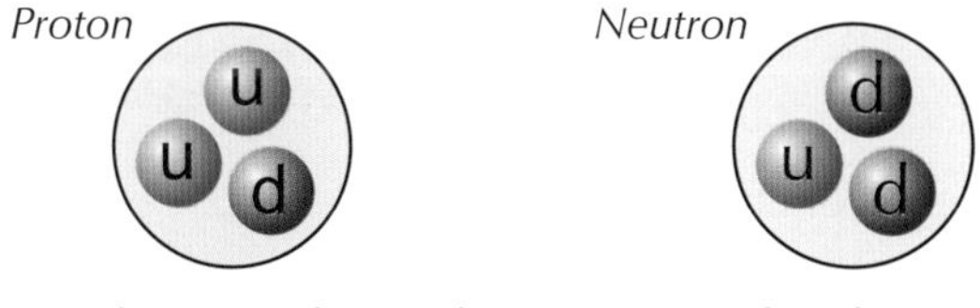

***Figure 2:** The quark composition of nucleons.*

Antiprotons are $\bar{u}\bar{u}\bar{d}$ and antineutrons are $\bar{u}\bar{d}\bar{d}$. Not all hadrons have three quarks though. Protons and neutrons are a type of hadron called baryons, which are made up of three quarks. There are also hadrons made up of a quark and an anti-quark, called mesons. You don't need to know anything more about them, but don't be surprised if you see one pop up in a question.

Tip: You only need to know the quark composition of protons and neutrons for your exam. If you're given any other hadrons, you'll be given their quark composition.

Quark confinement

It's not possible to get a quark by itself — this is called quark confinement. If you blasted a proton with a lot of energy, a single quark would not be removed. The energy that you supplied would just get changed into a quark-anti-quark pair.

Tip: Figure 3 just shows an example of pair production (see page 186). In this example, a meson (quark-anti-quark pair) is created.

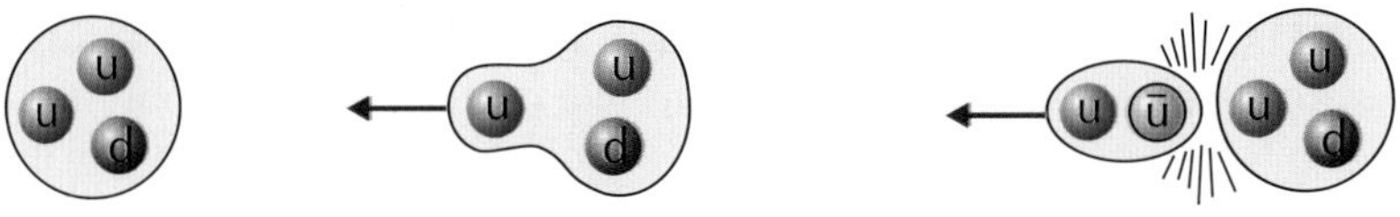

A proton.

Energy supplied to remove u quark.

When enough energy is supplied, a u and $\overline{u}$ pair is produced and the u quark stays in the proton.

Figure 3: *Quark confinement — the energy used trying to remove a u quark only creates a u and $\overline{u}$ pair in a pair production.*

Weak nuclear force

Hadrons can decay into other particles via the weak nuclear force (p.184). This is the only thing that can change one type of quark into another. A quark changing into another quark is known as changing a quark's flavour. In any other kind of interaction, the number of quarks of any type must be the same before the interaction as after it. In all interactions, charge must be conserved.

Beta-minus decay

Tip: A β^- particle is just an electron (see p.192).

Tip: You'll notice that charge is conserved in both of β^- and β^+ interactions — the total charge on the left of the arrow equals the total charge on the right.

Beta-minus (β^-) decay is a type of weak interaction. In β^- decay a neutron is changed into a proton — in other words udd changes into uud. It means turning a d quark into a u quark.

Beta-minus decay in terms of quarks: $d \rightarrow u + {}^{0}_{-1}e + \overline{\nu}$

Beta-minus decay in terms of charge: $(-\tfrac{1}{3}) \rightarrow (+\tfrac{2}{3}) + (-1) + 0$

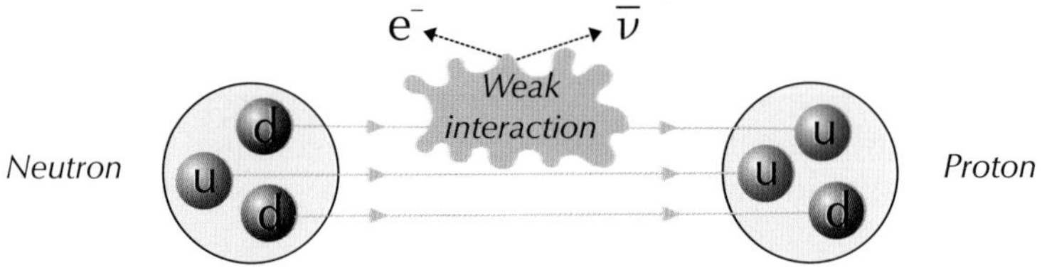

Figure 4: *A quark changing flavour in β^- decay.*

Beta-plus decay

Tip: Remember, ${}^{0}_{+1}e$ and β^+ are both symbols for a positron.

Tip: Don't get confused between the equations for beta-plus and beta-minus decay. There should always be one particle and one antiparticle emitted — so if an electron is emitted, an antineutrino is also emitted, and if a positron is emitted, a neutrino is also emitted.

Some unstable isotopes like carbon-11 decay by beta-plus (β^+) emission. β^+ is another type of weak interaction. β^+ decay just means a positron (a β^+ particle) is emitted. In this case a proton changes to a neutron, so a u quark changes to a d quark.

Beta-plus decay in terms of quarks: $u \rightarrow d + {}^{0}_{+1}e + \nu$

Beta-plus decay in terms of charge: $(+\tfrac{2}{3}) \rightarrow (-\tfrac{1}{3}) + (+1) + 0$

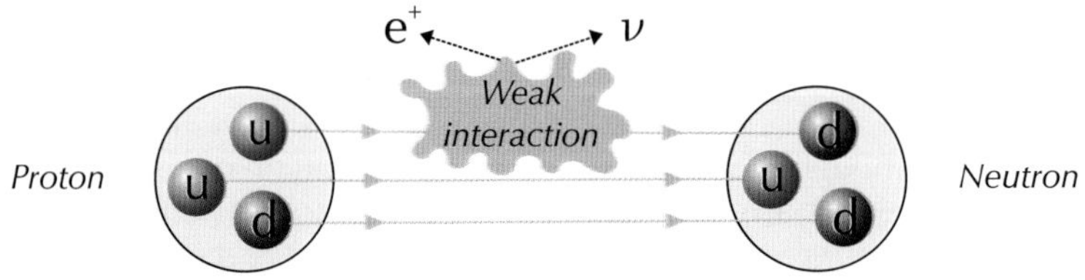

Figure 5: *A quark changing flavour in β^+ decay.*

Conservation of charge

In any particle reaction, the total charge after the reaction must equal the total charge before the reaction. You can use this rule to balance quark transformation equations.

Example — Maths Skills

Σ^+ and π^+ particles are both hadrons. A Σ^+ particle can decay to produce a π^+ particle and another particle A.

$$\Sigma^+ \rightarrow A + \pi^+$$
$$uus \rightarrow ?dd + u\bar{d}$$

By considering the charge on each quark, identify the missing quark (labelled '?') in this reaction, and hence state the name of particle A.

Write down the relative charge on each quark and then find the total charge on each particle.

$$\begin{array}{ccccccccccc} u & u & s & \rightarrow & ? & d & d & + & u & \bar{d} \\ +\frac{2}{3} & +\frac{2}{3} & -\frac{1}{3} & \rightarrow & ? & -\frac{1}{3} & -\frac{1}{3} & + & +\frac{2}{3} & +\frac{1}{3} \\ & +1 & & \rightarrow & & (? - \frac{2}{3}) & & + & & +1 \end{array}$$

For charge to be conserved, the missing quark must have a relative charge of $+\frac{2}{3}$. This means it must be an up quark.

This means particle A is made up of the quarks udd, so it is a neutron.

Tip: Remember, relative charge is the charge on a particle expressed as a multiple of the elementary charge e, 1.60×10^{-19} C.

Practice Questions — Application

Q1 Calculate the charge of the following quark compositions.

a) $u\bar{u}$ b) uus c) dss d) sss

Q2 A sigma-minus particle Σ^- is a hadron with –1 charge. It contains one strange quark, and two other quarks. Using u and/or d quarks, write down its quark composition.

Q3 A Xi-minus particle, Ξ^-, undergoes the following decay. The quark compositions of the particles are shown, with an unknown quark marked X.

$$\Xi^- \rightarrow \Lambda^0 + \pi^-$$
$$dss \rightarrow Xds + d\bar{u}$$

Using conservation of charge, work out the identity of quark X.

Practice Questions — Fact Recall

Q1 State the names and symbols of three quarks and three anti-quarks in the simple quark model.

Q2 State the charges of three anti-quarks in the simple quark model.

Q3 State the quark composition of a proton and a neutron.

Q4 Explain beta-minus decay in terms of the quark model.

Q5 Copy and complete the quark transformation equation for beta-plus decay: u → d + ☐ + ☐

Q6 Which quantity is always conserved in all particle interactions?

5. Radioactive Decay

Learning Objectives:

- Know what is meant by radioactive decay.
- Understand the spontaneous and random nature of radioactive decay.
- Know that α-particles, β-particles and γ-rays are types of nuclear radiation.
- Understand the nature, penetration and range of α-particles, β-particles and γ-rays.
- Understand techniques and procedures that can be used to investigate the absorption of α-particles, β-particles and γ-rays by appropriate materials (PAG7).

Specification Reference 6.4.3

Some nuclei are unstable, and decay to a more stable state by emitting nuclear radiation. There are four different types of nuclear radiation, and each has different properties.

Unstable nuclei

The nucleus is under the influence of the strong nuclear force holding it together and the electrostatic force pushing the protons apart. It's a very delicate balance, and it's easy for a nucleus to become unstable.

If a nucleus is unstable, it will break down to become more stable. Its instability could be caused by:

- too many neutrons
- too few neutrons
- too many nucleons in total (it's too heavy)
- too much energy in the nucleus

The nucleus decays by releasing energy and/or particles (nuclear radiation), until it reaches a stable form — this is called **radioactive decay**. An individual radioactive decay is spontaneous and random — the decay happens of its own accord (nothing triggers it) and it can't be predicted.

Although you can't predict the decay of an individual nucleus, if you take a very large number of nuclei, their overall behaviour shows a pattern. Any sample of a particular isotope (p.178) has the same rate of decay, i.e. the same proportion of nuclei will decay in a given time (p.198).

Types of nuclear radiation

There are four types of **nuclear radiation** — **alpha**, **beta-minus**, **beta-plus** and **gamma**. They are listed in Figure 1. The masses here have been given in unified atomic mass units (u) — one unified atomic mass unit = 1.661×10^{-27} kg, and is about the same as the mass of a proton or neutron (see page 177).

Radiation	Symbol	Constituent	Relative Charge	Mass (u)
Alpha	α	A helium nucleus — 2 protons & 2 neutrons	+2	4
Beta-minus (Beta)	β^- or β	Electron	−1	(negligible)
Beta-plus	β^+	Positron	+1	(negligible)
Gamma	γ	Short-wavelength, high-frequency electromagnetic wave.	0	0

Figure 1: *Types of nuclear radiation.*

When radiation hits an atom it can knock off electrons, creating an ion — so, radioactive emissions are also known as **ionising radiation**. The different types of radiation have different ionising powers as well as different speeds and penetrating powers — these are summarised in Figure 2.

Tip: Ionisation can be harmful. If it occurs in the cells of a living organism, it can do serious damage to it. This is what makes nuclear radiation so dangerous.

Radiation	Symbol	Ionising Power	Speed	Penetrating power	Affected by magnetic field
Alpha	α	Strong	Slow	Absorbed by paper or a few cm of air	Yes
Beta-minus (Beta)	β^- or β	Weak	Fast	Absorbed by ~3 mm of aluminium	Yes
Beta-plus	β^+	Annihilated by electron — so virtually zero range			
Gamma	γ	Very weak	Speed of light	Absorbed by many cm of lead, or several m of concrete	No

Figure 2: *A summary of the properties of the four types of nuclear radiation.*

Tip: Gamma radiation isn't made up of charged particles, so it won't be deflected by a magnetic field.

Tip: Any material that stops beta radiation will also stop alpha radiation, and any material that stops gamma radiation will also stop alpha and beta radiation.

The stronger the ionising power of radiation, the more energy it loses in a given distance, so the shorter the range of the radiation. As you can see from Figure 2, the more penetrating the type of radiation, the thicker or denser a material needs to be to absorb it — see Figure 3.

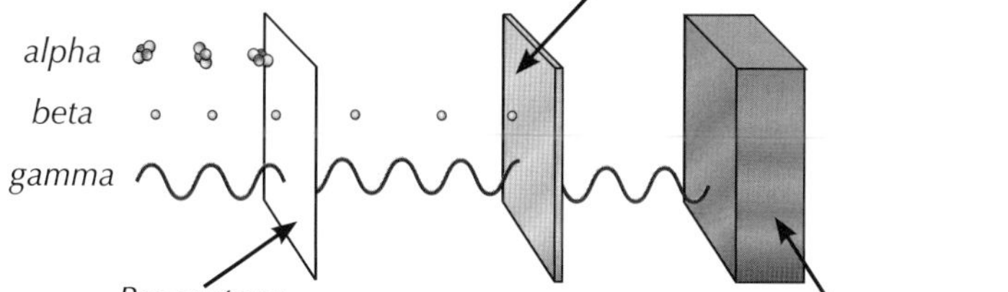
A few mm of aluminium stops beta radiation.
alpha
beta
gamma
Paper stops alpha radiation.
Several cm of lead stops gamma radiation.

Figure 3: *Examples of materials that stop each type of nuclear radiation.*

Both the thickness and the density of a material affect whether radiation will penetrate it — a very thick piece of aluminium could stop gamma radiation.

Figure 4: *The components inside a smoke alarm, including a source of alpha radiation. Alpha radiation's low penetrating power means it is easily absorbed by smoke particles, and so can be used to detect smoke.*

Background radiation

Background radiation is the low level of radiation that surrounds us at all times. It is made up of radiation from many different sources — these include naturally occurring isotopes in the environment, cosmic radiation from space, and man-made sources from industrial and medical radiation use. You'll need to take background radiation into account when you're investigating radiation (see the next page and page 203).

Figure 5: *Radioactive isotopes in granite rock contribute to background radiation.*

Investigating radiation

PRACTICAL ACTIVITY GROUP 7

You can investigate the penetration of different kinds of radiation by using different radioactive sources. These can be dangerous if you don't use them properly:

- Radioactive sources should be kept in a lead-lined box when they're not being used.
- They should only be picked up using long-handled tongs or forceps.
- Take care not to point them at anyone, and always keep a safe distance from them.

Tip: Lead will absorb all types of ionising radiation if it's thick enough.

Tip: Any experiments involving radiation can be dangerous if done incorrectly. As always, a full risk assessment must be carried out beforehand.

Measuring count rate

To begin, set up the equipment shown in Figure 6. A Geiger-Müller tube produces a 'count' in the form of an electrical pulse each time radiation enters it. The Geiger counter measures the total number of counts over a period of time. The count rate (counts per second) can be calculated by dividing the total number of counts recorded by the time (in seconds) over which this was measured.

Figure 6: *The experimental set-up used to investigate radiation.*

Tip: Wear gloves when performing this experiment, to help protect your hands from radiation.

Tip: You met background radiation on the previous page.

To measure the count rate of the radioactive source:

1. First of all, without the source present, measure the background counts. Radioactive decay is random, so to get an accurate and precise reading, the background counts need to be measured over a long enough time interval. For most experiments in the lab, around 30 seconds should do. Do this three times, and find the mean number of counts in this time interval.
2. Divide your counts by the number of seconds in the time interval to get the background count rate. You only need to do this once during a given experiment.
3. Place the source in front of the Geiger-Müller tube, as shown in Figure 6, making sure that it is close enough that the counter records a high number of counts (compared to the background counts).
4. Calculate the count rate of the radioactive source, using the same method you used for the background count rate. Again, take three sets of count measurements for three 30 second time intervals, and find the mean before you divide by the 30 seconds to find the count rate.
5. Subtract the background count rate from this value to find the count rate of the source.

Tip: Repeating your count measurement and taking an average helps make your results more precise.

Tip: You could instead attach the Geiger-Müller tube to a data-logger and a computer to reduce human error. This may be able to calculate the count-rate for you straight away. Or you could connect the tube to a ratemeter, which measures the count rate directly. In either case, measure the count rate three times, and then average the results.

Investigating penetration properties

To investigate the penetration of radiation through different materials, place a material between the source and the Geiger-Müller tube. Measure the count rate of the source both with and without the material. If the count rate remains about the same regardless of whether the material is present, then the radiation can penetrate the material. If the count rate drops by a large amount, then some of the radiation is being absorbed and blocked by the material. If the count rate drops to zero (after the background count rate is subtracted), the radiation is being completely absorbed. Repeat this for different materials and plot your results on a bar graph (since your results are categoric data).

You can also do this experiment with different sources to compare how different kinds of radiation are blocked by different materials. You'll probably need to change the distance between the source and the Geiger-Müller tube for this, as different kinds of radiation have different penetrating powers in air (see Figure 2 on p.193).

You could also adjust this experiment to investigate how the count rate for a particular source is affected by the thickness of a particular material — e.g. by using sheets of aluminium for beta radiation or different thicknesses of lead for gamma radiation. Or you could investigate how the count rate changes with distance between the source and the Geiger-Müller tube. For these variations, you can plot a scatter graph of your results as thickness and distance are continuous variables. This will allow you to see any trends in your results.

Tip: If you're comparing penetration through different materials, you should make sure they all have the same thickness. Thickness is a control variable here.

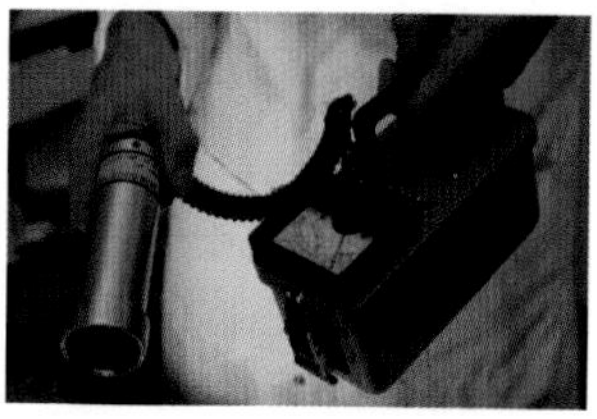

Figure 7: *A Geiger-Müller tube attached to a Geiger counter.*

Practice Question — Application

Q1 A radioactive source emits one unknown type of radiation. Using a Geiger-Müller tube placed in front of the source, a high count rate was measured. There was no decrease in the count rate recorded when a few sheets of paper were placed between the source and Geiger-Müller tube. The count rate dropped to almost zero when the sheets of paper were replaced with a thin sheet of aluminium.

a) Identify the type of radiation being emitted by the source.

b) The aluminium is replaced with a thick sheet of lead. The count rate is measured again. Explain why the count rate does not drop all the way to zero.

Practice Questions — Fact Recall

Q1 What is radioactive decay?

Q2 State the type of nuclear radiation that consists of:

a) 2 protons and 2 neutrons

b) a high frequency electromagnetic wave

c) an electron

Q3 List all the types of nuclear radiation which could be blocked by a thick sheet of lead.

6. Nuclear Decay Equations

Learning Objectives:

- Be able to balance nuclear transformation equations.
- Understand and be able to write nuclear decay equations for alpha, beta-minus and beta-plus decays.

Specification Reference 6.4.3

In any nuclear interaction, charge and nucleon number are always conserved. You need to use this fact to balance nuclear decay equations.

Representing nuclear decay

We can write equations to see exactly what's going on with a nucleus when it emits radiation. These equations are called nuclear decay equations (or nuclear transformation equations). To do this, we write the particles in standard notation (see page 178). This lets us clearly see how the numbers of protons and neutrons are changing.

Nuclear decay equations need to be balanced — in every nuclear reaction, including fission and fusion (p.210-213), charge and nucleon number must be conserved.

Tip: If you write out an equation for an interaction and the proton and nucleon numbers don't balance, that interaction cannot happen.

Tip: Remember, neutrons have no charge, so we just look at proton number for charge conservation.

Example

The nuclear decay equation for the decay of americium-241 to neptunium-237 by emission of an alpha particle is:

$$^{241}_{95}\text{Am} \longrightarrow {}^{237}_{93}\text{Np} + {}^{4}_{2}\alpha$$

There are 241 nucleons before the decay (in the americium-241 atom), and 241 nucleons after the decay (237 in the neptunium-237 atom and 4 in the alpha particle), so nucleon number is conserved.

You can see that charge is conserved by looking at the proton number (the blue numbers) — there are 95 protons before the decay and 95 after it.

Mass isn't conserved in nuclear reactions — the mass of an alpha particle is less than the individual masses of two protons and two neutrons. The difference in mass is called the mass defect (p.207), and the energy released when the nucleons bond together to form the alpha particle accounts for the missing mass. Once you have taken the conversion of mass to energy into account, then you can see that 'mass-energy' and momentum are conserved in all nuclear reactions. There's more on this on pages 207-209.

You need to know how to write balanced nuclear decay equations for the main types of radioactive decay.

Alpha emission

Alpha emission only happens in very heavy atoms, like uranium and radium. The nuclei of these atoms are too massive to be stable. When an alpha particle is emitted, the proton number decreases by two, and the nucleon number decreases by four.

Tip: An alpha particle is the nucleus of a helium atom, so you may see an alpha particle written with the symbol $^{4}_{2}\text{He}$.

Example

Uranium-238 decays to thorium-234 by emitting an alpha particle:

nucleon number decreases by 4

$$^{238}_{92}\text{U} \longrightarrow {}^{234}_{90}\text{Th} + {}^{4}_{2}\alpha$$

proton number decreases by 2

$238 = 234 + 4$ — *nucleon numbers balance*

$92 = 90 + 2$ — *proton numbers balance, so charge is conserved*

Beta-minus emission

Beta-minus decay happens in isotopes that are neutron-rich (i.e. have many more neutrons than protons in their nucleus). During beta-minus decay, one of the neutrons in the nucleus decays into a proton and ejects a beta-minus particle and an antineutrino. The proton number increases by one, and the nucleon number stays the same. Beta-minus particles have a negative charge, so they are written with a negative proton number ($^{0}_{-1}\beta$).

Example

Rhenium-188 decays to osmium-188 by emitting a beta-minus particle:

nucleon number stays the same

$$^{188}_{75}\text{Re} \longrightarrow {}^{188}_{76}\text{Os} + {}^{0}_{-1}\beta + {}^{0}_{0}\bar{\nu}$$

proton number increases by 1

188 = 188 + 0 + 0 — *nucleon numbers balance*

75 = 76 – 1 + 0 — *proton numbers balance, so charge is conserved*

Tip: A β^- particle is an electron, so you may see it written in equations as $^{0}_{-1}e$.

Beta-plus emission

Beta-plus decay happens in isotopes that are proton-rich (i.e. have a high proton to neutron ratio). A proton in the nucleus changes into a neutron, releasing a beta-plus particle and a neutrino. The proton number decreases by one, and the nucleon number stays the same.

Example

Sodium-22 decays to neon-22 by emitting a beta-plus particle:

nucleon number stays the same

$$^{22}_{11}\text{Na} \longrightarrow {}^{22}_{10}\text{Ne} + {}^{0}_{+1}\beta + {}^{0}_{0}\nu$$

proton number decreases by 1

22 = 22 + 0 + 0 — *nucleon numbers balance*

11 = 10 + 1 + 0 — *proton numbers balance, so charge is conserved*

Tip: A β^+ particle is a positron. You may see it written as $^{0}_{+1}e$.

Gamma emission

Even after a decay, the nucleus often has excess energy — it's in an excited state. This energy is lost by emitting a gamma ray. This often happens after an alpha or beta decay has occurred. During gamma emission, there is no change to the nuclear constituents — the nucleus just loses excess energy. We don't often write nuclear decay equations for just gamma decay, because the nuclear constituents don't change, but gamma ray photons may appear in equations showing other types of decay. The symbol for a gamma ray photon is $^{0}_{0}\gamma$.

Practice Questions — Application

Q1 Complete the following nuclear decay equation for the beta-minus decay of caesium-137: $^{137}_{55}\text{Cs} \longrightarrow {}^{?}_{?}\text{Ba} + ? + ?$

Q2 Write out the nuclear decay equation for the alpha decay of astatine-211 (Z = 85) to an isotope of bismuth. (The chemical symbols of astatine and bismuth are At and Bi respectively.)

Q3 Write out the nuclear decay equation for the beta-plus decay of aluminium-26 (Z = 13) to an isotope of magnesium. (The chemical symbols of aluminium and magnesium are Al and Mg respectively.)

Tip: Remember, Z is the proton number of a nucleus — see page 178.

7. Exponential Law of Decay

Learning Objectives:
- Understand what is meant by the activity of a radioactive source.
- Know what is meant by the decay constant, λ, of an isotope.
- Be able to use the equation $A = \lambda N$.
- Understand graphical methods and spreadsheet modelling of the equation $\frac{\Delta N}{\Delta t} = -\lambda N$ for radioactive decay.
- Be able to use the equation $N = N_0 e^{-\lambda t}$ where N is the number of undecayed nuclei.
- Understand the simulation of radioactive decay using dice.
- Be able to use the equation $A = A_0 e^{-\lambda t}$ where A is the activity.

Specification Reference 6.4.3

The number of unstable nuclei that decay each second in a radioactive sample depends on how many unstable nuclei are left in the sample.

The rate of radioactive decay

Radioactive decay is completely random — you can't predict which atom's nucleus will decay when. But although you can't predict the decay of an individual nucleus, if you take a very large number of nuclei, their overall behaviour shows a pattern and you can predict how many nuclei will decay in a given amount of time. Any sample of a particular isotope has the same rate of decay — i.e. the same proportion of atomic nuclei will decay in a given time. Each unstable nucleus within the isotope will also have a constant decay probability.

The decay constant and activity

The **activity**, A, of a sample is the number of nuclei that decay each second. It is proportional to the number of undecayed nuclei in the sample, N. For a given isotope, a sample twice as big would give twice the number of decays per second. The **decay constant**, λ, is the constant of proportionality. It is the probability of a specific nucleus decaying per unit time, and is a measure of how quickly an isotope will decay — the bigger the value of λ, the faster the rate of decay. The decay constant has units s^{-1} and the activity is measured in becquerels (1 Bq = 1 nucleus decaying per second).

$$A = \lambda N$$

A = *activity in Bq*; **N** = *number of undecayed nuclei in sample*; **λ** = *decay constant in* s^{-1}

Tip: Don't get the decay constant confused with wavelength — they both use the symbol λ.

Example — Maths Skills

A sample of a radioactive isotope contains 5.0 × 10^{-5} mol of undecayed nuclei. Its activity is measured to be 2.4 × 10^{12} Bq. Calculate the isotope's decay constant.

N = number of mol × N_A = 5.0 × 10^{-5} × 6.02 × 10^{23} = 3.01 × 10^{19}

Rearrange $A = \lambda N$ to give $\lambda = \frac{A}{N} = \frac{2.4 \times 10^{12}}{3.01 \times 10^{19}} = 7.973... \times 10^{-8}\ s^{-1}$

$= 8.0 \times 10^{-8}\ s^{-1}$ (to 2 s.f.)

Tip: Remember that 1 mol contains 6.02 × 10^{23} nuclei (from Avogadro's constant, N_A — see p.27).

Because the activity, A, is the number of nuclei that decay each second, you can write it as the change in the number of undecayed nuclei, ΔN, during a given time (in seconds) Δt:

$$A = -\frac{\Delta N}{\Delta t}$$

There's a minus sign in this equation because ΔN is always a decrease. Combining these two equations for the activity then gives the rate of change of the number of undecayed nuclei:

$$\frac{\Delta N}{\Delta t} = -\lambda N$$

$\frac{\Delta N}{\Delta t}$ = *rate of change of number of undecayed nuclei in* s^{-1}; **N** = *number of undecayed nuclei in sample*; **λ** = *decay constant in* s^{-1}

Tip: The minus sign has just been taken over to the other side by multiplying both sides by –1 here. This is how you'll get this equation in your data and formulae booklet.

Example — Maths Skills

A radioactive isotope has a decay constant of 1.2×10^{-4} s^{-1}.
What is the rate of change of N for a sample containing 7.5×10^{20} nuclei?

$$\frac{\Delta N}{\Delta t} = -\lambda N = -(1.2 \times 10^{-4}) \times (7.5 \times 10^{20}) = -9.0 \times 10^{16}\ \text{s}^{-1}$$

Tip: So N is currently decreasing by 9.0×10^{16} nuclei per second.

Modelling radioactive decay

HOW SCIENCE WORKS

Radioactive decay is an iterative process (the number of nuclei that decay in one time period determines the number that are available to decay in the next). You can use the formula on the last page to plot a graph of N against t, by working out how N changes for each small change in t. The easiest way to do this is to use a spreadsheet to model how a sample of an isotope will decay if you know the decay constant, λ, and the number of undecayed nuclei in the initial sample, N_0:

- Set up a spreadsheet with column headings for total time (t), ΔN and N, and a data input cell for each of Δt, λ and N_0.
- Decide on a value of Δt that you want to use — this is the time interval between the values of N that the spreadsheet will calculate.
  The most sensible time interval will depend on your decay constant.
- You can then enter formulas into the spreadsheet to calculate the number of undecayed nuclei left in the sample after each time interval. You'll need to use $\Delta N = -\lambda \times N \times \Delta t$ (rearranged from the equation on the previous page).

Data input cells		
	Δt	E.g. 100 s
	λ	E.g. 1×10^{-4} s^{-1}
	N_0	E.g. 7×10^{25}

t (s)	ΔN	N
$t_0 = 0$		N_0
$t_1 = t_0 + \Delta t$	$(\Delta N)_1 = -\lambda \times N_0 \times \Delta t$	$N_1 = N_0 + (\Delta N)_1$
$t_2 = t_1 + \Delta t$	$(\Delta N)_2 = -\lambda \times N_1 \times \Delta t$	$N_2 = N_1 + (\Delta N)_2$

***Figure 1:** An example of the formulas that can be used to create an iterative spreadsheet of the number of undecayed nuclei over time in a radioactive sample.*

- Plot a graph of the number of undecayed nuclei against time — see Figure 2.

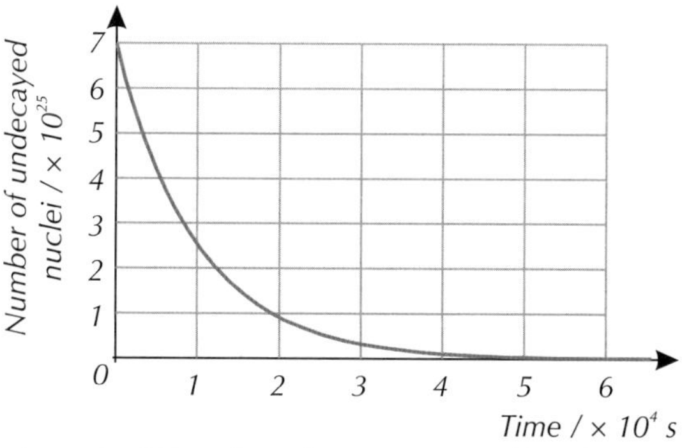

***Figure 2:** A graph of N against t for an isotope with a decay constant of 1×10^{-4} s^{-1} that originally contained 7×10^{25} undecayed nuclei. Plotted from an iterative spreadsheet using a time interval of 100 s.*

Tip: You could also do this by hand, using the same equations shown in Figure 1. However, this can take a long time, so using a spreadsheet is usually much faster.

Tip: If you write the formulas properly, the spreadsheet can automatically fill them in for as many rows (iterations) as you want.

Tip: You may have to fiddle with your value for Δt to get a graph with a nice shape.

Tip: Make sure you iterate enough times, so that you have enough data points to form a graph with a clear curve.

The decay equation

The graph produced by the spreadsheet model (see Figure 2) shows that radioactive decay is a form of **exponential** decay — the number of undecayed nuclei decreases exponentially with time.

It is known that the number of undecayed nuclei remaining, N, depends on the number originally present, N_0, the decay constant, λ, and how much time has passed, t. This is described by the equation:

$$N = N_0 e^{-\lambda t}$$

N = *the number of undecayed nuclei remaining*

$\mathbf{N_0}$ = *the original number of undecayed nuclei*

t = *time in s*

λ = *the decay constant in* s^{-1}

Tip: Remember to always round your final answer to the same no. of significant figures (s.f.) as the least precise data you use in your calculation. Here the 800 seconds is an exact value, and the other pieces of data are to 3 s.f. — so the answer to this calculation should be given to 3 s.f.

Example — Maths Skills

A sample of the radioactive isotope ^{13}N contains 5.00×10^6 nuclei. The decay constant for this isotope is 1.16×10^{-3} s^{-1}.

How many nuclei of ^{13}N will remain after exactly 800 seconds?

$$N = N_0 e^{-\lambda t} = (5.00 \times 10^6) \times e^{-(1.16 \times 10^{-3}) \times 800}$$
$$= 1.976... \times 10^6$$
$$= 1.98 \times 10^6 \text{ nuclei (to 3 s.f.)}$$

Plotting the natural log (ln) of the number of undecayed nuclei (or the activity, see next page) against time gives a straight-line graph (see Figure 3).

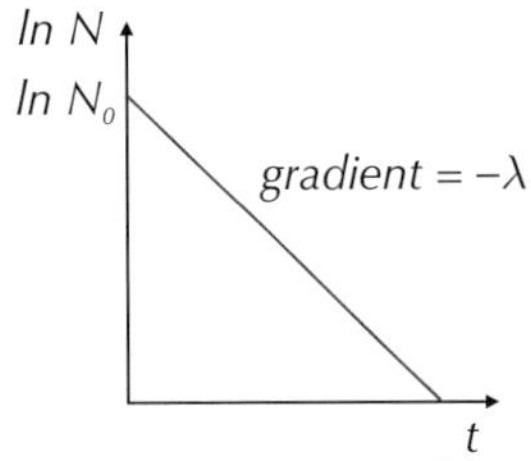

Figure 3: *A graph showing the natural log of the number of undecayed nuclei against time.*

Tip: Taking natural log of both sides of $N = N_0 e^{-\lambda t}$ and using the log rules on p.237 gives $\ln N = \ln N_0 - \lambda t$. Comparing this to the straight-line graph equation $y = mx + c$, you can see that a plot of $\ln N$ against t will be a straight line with a y-intercept of $\ln N_0$ and a gradient of $-\lambda$ (see p.244).

Dice simulation of radioactive decay

HOW SCIENCE WORKS

Radioactive decay is a random process where there is a constant probability that an undecayed nucleus will decay. The probability is given by the decay constant. The result of rolling a fair 6-sided dice is also random with a constant probability — the probability of rolling any one number is 1/6. These similarities mean that you can simulate radioactive decay using dice, where each dice represents an undecayed nucleus.

You need at least 100 dice for a good simulation of the undecayed nuclei in a small radioactive sample. Roll all the dice and count how many of them landed on a 6 — these dice represent the nuclei that have decayed. Record the total number of dice rolled and the number of dice that have 'decayed' in a table. Remove the 'decayed dice' and roll the remaining dice again.

Repeat this process until all of the dice have 'decayed'. Each roll counts as 1 unit of time passing in the 'lifespan' of the radioactive sample. If you plot a graph of the number of dice rolled each time (i.e. the number of undecayed nuclei left in the sample, N) against time, then you'll see the same exponential relationship as shown in Figure 2 for radioactive decay.

Figure 4: *Wooden cubes being used to simulate radioactive decay. Each cube represents an undecayed nucleus that 'decays' when it lands with the black side facing upwards.*

Activity and the decay equation

The number of unstable nuclei decaying per second (the activity) is proportional to the number of nuclei remaining. As a sample decays, its activity goes down — there's an equation for that too:

$$A = A_0 e^{-\lambda t}$$

A = the activity of the sample at time t in Bq

A_0 = the original activity of the sample in Bq

t = time in s

λ = the decay constant in s^{-1}

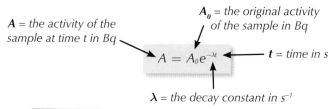

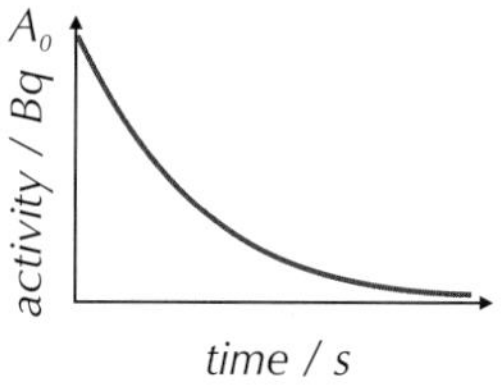

Figure 5: *Graph showing the exponential decay of the activity against time.*

Example — Maths Skills

The isotope radon-220 has a decay constant of 1.25×10^{-2} s^{-1}. How long will it take for the activity of a sample of radon-220 to fall from 85 Bq to 55 Bq?

First rearrange $A = A_0 e^{-\lambda t}$ to make t the subject.
Divide by A_0, take the natural log (ln) of both sides then divide by $-\lambda$.

$$\frac{A}{A_0} = e^{-\lambda t} \Rightarrow \ln\left(\frac{A}{A_0}\right) = -\lambda t \Rightarrow t = \frac{\ln\left(\frac{A}{A_0}\right)}{-\lambda}$$

Now just plug the numbers in and solve for t.

$$t = \frac{\ln\left(\frac{A}{A_0}\right)}{-\lambda} = \frac{\ln\left(\frac{55}{85}\right)}{-1.25 \times 10^{-2}} = 34.825... = 35 \text{ s (to 2 s.f.)}$$

Tip: Remember, in all the equations for radioactivity the time needs to be in seconds.

Practice Questions — Application

Q1 A sample contains 4.5×10^{18} atoms of a radioactive isotope. If it has a decay constant of 1.1×10^{-13} s^{-1}, what is the activity of the sample?

Q2 A sample of a radioactive isotope has an activity of 3.2 kBq. If the isotope has a decay constant of 1.3×10^{-4} s^{-1}, what will the sample's activity be 6.5 hours later?

Q3 An isotope of protactinium has a decay constant of 9.87×10^{-3} s^{-1}. If a sample initially contains 2.5×10^{15} unstable nuclei, how many will there be after 35 minutes?

Practice Questions — Fact Recall

Q1 a) What is the decay constant of a radioactive isotope?

b) What is meant by the activity of a radioactive sample?

Q2 a) Write down an equation for the rate of change in the number of unstable nuclei in a radioactive sample. Define all symbols used.

b) State one method you could use to model the decay of an isotope over time using the formula for rate of change given in part a).

Q3 Sketch a graph to show how the number of undecayed nuclei in a radioactive sample changes over time.

Q4 A student rolls a number of ordinary dice. She records how many dice show a 6, removes those showing 6, and rolls the remaining dice again. She repeats the process until she has removed all of the dice. Explain why this process is used to demonstrate radioactive decay.

Learning Objectives:

- Know what is meant by the half-life of an isotope.
- Understand techniques and procedures used to determine the half-life of an isotope such as protactinium (PAG7).
- Be able to calculate the half-life of an isotope using the equation $\lambda t_{½} = \ln(2)$.
- Know how isotopes can be used in radioactive dating, e.g. carbon-dating.

Specification Reference 6.4.3

Tip: Make sure you check your units carefully when you're calculating half-life.

8. Half-life and Radioactive Dating

The time it takes the number of unstable nuclei in a sample to halve is constant for a given substance. This quantity is called half-life, and you need to be able to calculate it...

Half-life of radioactive isotopes

The **half-life** ($t_{½}$) of an isotope is the average time it takes for the number of undecayed nuclei to halve.

Measuring the number of unstable nuclei isn't the easiest job in the world. In practice, half-life isn't measured by counting nuclei, but by measuring the time it takes the activity to halve. The longer the half-life of an isotope, the longer it takes for the radioactivity level to fall.

Calculating half-life from decay curves

You can calculate the half-life of an isotope from a decay curve — see Figure 1.

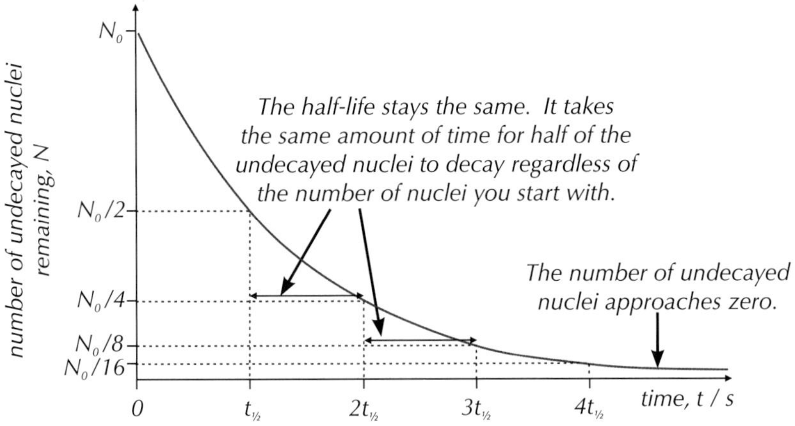

Figure 1: *A graph showing that the time taken for the number of undecayed nuclei remaining to halve is always equal to the half-life.*

- Read off the value for the number of unstable nuclei when $t = 0$.
- Go to half the original number of unstable nuclei on the y-axis.
- Draw a horizontal line to the curve, then a vertical line down to the x-axis. Read off the half-life where the line crosses the x-axis.
- It's always a good idea to check your answer — repeat these steps for a quarter of the original value and divide your answer by two. That will also give you the half-life. You can do the same for an eighth of the original value (divide the time by 3), and a sixteenth of the original value. Check that you get the same answer each time.

You can calculate half-life from an activity-time or count rate-time graph in exactly the same way, see Figure 2. N, A and count rate for an isotope will halve in the same amount of time, giving the half-life. (You met count rate on page 194 — it's the number of decays detected by a detector in 1 second.)

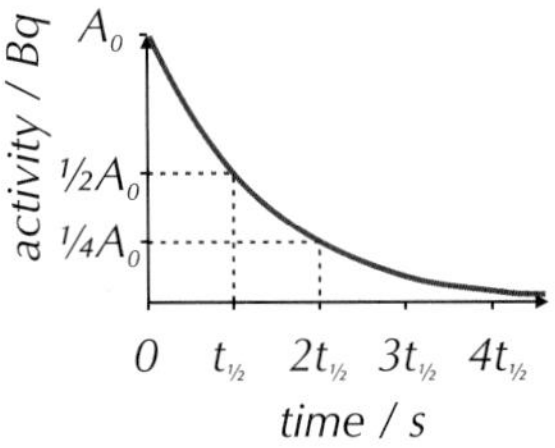

Figure 2: *A graph of activity against time. The activity halves after each half-life $t_{½}$.*

Tip: Be careful — the count rate is the number of decays detected per second, but activity is the total number of decays per second. So the count rate of a source will always be lower than the activity.

Example — Maths Skills

The graph on the right shows the activity-time graph of a sample of a radioisotope. Calculate the half-life of the isotope.

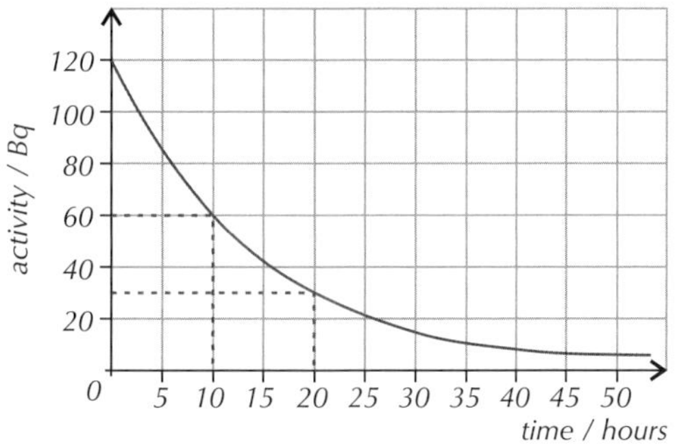

The starting activity of the sample is 120 Bq. So to find the half-life, read off the time at the point when the activity has dropped to half this value.

$120 \div 2 = 60$. When the activity is 60 Bq, time = 10 hours.

Check this by finding the time taken for the activity to fall to a quarter of its initial value and halving it. When the activity is 30 Bq, time = 20 hours. $20 \div 2 = 10$ hours. So, the half-life is 10 hours.

Investigating half-life

PRACTICAL ACTIVITY GROUP 7

You can determine the half-life of an isotope for yourself by measuring how the count rate detected from a sample decreases with time. You're most likely to do this using the isotope protactinium-234. Protactinium-234 is formed when uranium decays (via another isotope). You can measure protactinium-234's decay rate using a protactinium generator — a bottle containing a uranium salt and two solvents, which separate out into layers, as shown in Figure 3. The uranium salt is only soluble in the bottom layer. The uranium salt decays into protactinium-234 (as well as some other isotopes, but you don't need to worry about those).

When the generator is shaken, the solvents are mixed. Only the protactinium-234 can form a solution with the top layer solvent, so when the layers in the bottle are allowed to separate, the top layer will only contain the protactinium-234 in solution, while the uranium salt and other decay products are still left in the bottom layer.

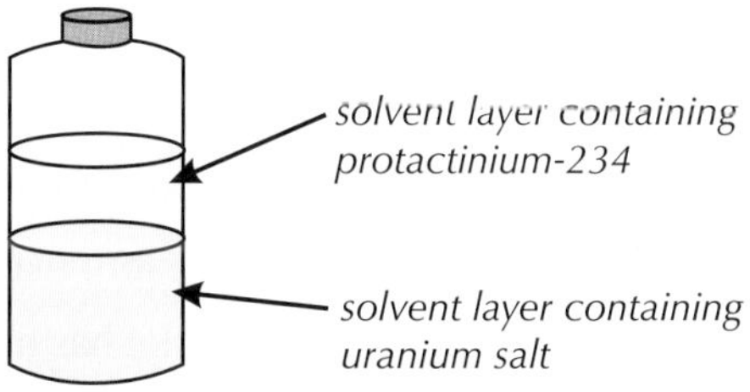

Figure 3: *A protactinium generator.*

So if a Geiger-Müller tube is directed towards the top layer only, the activity of protactinium-234 can be monitored without confusing its counts with those from other decaying isotopes.

1. Start by setting up the equipment shown in Figure 5. Don't shake the protactinium generator yet. Find the background counts by measuring the number of counts detected over a time period (at least 30 s). Repeat this two more times, find the mean of your three results, and divide the mean counts by the number of seconds. This is the background count rate you will need to subtract from your future results.

Tip: Make sure you do a risk assessment before carrying out this experiment.

Figure 4: *Uranyl nitrate, a uranium salt. It can be dissolved in water and used in protactinium generators.*

Tip: The uranium in the salt is always decaying into protactinium-234, so you can reuse the protactinium generator by shaking the bottle again to dissolve more protactinium-234 in the top layer.

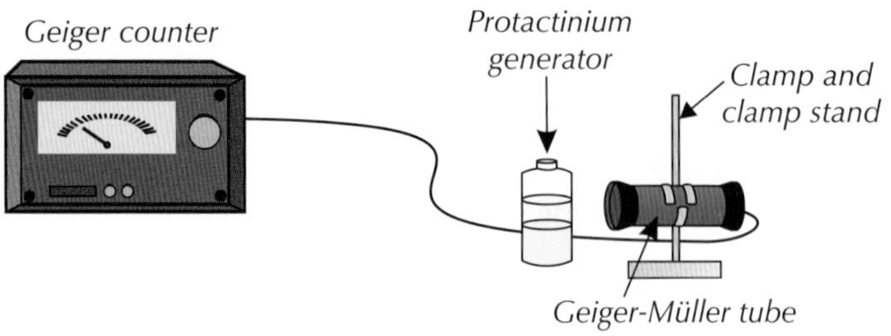

Figure 5: *The experimental set-up for an investigation into the half-life of protactinium-234.*

Tip: Make sure the Geiger-Müller tube is directed at the top layer of the protactinium generator, and that it is close enough that it records a large number of counts.

2. Shake the protactinium generator to mix the solvents together, then place it back in its position shown in Figure 5. Wait for the liquids to separate.
3. As soon as the liquids separate, record the count rate of the protactinium-234 in the top layer (e.g. how many counts you get in 10 seconds, divided by 10). Measure the count rate at sensible intervals (e.g. every 30 seconds) for at least 5 minutes.
4. Subtract your background count rate from each of your results, then plot a graph of count rate against time. It should look like the graph in Figure 6. You can use this graph to find the half-life in exactly the same way as in Figure 1. In this case the half-life is the time taken for the count rate (or activity) to halve.

Tip: If you have enough data, calculate the half-life from the time taken for the count rate to drop to a quarter, and an eighth of the original count rate, as shown in Figure 1. Take an average of these values to get your final half-life result.

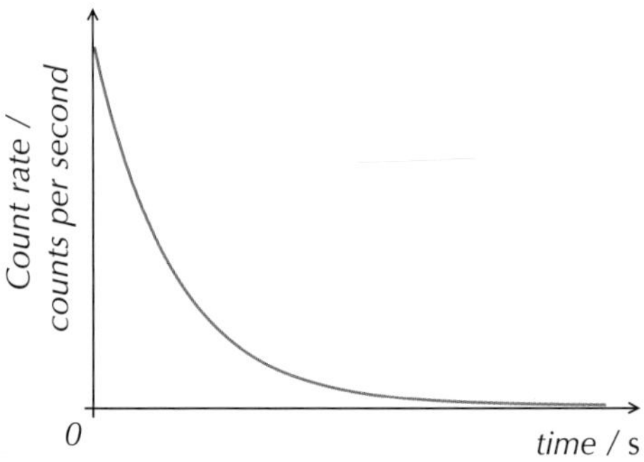

Figure 6: *A graph of count rate against time for a decaying sample of protactinium-234.*

The half-life equation

You can derive an equation for half-life from the formula $N = N_0 e^{-\lambda t}$ on p.200.

- When $t = t_{1/2}$, the number of undecayed nuclei has halved, so $N = \frac{1}{2}N_0$.
- Substituting these values into the equation for N:

$$\frac{1}{2}N_0 = N_0 e^{-\lambda t_{1/2}}$$

- Cancelling N_0 and taking the natural log of both sides:

$$\ln\left(\frac{1}{2}\right) = -\lambda t_{1/2}$$

- $\frac{1}{2} = 2^{-1}$ so $\ln\left(\frac{1}{2}\right) = \ln(2^{-1}) = -\ln(2)$ using the rules on page 237. So:

Tip: You'll need to use the power law $\frac{1}{a^n} = a^{-n}$ here.

$$\lambda t_{1/2} = \ln(2)$$

$\boldsymbol{\lambda}$ = the decay constant, s^{-1}

$\boldsymbol{t_{1/2}}$ = the half-life, s

Example — Maths Skills

A radioactive isotope has a decay constant of $1.16 \times 10^{-3}\ s^{-1}$. What is the half-life for this isotope?

Divide both sides of $\lambda t_{1/2} = \ln(2)$ by λ, and substitute in the values.

$$t_{1/2} = \frac{\ln(2)}{\lambda} = \frac{\ln(2)}{1.16 \times 10^{-3}} = 597.54... = 598 \text{ s (to 3 s.f.)}$$

Tip: Here it's fine to give the half-life in seconds. Some isotopes have much longer half-lives though, which are more sensible to give in years.

Radioactive dating

HOW SCIENCE WORKS

The radioactive isotope carbon-14 is used in **radioactive dating**. Living plants take in carbon dioxide from the atmosphere as part of photosynthesis, including some molecules containing the radioactive isotope carbon-14. Animals then take this carbon-14 in when they eat the plants. All living things contain the same percentage of carbon-14.

When they die, and so stop taking in any more carbon-14, the activity of carbon-14 in the plant or animal starts to fall, with a half-life of around 5730 years. Archaeological finds of once-living material (like wood or animal bone) can be tested to find the current amount of carbon-14 in them. This can be used to calculate how long the organism has been dead, and so how old the material is.

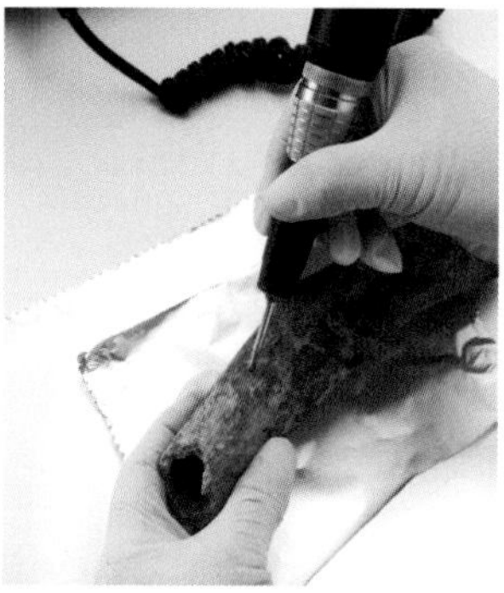

Figure 7: *The carbon-14 in living tissue decays after an organism dies. By measuring how much carbon-14 remains in dead tissue, the age of a sample can be estimated.*

Example — Maths Skills

A sample of 6.5×10^{23} carbon atoms is taken from a spear, and found to have an activity of 0.052 Bq. The ratio of radioactive carbon-14 to stable carbon-12 in living wood is $1 : 1.4 \times 10^{12}$, and the half-life of carbon-14 is 5730 years. How old is the wood the spear is made from?

The half-life of carbon-14 in seconds is

$$t_{1/2} = 5730 \times (365 \times 24 \times 3600) = 1.8070... \times 10^{11} \text{ s}$$

So the decay constant, $\lambda = \frac{\ln 2}{t_{1/2}} = \frac{\ln 2}{1.8070... \times 10^{11}} = 3.8358... \times 10^{-12}\ s^{-1}$

Rearrange and use the equation for activity to find the number of carbon-14 nuclei in the wood:

$$N = \frac{A}{\lambda} = 0.052 \div (3.8358... \times 10^{-12}) = 1.3556... \times 10^{10}$$

Use the ratio given in the question to calculate the expected number of carbon-14 nuclei in a sample of 6.5×10^{23} carbon atoms from living wood:

$$N_0 = (1 \div (1.4 \times 10^{12})) \times 6.5 \times 10^{23} = 4.6428... \times 10^{11}$$

N and N_0 are related by $N = N_0 e^{-\lambda t}$. You can rearrange this by dividing by N_0 and taking the natural log (ln) of both sides to make t the subject and find the age of the wood:

$$N = N_0 e^{-\lambda t} \Rightarrow \frac{N}{N_0} = e^{-\lambda t} \Rightarrow \ln\left(\frac{N}{N_0}\right) = \ln(e^{-\lambda t}) \Rightarrow \ln\left(\frac{N}{N_0}\right) = -\lambda t$$

$$\text{So } t = -\frac{1}{\lambda} \times \ln\left(\frac{N}{N_0}\right) = -\frac{1}{3.8358... \times 10^{-12}} \times \ln\left(\frac{1.3556... \times 10^{10}}{4.6428... \times 10^{11}}\right)$$

$$= 9.2121... \times 10^{11} \text{ s}$$

$$= 29\,000 \text{ years (to 2 s.f.)}$$

Tip: You'll need to use the equations for N back on pages 198 and 200 to answer this question.

Tip: $\ln(e^a) = a$, which is why $\ln(e^{-\lambda t})$ becomes $-\lambda t$ when you're rearranging to find t. For more help on logs, skip on over to pages 236-237.

Practice Questions — Application

Q1 The activity of a radioactive sample fell from 2400 Bq to 75 Bq over a period of 24 hours. How long is the isotope's half-life?

Q2 An isotope has a half-life of 483 seconds. What is its decay constant?

Q3 A sample of carbon atoms from an ancient bone has an activity of 0.45 Bq. The same number of carbon atoms from a piece of living tissue has an activity of 1.2 Bq. Given that the half-life of carbon-14 is 5730 years, how old is the bone?

Tip: You'll need to use the activity equation from page 201 to answer Q3.

Practice Questions — Fact Recall

Q1 What is the half-life of an isotope?

Q2 Describe how you would find an isotope's half-life from a graph showing its activity against time.

Q3 Nobelium-255 is a radioactive alpha-emitter with a half-life of a few minutes. Describe an experiment you could perform to find the half-life of a sample of Nobelium-255.

Q4 Give an example of an isotope used in radioactive dating.

9. Binding Energy

The binding energy is a measure of how strongly a nucleus is held together — the greater the binding energy per nucleon, the more stable the nucleus.

Learning Objectives:

- Know what is meant by the mass defect.
- Know what is meant by the binding energy of a nucleus.
- Be able to calculate the binding energy of nuclei using the equation $\Delta E = \Delta mc^2$ and the masses of nuclei.
- Be able to calculate the binding energy per nucleon for a nucleus.
- Know the shape of the curve for the graph of binding energy per nucleon against nucleon number.

Specification Reference 6.4.4

Mass defect and binding energy

The mass of a nucleus is less than the mass of its constituent nucleons — the difference is called the **mass defect**.

Example — Maths Skills

The mass of a nucleus of potassium, $^{40}_{19}K$, is 39.9536 u. The mass of a proton is 1.00728 u and the mass of a neutron is 1.00867 u. Calculate the mass defect of the nucleus in u.

Number of protons = 19, number of neutrons = (40 – 19) = 21
Mass of nucleons = (19 × 1.00728 u) + (21 × 1.00867 u) = 40.32039 u

So mass defect = mass of nucleons – mass of nucleus
= 40.32039 u – 39.9536 u = 0.36679 u

Einstein's equation says that mass and energy are equivalent:

$$\Delta E = \Delta mc^2$$

E *= energy in J*
c *= the speed of light in a vacuum in ms⁻¹*
m *= mass in kg*

Tip: You've already met this equation on page 185.

As nucleons join together, the total mass decreases — this 'lost' mass is converted into energy and released. You can calculate this energy using the equation above. The amount of energy released is equivalent to the mass defect.

If you pulled the nucleus completely apart, the energy you'd have to use to do it would be the same as the energy released when the nucleus formed. The energy needed to separate all of the nucleons in a nucleus is called the **binding energy** (measured in MeV), and it is equivalent to the mass defect.

Example — Maths Skills

Calculate the binding energy in MeV of the nucleus of a lithium-6 atom, $^{6}_{3}Li$, given that its mass defect is 0.0343 u.

Convert the mass defect into kg:
Mass defect = $0.0343 \times (1.661 \times 10^{-27}) = 5.697... \times 10^{-29}$ kg

Use $\Delta E = \Delta mc^2$ to calculate the binding energy:
$\Delta E = (5.697... \times 10^{-29}) \times (3.00 \times 10^{8})^2 = 5.127... \times 10^{-12}$ J
Convert to MeV:
$\Delta E = ((5.127... \times 10^{-12}) \div (1.60 \times 10^{-19})) \div 10^6 = 32.046... = 32.0$ MeV (to 3 s.f.)

Tip: Remember — 1 u = 1.661×10^{-27} kg. You'll be given this in the data and formulae booklet in the exams.

Tip: To convert from J to eV, you divide by the magnitude of the charge on an electron, $e = 1.60 \times 10^{-19}$ C. Then to convert to MeV you just divide by 10^6.

The binding energy per unit of mass defect can also be calculated. Using the fact that 1 u = 1.661×10^{-27} kg:

$$\Delta E = \Delta mc^2 = 1.661 \times 10^{-27} \times (3 \times 10^8)^2$$
$$= 1.4949 \times 10^{-10} \text{ J}$$
$$\approx 930 \text{ MeV (to 2 s.f.)}$$

This means that a mass defect of 1 u is equivalent to about 930 MeV of binding energy.

$$1\ \text{u} \approx 930\ \text{MeV}$$

Tip: This is good for checking your answers, or making estimates, but it's not very accurate. If you're asked to find the binding energy, you should always use $\Delta E = \Delta mc^2$.

Example — Maths Skills

Using the binding energy per unit mass defect, the binding energy in MeV of lithium-6 from the last example on the previous page is:

$\Delta E \approx 0.0343 \times 930 = 31.899$ MeV = 32 MeV (to 2 s.f.)

Binding energy per nucleon

A useful way of comparing the binding energies of different nuclei is to look at the binding energy per nucleon.

$$\text{Binding energy per nucleon} = \frac{\text{Binding energy } (B)}{\text{Nucleon number } (A)}$$

Figure 1: Some blocks of pure iron. Iron has the most stable nucleus of all the elements.

Example — Maths Skills

What is the binding energy per nucleon for a $^6_3$Li nucleus?

You know from the second example on the previous page that binding energy = 32.046... MeV. Nucleon number = A = 6

Binding energy per nucleon = $\frac{B}{A}$ = 32.046... ÷ 6 = 5.341...
= 5.34 MeV (to 3 s.f.)

If you plot a graph of binding energy per nucleon against nucleon number, for all elements, the line of best fit is a curve — see Figure 2. A high binding energy per nucleon means that more energy is needed to remove nucleons from the nucleus. In other words the most stable nuclei occur around the maximum point on the graph — which is at nucleon number 56 (i.e. iron, Fe).

Exam Tip
Make sure you know the shape of this graph, as well as the axes labels — you could be asked to sketch it in an exam.

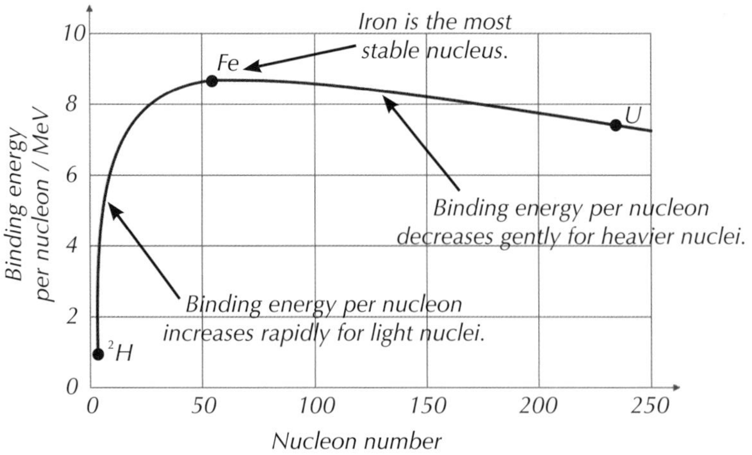

Figure 2: Graph showing how the binding energy per nucleon varies with nucleon number. The red line shows the line of best fit.

Tip: You can use this graph to find the energy released in nuclear reactions — see page 212.

Practice Questions — Application

Q1 The mass defect of a carbon-12 nucleus is 0.0989 u. Calculate the binding energy of the nucleus in MeV.

Q2 Calculate the mass defect of a nucleus of $^{16}_{8}O$ (mass = 15.994915 u) in atomic units, u.

Q3 The binding energy per nucleon of a nucleus of iron-56 is 8.79 MeV. What is its mass defect in u?

Q4 A student finds a graph of binding energy per nucleon against nucleon number in her textbook. A section of the graph is shown below.

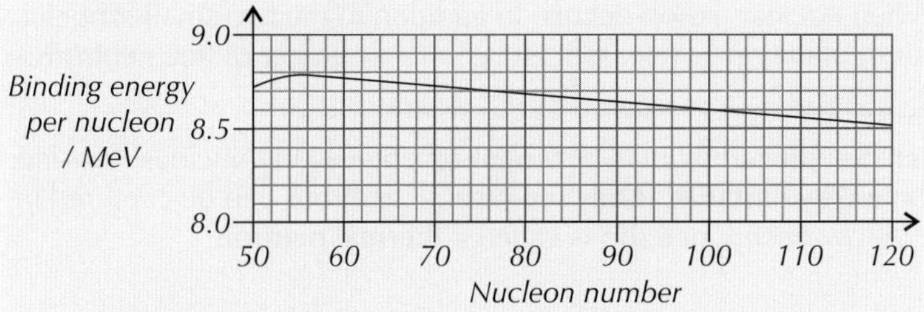

a) Using the graph, find the binding energy per nucleon of a molybdenum-100, $^{100}_{42}Mo$, nucleus.

b) Calculate the total binding energy of a molybdenum-100 nucleus.

Practice Questions — Fact Recall

Q1 What is meant by the binding energy of a nucleus?

Q2 Sketch the graph of the binding energy per nucleon against nucleon number. Label the positions of ^{2}H and ^{56}Fe on the curve.

Tip: For a reminder on standard notation, see page 178.

Tip: The mass of a proton is 1.00728 u, and the mass of a neutron is 1.00867 u.

10. Nuclear Fission and Fusion

Learning Objectives:

- Understand what is meant by induced nuclear fission.
- Understand what is meant by nuclear fusion.
- Know why fusion reactions require very high temperatures.
- Be able to balance nuclear transformation equations.
- Be able to use the graph of binding energy per nucleon against nucleon number to calculate energy changes in reactions.
- Be able to calculate the energy released (or absorbed) in simple nuclear reactions.

Specification Reference 6.4.4

Radioactive decay isn't the only way that nuclei can change — they can also split into two smaller nuclei, or fuse with other nuclei to form larger ones.

Nuclear fission

Heavy nuclei (e.g. uranium) are unstable, and some can randomly split into two smaller nuclei — this is called **nuclear fission**. This process is called spontaneous if it just happens by itself, or induced if we encourage it to happen. When nuclear fission occurs, in addition to the two smaller nuclei, a large amount of energy is released, along with a number of free neutrons.

Example

Fission can be induced by making a neutron enter a ^{235}U nucleus, causing it to become very unstable. Only low-energy neutrons can be captured in this way. A low-energy neutron is called a **thermal neutron**.

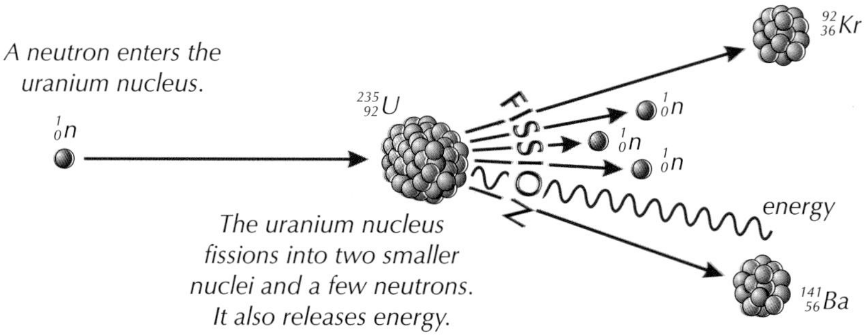
A neutron enters the uranium nucleus.
$^{1}_{0}n$
$^{235}_{92}U$
FISSION
$^{1}_{0}n$
$^{1}_{0}n$
$^{1}_{0}n$
$^{92}_{36}Kr$
energy
$^{141}_{56}Ba$
The uranium nucleus fissions into two smaller nuclei and a few neutrons. It also releases energy.

Figure 1: *A possible fission of a uranium-235 nucleus.*

Energy is released during nuclear fission because the new, smaller nuclei have a higher average binding energy per nucleon (see page 208). The larger the nucleus, the more unstable it will be — so large nuclei are more likely to spontaneously fission. This means that spontaneous fission limits the number of nucleons that a nucleus can contain — in other words, it limits the number of possible elements.

Tip: Heavy nuclei don't always fission to form the same daughter nuclei — ^{235}U can split into lots of different pairs of nuclei (with nucleon numbers around 90 and 140).

Nuclear fusion

Two light nuclei can combine to create a larger nucleus. This is called **nuclear fusion**.

Example

In the Sun, hydrogen nuclei fuse in a series of reactions to form helium. One of the reactions is: $^{2}_{1}H + ^{1}_{1}H \longrightarrow ^{3}_{2}He$ + energy.

$^{2}_{1}H$
$^{1}_{1}H$
fusion
energy
$^{3}_{2}He$

Figure 2: *Two isotopes of hydrogen fuse to form helium.*

Tip: Just like decay equations, equations showing nuclear fission and fusion need to be balanced — see page 196.

Nuclei can only fuse if they are moving very fast, and so have enough energy to overcome the electrostatic (Coulomb) repulsion between them (p.134) and get close enough for the strong interaction to bind them. This means fusion reactions require much higher temperatures than fission, as well as high pressures (or high densities). Under such conditions, generally only found inside stars, matter turns into a state called a plasma.

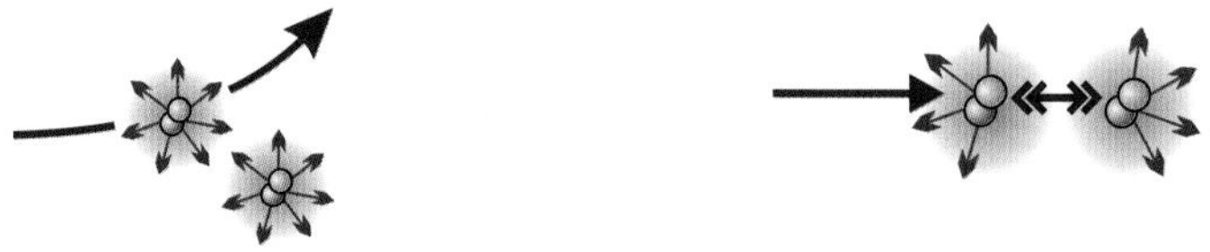

Low-energy nuclei are deflected by electrostatic repulsion.
High-energy nuclei overcome electrostatic repulsion and are attracted by the strong interaction.

Figure 3: *Nuclei must overcome their mutual electrostatic repulsion to fuse together.*

Tip: The strong interaction can only attract nucleons together when they're less than a few femtometres apart (see p.182) — so the fusing nuclei must have enough energy to get very close to each other.

A lot of energy is released during nuclear fusion because the new, heavier nucleus has a much higher binding energy per nucleon (and so a lower total mass, see p.208). The energy released helps to maintain the high temperatures needed for further fusion reactions.

Nuclear equations

Just like other nuclear transformations (see p.196-197), we can express nuclear fission and fusion reactions as nuclear equations. Just like all equations though, we have to make sure these balance, since charge must be conserved. The total nucleon number and proton number on each side must be equal.

Tip: 2^1_0n just means two free neutrons.

Examples

The nuclear equation for the spontaneous fission of californium-238 into cadmium-115 and tin-121 is:

$$^{238}_{98}\text{Cf} \longrightarrow {}^{115}_{48}\text{Cd} + {}^{121}_{50}\text{Sn} + 2^1_0\text{n}$$

The nuclear equation for the fusion of lithium-6 and deuterium into beryllium-7 is:

$$^{6}_{3}\text{Li} + {}^{2}_{1}\text{H} \longrightarrow {}^{7}_{4}\text{Be} + {}^{1}_{0}\text{n}$$

Tip: You can often get another small particle produced in nuclear fusion, alongside the main fusion product. This is usually a proton or neutron. If you're fusing larger nuclei (e.g. carbon, oxygen, etc) you may even get a small nucleus, like a hydrogen or helium isotope.

You may be expected to solve problems involving balancing nuclear fission and fusion equations in your exam.

Example — Maths Skills

The nuclear equation below shows the induced fission of a uranium-235 nucleus.

$$^{235}_{92}\text{U} + {}^{1}_{0}\text{n} \longrightarrow {}^{96}_{37}\text{Rb} + {}^{A}_{Z}\text{Cs} + 3^1_0\text{n}$$

Tip: Cs is the chemical symbol for caesium.

Find the nucleon number, A, and proton number, Z, of the Cs nucleus.

To find the nucleon number, subtract the total nucleon number on the right from the total on the left.

$A = (235 + 1) - (96 + (3 \times 1)) = 236 - 99 = 137$

Do the same for proton number.

$Z = (92 + 0) - (37 + 0) = 92 - 37 = 55$

So the Cs nucleus has a nucleon number of 137 and a proton number of 55.

Tip: Energetically favourable means that the fusion or fission will give out energy. Reactions that aren't energetically favourable take in energy rather than release it.

Energy released by fission and fusion

You can tell whether it is energetically favourable for an element to undergo fission or fusion by looking at the graph of binding energy per nucleon against nucleon number. Only elements to the right of ^{56}Fe can release energy through nuclear fission. Similarly only elements to the left of ^{56}Fe can release energy through nuclear fusion. This is because energy is only released when the binding energy per nucleon increases.

Tip: For a reminder on this graph, take a look back at page 208.

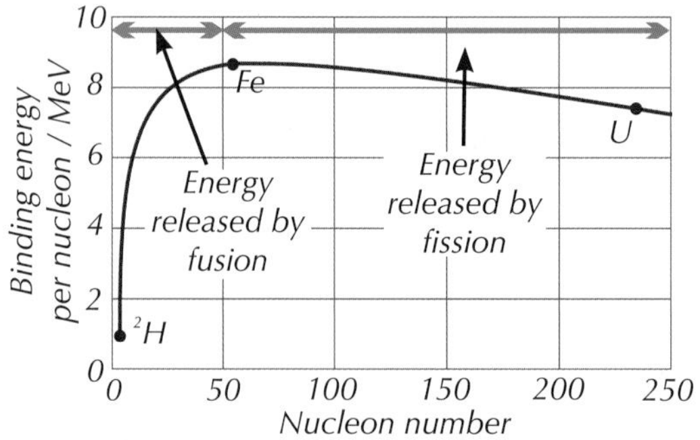

Figure 4: *Graph showing the regions where fusion and fission reactions are energetically favourable.*

The change in binding energy gives the energy released. The binding energy per nucleon graph can be used to estimate the energy released in nuclear reactions.

Example — Maths Skills

Calculate the energy released through the fusion of ^2H and ^3H into ^4He.

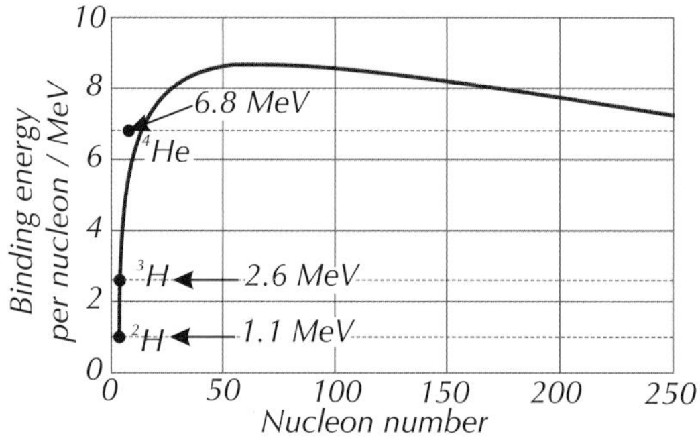

Binding energy of $^4He = 4 \times 6.8 = 27.2$ MeV

Binding energy of ^2H and $^3H = (2 \times 1.1) + (3 \times 2.6)$
$= 10.0$ MeV

Therefore total energy released $= 27.2 - 10.0$
$= 17.2$
$= 17$ MeV (to 2 s.f.)

Tip: There would be some free neutrons produced in these interactions as well, but you don't need to include them in binding energy calculations. The binding energy of a free neutron is 0 MeV.

Example — Maths Skills

Calculate the energy released in the induced nuclear fission of ^{235}U into ^{92}Rb and ^{140}Cs.

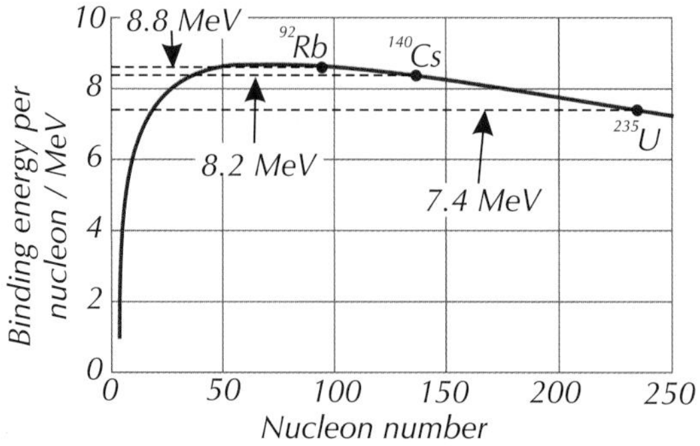

Binding energy of ^{235}U
$= 235 \times 7.4 = 1739$ MeV

Binding energy of ^{92}Rb and ^{140}Cs
$= (92 \times 8.8) + (140 \times 8.2)$
$= 1957.6$ MeV

So the total energy released $= 1957.6 - 1739$
$= 218.6$
$= 220$ MeV (to 2 s.f.)

The energy released per reaction is generally lower in nuclear fusion than fission. But the nuclei used in fusion have a lower mass, so one mole of the reactants in a fusion reaction weighs less than one mole of the reactants in a fission reaction. Gram for gram, fusion can release more energy than fission.

In both fission and fusion reactions, the total mass decreases, but mass-energy is conserved. So you can calculate the energy released during a fission or fusion reaction using the equation $\Delta E = \Delta mc^2$, where Δm is the total difference in mass between the reactants and products.

Example — Maths Skills

Calculate the energy released (in MeV) by the following fission reaction using the data in Figure 5: $^{235}_{92}\text{U} + ^{1}_{0}\text{n} \longrightarrow ^{89}_{36}\text{Kr} + ^{144}_{56}\text{Ba} + 3^{1}_{0}\text{n}$

$\Delta m = m_{Kr\text{-}89} + m_{Ba\text{-}144} + 3m_n - m_{U\text{-}235} - m_n$

$= 88.89783 + 143.89215 + (3 \times 1.00867) - 234.99333 - 1.00867$

$= -0.18601 \text{ u}$

$= (-0.18601 \times 1.661 \times 10^{-27}) \text{ kg} = -3.0896... \times 10^{-28} \text{ kg}$

$\Delta E = \Delta mc^2 = -3.0896... \times 10^{-28} \times (3.00 \times 10^8)^2$

$= -2.7806... \times 10^{-11} \text{ J}$

$= ((-2.7806... \times 10^{-11}) \div (1.60 \times 10^{-19})) \div 10^6 \text{ MeV}$

$= -173.79... \text{ MeV}$

$= -174 \text{ MeV (to 3 s.f.)}$

So the fission reaction releases 174 MeV of energy.

Nucleus / Particle	Mass (u)
^{235}U	234.99333
^{89}Kr	88.89783
^{144}Ba	143.89215
^{139}Te	138.90613
^{94}Zr	93.88431
^{2}H	2.01355
proton	1.00728
neutron	1.00867
positron	0.00055
neutrino	0

Figure 5: *The masses of some nuclei and particles.*

In extreme conditions, nuclei can undergo transformations which are not energetically favourable. For example, the large amount of energy released in supernovae allows iron in an exploding star to fuse into bigger nuclei, absorbing some of that energy. Scientists think this is how almost all naturally occurring elements heavier than iron came to exist.

Tip: There's more on supernovae on page 87.

You can calculate the energy absorbed by an energetically unfavourable reaction using the same method as in the example above. In this case the binding energy per nucleon decreases during the transformation, so the total mass increases, and you get a positive values for Δm and ΔE. This indicates that energy was absorbed instead of emitted.

Tip: The peak of the binding energy per nucleon against nucleon number graph is at a nucleon number of 56.

Practice Questions — Application

Q1 For each of the nuclei below, state whether it is more energetically favourable for them to undergo nuclear fission or nuclear fusion:
a) $^{28}_{14}\text{Si}$ b) $^{227}_{89}\text{Ac}$ c) $^{50}_{24}\text{Cr}$ d) $^{120}_{52}\text{Te}$

Q2 A deuterium nucleus, $^{2}_{1}\text{H}$, fuses with a tritium nucleus, $^{3}_{1}\text{H}$, to form a helium-4 nucleus, $^{4}_{2}\text{He}$, and another particle, X. Write a balanced nuclear equation for this reaction and identify particle X.

Q3 Using the data in Figure 5, calculate the energy (in MeV) released during the fusion of two protons: $^{1}_{1}\text{p} + ^{1}_{1}\text{p} \longrightarrow ^{2}_{1}\text{H} + ^{0}_{1}\text{e} + ^{0}_{0}\nu$.

Q4 Using the data in Figure 5, calculate the energy (in MeV) released when a ^{235}U nucleus is hit by a neutron and fissions into ^{94}Zr and ^{139}Te and a number of neutrons.

Tip: Remember, if you're using $\Delta E = \Delta mc^2$, you need to account for all the particles that you started with, and all the particles produced.

Practice Questions — Fact Recall

Q1 What is nuclear fission? How can it be induced?

Q2 What is nuclear fusion?

Q3 Explain why light nuclei require high temperatures to fuse together.

11. Fission Reactors

You need to know about the different bits and pieces that make up a fission reactor, as well as the potential environmental problems that come from them.

Learning Objectives:
- Understand what is meant by a chain reaction.
- Know the basic structure of a fission reactor.
- Know and be able to explain the roles of components of a fission reactor, including fuel rods, control rods and the moderator.
- Understand the environmental impact of nuclear waste.

Specification Reference 6.4.4

Tip: In Figure 1, the moderator also acts as a coolant (see page 215).

Tip: Make sure you're familiar with the basic structure and features of a fission reactor.

Tip: The fuel rods are placed into the reactor remotely, which keeps workers as far away from the radiation as possible.

Structure of fission reactors

We can harness the energy released during nuclear fission reactions in a thermal fission reactor (see Figure 1), but it's important that these reactions are very carefully controlled.

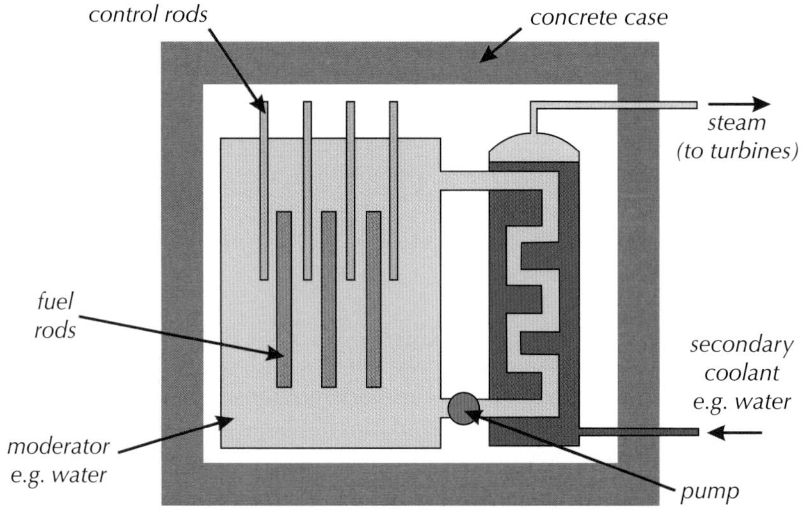

Figure 1: *The key features of a thermal fission reactor.*

Chain reactions

Fission reactors use rods of uranium that are rich in ^{235}U (or sometimes plutonium rods rich in ^{239}Pu) as 'fuel' for fission reactions. (The rods also contain other isotopes, but they don't undergo fission.)

These fission reactions produce neutrons which then induce other nuclei to fission, which produce more neutrons which go on to induce more fission. This is called a **chain reaction**. The neutrons will only cause a chain reaction if they are slowed down, which allows them to be absorbed by the uranium nuclei — these slowed down neutrons are called thermal neutrons (see page 210).

If the chain reaction in a fission reactor is left to continue unchecked, large amounts of energy are released in a very short time. Many new fission reactions will follow each fission, causing a runaway reaction which could lead to an explosion. This is what happens in a nuclear fission bomb.

Figure 2: *Fuel rods being lowered into the reactor at a nuclear power station. The core is surrounded by water, which acts as a moderator.*

Moderator

Fuel rods need to be placed in a **moderator** (for example, water) to slow down and/or absorb neutrons. You need to choose a moderator that will slow down some neutrons enough so they can cause further fission, keeping the reaction going at a steady rate.

Control rods

You want the chain reaction to continue on its own at a steady rate, where one fission follows another. The amount of 'fuel' you need to do this is called the **critical mass** — any less than the critical mass (sub-critical mass) and the reaction will just peter out. Fission reactors use a supercritical mass of fuel (i.e. more than the critical mass, so that several new fissions normally follow each fission) and control the rate of fission using **control rods**.

Control rods control the chain reaction by limiting the number of neutrons in the reactor. They absorb neutrons so that the rate of fission is controlled. Control rods are made up of a material that absorbs neutrons (e.g. boron), and they can be inserted by varying amounts to control the reaction rate. In an emergency, the reactor will be shut down automatically by the release of the control rods into the reactor, which will stop the reaction as quickly as possible.

Figure 3: *Cooling towers at a nuclear power plant. The steam used to drive the turbines is passed through the cooling tower to cool it down.*

Coolant

A coolant is a substance which transfers heat in a reactor. The material used should be a liquid or gas at room temperature, so that it can be pumped around the reactor, and be efficient at transferring heat. The moderator can serve as a primary coolant, which transfers the heat produced by fission to another coolant (the secondary coolant), such as water, which is converted to steam as it's heated. This steam then passes through and powers electricity-generating turbines.

Tip: When we say 'nuclear power station', we mean a power station which generates electricity using a fission reactor.

Environmental impact of fission reactors

HOW SCIENCE WORKS

Deciding whether or not to build a nuclear power station (and if so, where to build it) is a tricky business.

Nuclear fission doesn't produce carbon dioxide, unlike burning fossil fuels, so it doesn't contribute to global warming. It also provides a continuous energy supply, unlike many renewable sources (e.g. wind/solar).

Tip: Fission doesn't produce any greenhouse gases, but parts of the process do, e.g. transporting uranium fuel rods to a power station.

However, some of the waste products of nuclear fission are highly radioactive and difficult to handle and store. When material is removed from the reactor, it is initially very hot, so it is placed in cooling ponds until the temperature falls to a safe level. The radioactive waste is then stored in sealed containers in specialist facilities. These facilities often have to be built deep underground. Digging into the ground to build these facilities can have a significant effect on the local landscape and environment, and damage local wildlife habitats. The waste is kept there until its activity has fallen sufficiently for it to be considered safe. This can take many years, and there's a risk that material could escape from these containers.

A leak of radioactive material could be harmful to the environment and local human populations both now and in the future, particularly if the material contaminated water supplies. This would allow the isotopes to contaminate plants and animals, and so enter the food chain, exposing organisms (including humans) to radiation. Radiation sources are very dangerous to living cells — if radiation reaches a cell it can ionise atoms inside it. This can kill the cell, immediately harming the organism. Ionisation can also cause a mutation in the cell's DNA, which can result in cancer in animals, or produce a harmful genetic mutation which could be passed on to offspring.

Tip: Alpha radiation is particularly dangerous inside the body, as its short penetration range (see p.193) means it will almost certainly be absorbed by cells, which can damage them.

Accidents or natural disasters pose a risk to fission reactors. In 2011 an earthquake and subsequent tsunami in Japan caused a meltdown at the Fukushima nuclear power plant. Over 100 000 people were evacuated from the area, and many tonnes of contaminated water leaked into the sea.

Tip: Scientists are trying to develop fusion reactors so that we can generate nuclear electricity without the waste you get from fission reactors, but they haven't yet succeeded in creating one that makes more electricity than it uses.

The perceived risk of this kind of disaster leads many people to oppose the construction of nuclear power plants near their homes. Because of all of the necessary safety precautions, building and decommissioning nuclear power plants is very time-consuming and expensive.

Practice Questions — Fact Recall

Q1 Describe a fission chain reaction.

Q2 State why neutrons need to be slowed down by a moderator in a fission reactor.

Q3 How are control rods used to control the rate of reaction in a fission reactor? Give an example of a material used to make control rods.

Q4 What properties does the material used for the coolant in a fission reactor need to have? Give an example of a material that could be used as a coolant.

Q5 Give one way that nuclear waste can impact the environment.

Section Summary

Make sure you know...

- How the alpha-particle scattering experiment disproved the Thomson model of the atom, and provided evidence for the atom having a small, charged nucleus.
- That the nuclear model of the atom states that the atom consists of a small, positively-charged nucleus, made up of protons and neutrons, orbited by negatively-charged electrons.
- What the proton number, Z, and nucleon number, A, of a nucleus are.
- How to represent an atom using standard notation, ${}^{A}_{Z}X$, where X is the element's chemical symbol.
- That isotopes are forms of an element with different numbers of neutrons.
- That atoms have a diameter of around 1×10^{-10} m and a nucleus has a typical diameter of a few fm.
- How to use the equation $R = r_0A^{1/3}$ to calculate the radius of a nucleus, R, where r_0 is constant.
- How to calculate the mean densities of atoms and nuclei by approximating them as spheres.
- That the strong nuclear force is responsible for holding nuclei together, and its nature at different ranges.
- That hadrons, such as protons and neutrons, are particles that feel the strong nuclear force.
- That leptons, such as electrons and neutrinos, are fundamental particles that do not feel the strong nuclear force but are subject to the weak nuclear force.
- That every particle has an equivalent antiparticle which has the same mass but opposite charge.
- That the antiparticles of the electron, proton, neutron and neutrino are the positron, antiproton, antineutron and antineutrino respectively.
- That Einstein's mass-energy equation, $\Delta E = \Delta mc^2$, states that mass and energy are equivalent.
- That pair production is when energy changes into matter in the form of a particle-antiparticle pair.
- That when a particle comes into contact with its corresponding antiparticle, they annihilate and form a pair of photons.
- How to use Einstein's mass-energy equation to make calculations for pair production and annihilation.
- That hadrons are made up of fundamental particles called quarks.
- The basic properties of the up quark, u, the down quark, d, and the strange quark, s, as well as their respective anti-quarks.
- That a proton is made of two up quarks and a down quark, uud, and a neutron is made of one up quark and two down quarks, udd.
- How to describe the decay of particles in terms of the quark model.

cont...

- That in beta-minus decay, a d quark turns into a u quark and emits an electron and an antineutrino.
- That in beta-plus decay, a u quark turns into a d quark and emits a positron and a neutrino.
- That in any particle interaction charge must be conserved, and how to use charge conservation to balance particle interaction equations.
- That unstable nuclei will undergo radioactive decay to become more stable by emitting a particle or energy, and that this process is spontaneous and random.
- The four types of nuclear radiation: alpha particles (helium nuclei), beta-minus particles (electrons), beta-plus particles (positrons) and gamma rays, and their range and penetration properties.
- How to perform techniques and procedures to investigate the absorption of alpha particles, beta particles and gamma rays by different materials.
- How to write and balance nuclear decay equations for alpha, beta-minus and beta-plus decays.
- That the activity of a radioactive source is the number of nuclei in the source that decay each second.
- That the decay constant, λ, of a radioactive isotope is a measure of the rate of decay of an isotope.
- How to use the equation $A = \lambda N$ to calculate the activity of a sample of a radioactive isotope, where N is the number of nuclei of that isotope in the sample.
- How to use the equation $\frac{\Delta N}{\Delta t} = -\lambda N$ to calculate the rate of change of the number of undecayed nuclei of an isotope at a given time, and model radioactive decay using a spreadsheet.
- How to calculate the number of undecayed nuclei of an isotope at time t using the equation $N = N_0 e^{-\lambda t}$, where N_0 is the initial number of undecayed nuclei.
- How to use dice to simulate radioactive decay.
- How to calculate the activity of a source at time t using the equation $A = A_0 e^{-\lambda t}$, where A_0 is the initial activity of the source.
- That the half-life, $t_{1/2}$, of an isotope is the average time taken for the number of undecayed nuclei of the isotope in a sample to halve, and how to find the half-life of an isotope from its decay graph.
- How to perform techniques and procedures to determine the half-life of protactinium.
- How to calculate the half-life of an isotope using the equation $\lambda t_{1/2} = \ln(2)$.
- How radioactive isotopes such as carbon-14 can be used in radioactive dating.
- That the 'mass defect' is the difference between the mass of a nucleus and the total mass of its constituent nucleons.
- That the binding energy of a nucleus is the energy required to separate all the nucleons in a nucleus, and is equivalent to its mass defect converted into energy.
- How to calculate the binding energy of a nucleus from its mass defect using $\Delta E = \Delta mc^2$, and how to calculate the binding energy per nucleon for a given nucleus.
- The shape of the graph of binding energy per nucleon against nucleon number.
- That nuclear fission is when a large nucleus splits into two smaller nuclei with a higher binding energy per nucleon, and that this process releases energy and free neutrons.
- That induced nuclear fission is when fission is caused by the absorption of a low energy neutron.
- That nuclear fusion is when two nuclei combine to form a larger nucleus with higher binding energy per nucleon, and that this process releases energy, but requires very high temperatures and pressures.
- How to write and balance nuclear equations for fission and fusion reactions.
- How to calculate the energy released (or absorbed) in nuclear reactions from the binding energy per nucleon against nucleon number graph and by using $\Delta E = \Delta mc^2$.
- That a chain reaction is where neutrons produced by a fission reaction induce more fission reactions.
- The basic structure of a fission reactor, including details of the moderator, control rods, fuel rods and coolant.
- The environmental impact of nuclear waste.

Exam-style Questions

1 At what range does the strong nuclear force lead to the greatest attraction between nucleons?

A Less than 0.5 fm.

B Between 0.5 fm and 3 fm.

C Between 3 and 6 fm.

D More than 6 fm.

(1 mark)

2 Which of the following interactions is possible?

A $e^+ \rightarrow e^- + \nu$

B $p \rightarrow n + e^- + \nu$

C $n \rightarrow p + e^- + \overline{p}$

D $p \rightarrow n + e^+ + \nu$

(1 mark)

3 A sample contains 8.04×10^{21} atoms of the radioisotope ^{60}Co. It takes 10.4 years for the number of ^{60}Co atoms remaining in the sample to fall to 2.01×10^{21}.
What is the decay constant of ^{60}Co?

A 4.23×10^{-9} s^{-1}

B 2.11×10^{-9} s^{-1}

C 2.36×10^{8} s^{-1}

D 1.64×10^{8} s^{-1}

(1 mark)

4 Subatomic particles can be classified into categories such as hadrons and leptons.

(a) Give one similarity and one difference between hadrons and leptons.

(2 marks)

The Δ^{++} particle is a hadron. It has a charge of $+2e$, and a mass of 2.19×10^{-27} kg.

(b) The quark composition of the Δ^{++} particle is u u x.
Identify quark x.

(1 mark)

(c) All particles have a corresponding antiparticle.
State the mass and charge of the antiparticle of the Δ^{++} particle.

(1 mark)

(d) Quarks cannot be found in isolation. When energy is supplied to a hadron to remove a quark, it instead pair-produces a quark-anti-quark pair. Given that the mass of a strange quark is approximately 1.71×10^{-28} kg, calculate the minimum energy, in MeV, required to produce a strange-anti-strange pair.

(3 marks)

5 A plutonium isotope can be represented by $^{240}_{94}\text{Pu}$.

(a) State how many neutrons and protons are in the nucleus of an atom of $^{240}_{94}\text{Pu}$.

(1 mark)

(b) Name the force that stabilises an atomic nucleus by acting against the repulsive electrostatic force between the protons.

(1 mark)

(c) $^{240}_{94}\text{Pu}$ decays by α decay to form an element with symbol U.
Write an equation for this decay using standard notation.

(2 marks)

Another isotope of plutonium, with nucleon number 241, decays by emitting α and β^- particles in stages (a decay chain), eventually forming $^{205}_{81}\text{Tl}$.

(d) There are nine α decays in the decay chain from the plutonium isotope to $^{205}_{81}\text{Tl}$.
Calculate the number of β^- decays in the decay chain.

(2 marks)

(e) Describe the process of β^- decay in terms of quark flavour change.

(2 marks)

6 **Fig. 6.1** shows the activity-time graph for a sample of a radioactive isotope.

Fig. 6.1

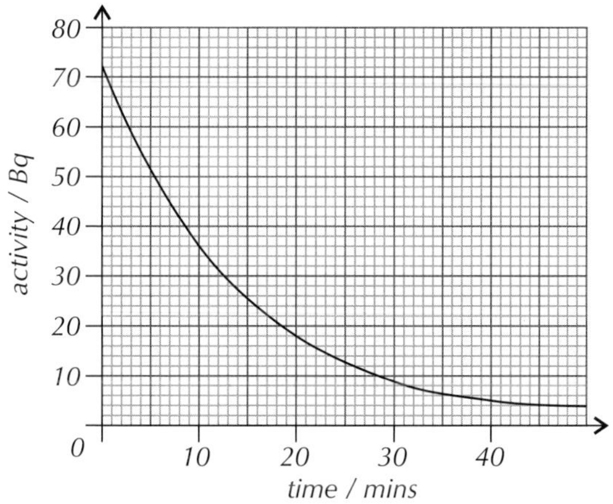

(a) Find the half-life of the isotope in minutes.

(1 mark)

(b) Calculate the number of undecayed nuclei in the sample after 17 minutes.

(3 marks)

When the isotope decays, the nucleus emits a positron and a neutrino.

(c) The positron goes on to annihilate with an electron. Calculate the minimum frequency of one of the photons produced in this annihilation.

(3 marks)

7 ^{33}P is an isotope of phosphorus.

(a) Calculate the radius of a nucleus of ^{33}P. Use the value $r_0 = 1.4 \times 10^{-15}$ m.

(2 marks)

(b) A nucleus of ^{33}P has a mass of 32.97 u. Calculate the nuclear density of ^{33}P. You may assume the nucleus is spherical.

(3 marks)

^{33}P emits β^- radiation as it decays to an isotope of sulfur.
^{33}P has a half-life of 25.4 days, and a proton number of 15.

(c) Write the nuclear equation for the decay of ^{33}P. (The chemical symbol of sulfur is S.)

(2 marks)

(d) Calculate the decay constant of ^{33}P.

(2 marks)

(e) A sample of ^{33}P contains 1.6×10^{15} undecayed atoms. How long will it take for the number of undecayed atoms to fall to 7.0×10^{13}? Give your answer in days.

(2 marks)

8 This question is about binding energy.

(a) What is meant by the term *binding energy*?

(1 mark)

(b) Copy the axes in **Fig 8.1** and sketch the graph of binding energy per nucleon against nucleon number. Indicate which nucleus is found at the peak of the graph.

Fig. 8.1

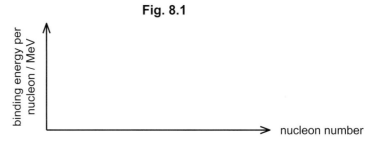

(2 marks)

(c) Calculate the binding energy per nucleon of zinc-66, given that it has a mass defect of 0.62065 u. Give your answer in MeV.

(2 mark)

After absorbing a neutron, ^{235}U can fission into ^{94}Sr and ^{140}Xe, along with a number of neutrons:

$$^{235}_{92}U + ^{1}_{0}n \longrightarrow ^{140}_{54}Xe + ^{94}_{a}Sr + b^{1}_{0}n$$

(d) Calculate the proton number of ^{94}Sr, a, and the number of neutrons produced, b.

(2 marks)

(e) Explain how the fission of a heavy nucleus releases energy.

(2 marks)

(f) Calculate the energy released by the above reaction in MeV.
(Nuclear masses: ^{235}U = 234.99333 u, ^{94}Sr = 93.89446 u,
^{140}Xe = 139.89194 u, ^{1}n = 1.00867 u.)

(3 marks)

1. X-ray Imaging

Non-invasive medical imaging techniques let doctors see what's going on (or going wrong) inside the body, so they won't need to resort to surgery to have a look. X-ray imaging is one of these techniques.

Learning Objectives:
- Know the basic structure of an X-ray tube and that its components include a heater (cathode), an anode, a target metal and a high voltage supply.
- Understand how X-ray photons are produced from an X-ray tube.
- Know what is meant by the attenuation of X-rays and be able to calculate the intensity of an X-ray beam using $I = I_0 e^{-\mu x}$, where μ is the attenuation (absorption) coefficient.
- Know that the X-ray attenuation mechanisms are simple scatter, photoelectric effect, Compton effect and pair production.
- Explain how barium and iodine can be used as contrast media in X-ray imaging.
- Know what computerised axial tomography (CAT) scanning is, and that the components of a CAT scanner include a rotating X-tube producing a thin fan-shaped beam, a ring of detectors, and computer software and display.
- Know advantages of a CAT scan over an X-ray image.

Specification Reference 6.5.1

X-ray tubes

The X-rays used for diagnostic imaging are produced in X-ray tubes. An **X-ray tube** is a glass tube containing an electrical circuit, with a **cathode** where electrons are emitted, and an **anode** (called the target metal) that the electrons are directed towards. X-ray tubes are surrounded by a lead housing (to keep the X-rays contained), which has a small window in it through which X-rays can pass. To prevent emitted electrons from colliding with gas particles while still in the tube, the X-ray tube is evacuated (i.e. the air is removed).

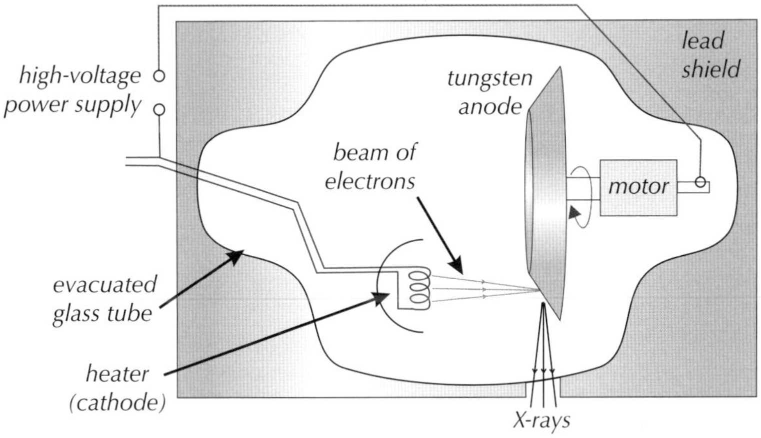

Figure 1: *An X-ray tube.*

At the cathode, electrons are emitted (boiled off) by the hot filament. This filament is heated by passing a current through it — this current is separate from the current that is flowing to the anode. The cathode filament is usually contained in cup-shaped housing, which focuses the beam of electrons onto the target metal.

The target metal (tungsten) acts as the anode of the circuit. The high potential difference across the tube (the 'tube voltage') causes the electrons to accelerate towards it. When the electrons smash into the anode, they decelerate and some of their kinetic energy is converted into electromagnetic energy, in the form of X-ray photons. Whatever energy is lost by the electron is gained by the photon — the result is that the tungsten anode emits a continuous spectrum of X-ray radiation.

The maximum kinetic energy of the electrons (and therefore the maximum energy of the X-ray photons) is equal to the potential difference of the X-ray tube multiplied by the charge of an electron:

$$E = eV$$

E *= maximum kinetic energy of the electrons in J*
e *= charge of an electron in C*
V *= potential difference of the X-ray tube in V*

Figure 2: William David Coolidge, American physicist, holding an early version of the X-ray tube he invented.

Tip: The X-ray photon is released when the outer electron moves into a lower energy level, not when the inner electron is ejected. This means these photons can only be emitted with very specific energies (which depend on the energies of the shells the electron moves between — see p.94 for more information).

Tip: Because X-rays are harmful, the intensity of medical X-rays needs to be carefully controlled — you need the lowest possible intensity that will give a clear image.

Tip: Current is a measure of the number of electrons passing a point, not of their energy. So increasing the current flowing through the filament increases the number of X-ray photons produced, but doesn't change the maximum photon energy.

So, if a potential difference of 50 kV is used in the tube, the maximum X-ray energy will be 50 keV.

On average, only about 1% of the electrons' kinetic energy is converted into X-rays, while the rest is converted into heat. To avoid overheating, the tungsten anode is rotated at about 3000 rpm (revolutions per minute) to spread heat generated around the whole anode. The anode is also mounted on copper — this conducts the heat away effectively.

Example — Maths Skills

Electrons hitting an anode in an X-ray tube produce X-ray photons whose maximum energy corresponds to a wavelength of 5.0×10^{-11} m. The tube voltage is then halved but all other factors are kept the same. Calculate the wavelength of the most energetic X-ray photons produced at this lower voltage.

The maximum energy of the electrons produced in the X-ray tube is $E = e \times V$. Halving the voltage would halve the maximum energy of the electrons and hence halve the maximum energy of the X-ray photons.

Since the energy of a photon (E) is given by $E = \frac{hc}{\lambda}$ (where h and c are constant), halving the energy would mean the wavelength is doubled.

So the new wavelength of the most energetic photons is:

$$2 \times 5.0 \times 10^{-11} = 1.0 \times 10^{-10} \text{ m}$$

X-rays are also produced when beam electrons knock out electrons from the inner shells of the tungsten atoms. Electrons in the atoms' outer shells fall into the vacancies in the inner energy levels, and release energy in the form of X-ray photons — see Figure 3.

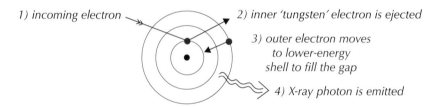

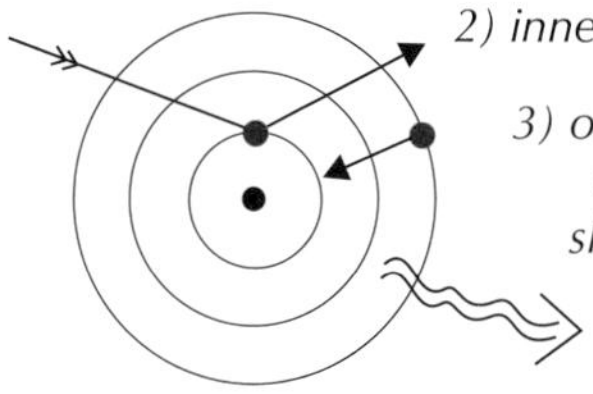
1) incoming electron
2) inner 'tungsten' electron is ejected
3) outer electron moves to lower-energy shell to fill the gap
4) X-ray photon is emitted

Figure 3: X-rays are emitted when outer electrons move to inner energy shells to fill vacancies in tungsten atoms.

Varying beam intensity

The intensity of an X-ray beam is the energy per second that passes through a unit area at right angles to the beam. There are two ways to increase the intensity of the X-ray beam:

- Increase the tube voltage. Increasing the potential difference between the cathode and anode gives the electrons more kinetic energy — meaning they have more energy available to be converted into photons on deceleration. Higher energy electrons can also knock out electrons from shells deeper within the tungsten atoms. Both these factors lead to higher energy X-ray photons being produced.
- Increase the current supplied to the filament. This liberates more electrons per second from the cathode, which then produce more X-ray photons per second. Individual photons have the same energy as before.

X-ray attenuation

When X-rays pass through matter (e.g. a patient's body), they are absorbed and scattered, which leads to a fall in their intensity. The intensity of the X-ray beam decreases (attenuates) exponentially as they travel deeper into the material — the intensity at a depth of x cm depends on the material's **attenuation (absorption) coefficient**:

$$I = I_0 e^{-\mu x}$$

I = intensity of the attenuated X-ray beam in Wm^{-2}

I_0 = incident intensity of the X-ray beam in Wm^{-2}

μ = the material's attenuation coefficient in cm^{-1}

x = the depth to which the beam has penetrated in cm

Tip: You might also see the attenuation coefficient given in m^{-1}. In this case you'll need to either convert the attenuation coefficient to cm^{-1} or make sure the penetration depth is in m before doing any calculations.

Example — Maths Skills

The attenuation coefficient of a patient's body tissue is 0.2 cm^{-1}. At what depth inside the tissue will an X-ray be attenuated to 40% of the intensity of the incident beam?

You know that the intensity is 40% of the original intensity, so:

$$\frac{I}{I_0} = 0.4 \Rightarrow e^{-\mu x} = 0.4$$

Take the natural logarithm of both sides and rearrange to make x the subject:

$$-\mu x = \ln 0.4 \;\Rightarrow\; x = -\frac{\ln 0.4}{\mu} = -\frac{\ln 0.4}{0.2} = 4.58... = 5\text{ cm (to 1 s.f.)}$$

Tip: 'ln' is the natural logarithm and you can use it to get rid of exponentials because $\ln e^x = x$. You might see 'ln' written as '$\log_e$'. There's more on this on page 236.

X-rays are attenuated as they pass through a material in four main ways:

- Simple scattering — at low energies a photon may be deflected by atoms in the material, but without losing energy (i.e. it's elastically scattered).
- The photoelectric effect — a photon with around 30 keV of energy can be absorbed by an electron — the electron is then ejected from its atom. The gap in the electron shell is then filled by another electron, which results in a photon being emitted.
- The Compton effect — a photon with around 0.5-5 MeV of energy loses some of its energy as it interacts with an electron in an atom. The electron is knocked out of the atom and the photon is deflected (i.e. scattered).
- Pair production — a high (> 1.1 MeV) energy photon decays into an electron and a positron (i.e. photon energy is converted to mass) — see page 186.

Tip: You learnt about the photoelectric effect in year 1 of A-level.

One factor that affects the amount of energy absorbed by a material is the atomic number of atoms within it. So different tissues (e.g. soft tissue and bone) containing atoms with different atomic numbers will appear different in an X-ray image because they're absorbing different amounts of energy. For example, X-rays are absorbed more by bone than soft tissue, so bones show up more brightly.

If the tissues in the region of interest have similar attenuation coefficients then artificial **contrast media** can be used so these different tissues appear more distinct in an X-ray image. Iodine and barium are both relatively harmless and have high atomic numbers, so they show up clearly in X-ray images. They can also be followed as they move through a patient's body. For example, a barium meal can be swallowed and its path through the digestive system imaged, while iodine is usually injected into blood vessels or tissues so that they can be viewed more clearly.

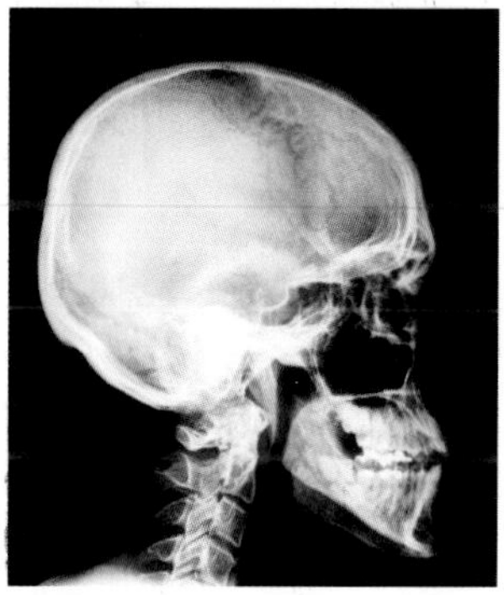

***Figure 4:** An X-ray image of a human skull. The white parts of the image are where the film behind the subject has been exposed to X-rays of lower intensity due to their attenuation as they pass through the subject's body. The more intense an X-ray beam is when it hits the film, the darker that part of the film will be.*

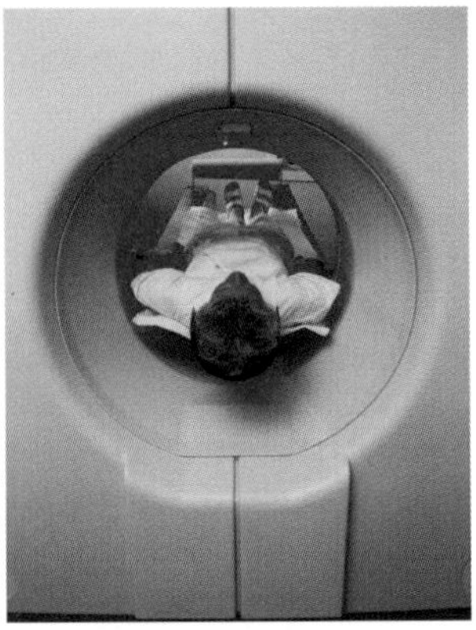

Figure 5: A patient undergoing an upper-body CAT scan.

CAT scanning

Computed axial tomography (CT or CAT) scans produce an image of a two-dimensional slice through the body. The patient lies on a table, which slides in and out of a ring containing a rotating X-ray tube (or X-tube) and X-ray detectors.

A thin X-ray beam in the shape of a fan is emitted by the X-ray tube, and the tube is rotated around the body. This beam is picked up by the detectors. Computer software works out how much attenuation has been caused by each part of the body, and converts this information into a high quality image of the patient's tissues. CAT scans produce more detailed images than regular X-rays, especially for soft tissue. The data can also be manipulated to generate a 3D image.

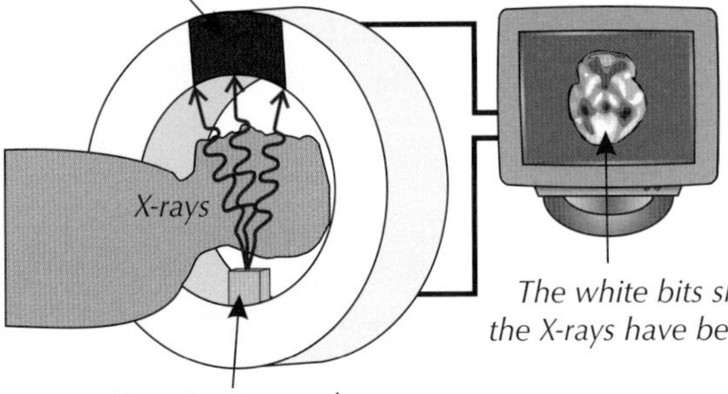
X-rays detected on other side.
Image of a 'slice' through patient's head.
X-rays
The white bits show where the X-rays have been absorbed.
Rotating X-ray tube emits beam of X-rays in a fan shape.

Figure 6: Diagram showing how a CAT scan works.

Practice Questions — Application

Q1 In an X-ray tube, electrons are accelerated through a potential difference of 85.0 kV towards an anode. What's the maximum energy an emitted X-ray photon could have in joules?

Q2 X-rays are used to image a bone in a patient's leg. The X-rays have an intensity of 30.0 Wm^{-2} when incident on the bone. The attenuation coefficient of bone (μ) is given by $\mu = 2.5\ cm^{-1}$. Find the intensity of the X-rays at a depth of 0.69 cm beneath the surface of the bone.

Practice Questions — Fact Recall

Q1 a) Draw and label a diagram of an X-ray tube.

b) Explain why the anode in the X-ray tube rotates.

Q2 State the names of four mechanisms by which X-rays can be attenuated.

Q3 Explain how a barium meal can help to improve the contrast between tissues in an X-ray image of the digestive tract.

Q4 Describe how a CAT scanner works.

Q5 Give two advantages of CAT scans over standard X-ray images.

2. Medical Uses of Nuclear Radiation

Medical tracers move through the patient's body to the region of interest and emit radiation. This radiation can be detected with a gamma camera or a PET scanner to map and study the function and structure of tissues or organs.

Learning Objectives:

- Know how medical tracers, including technetium-99m and fluorine-18, are used.
- Know how a gamma camera can be used to aid diagnosis.
- Be able to describe the components of a gamma camera and their functions, including the collimator, scintillator, photomultiplier tubes, computer and display.
- Be able to describe how an image is formed in a gamma camera.
- Know that a positron emission tomography (PET) scanner relies on the annihilation of positron-electron pairs, and describe how an image is formed.
- Know how PET scanning can be used to aid diagnosis.

Specification Reference 6.5.2

Medical tracers

Many types of imaging, e.g. X-rays (pages 223-224), only show the structure of organs. **Medical tracers** (or radiotracers) are radioactive substances that are used to show the function, as well as the structure, of tissues or organs.

Medical tracers usually consist of a radioactive isotope (e.g. fluorine-18 or technetium-99m) bound to a substance that is used by the body (e.g. glucose or water). The tracer is injected into or swallowed by the patient and then moves through the body to the region of interest. Where the tracer goes depends on the substance the isotope is bound to — i.e. it goes anywhere that the substance would normally go as it's used by the body. The radiation emitted due to the isotope is detected (e.g. by a gamma camera or PET scanner, see below and page 226) and used to produce an image of inside the patient.

This can be useful in different areas — for example:

- Tracers can show areas of damaged tissue in the heart by indicating areas with decreased blood flow. This can reveal coronary artery disease and damaged or dead heart muscle caused by heart attacks.
- They can help identify active cancer tumours. Cancer cells have a much higher metabolism than healthy cells because they're growing fast, so they take up more tracer.
- Tracers can show blood flow and activity in the brain. This helps to research and treat neurological conditions such as Parkinson's, Alzheimer's disease, epilepsy and depression.

Technetium-99m is widely used in medical tracers because the gamma radiation it emits can easily pass out of the body to reach the detector, and it has a half-life of 6 hours (long enough for data to be recorded, but short enough to limit the radiation to which the patient is exposed to an acceptable level). It also decays to a much more stable (less radioactive) isotope.

Tip: Don't worry, you don't need to know what the 'm' at the end of technetium-99m means.

Fluorine-18 is used in PET scans as it usually undergoes beta-plus decay (which is necessary for a PET scan). It also has a short half-life (110 minutes), meaning the patient's exposure to dangerous radioactivity is kept as low as possible.

Tip: The beta-plus decay of fluorine-18 in a PET scan produces positrons. These annihilate with electrons in the body, producing high-energy gamma rays — see the next page.

Gamma cameras

The gamma rays emitted by some medical tracers in a patient's body are detected using a gamma camera. Once the medical tracer has had time to move to the tissue of interest in the patient, areas where the tracer has been taken up will emit gamma radiation. The patient can then lie on a table and a gamma camera positioned above or below the body can be used to detect the emitted gamma rays.

Tip: The scintillator turns the energy of a gamma photon into a large number of visible light photons, which are easier to detect.

Tip: Photomultiplier tubes are also used in PET scanners (see below).

Gamma cameras consist of five main parts (see Figure 1):

- Lead shield — this surrounds most of the camera and stops radiation from other sources being detected.
- Lead collimator — a piece of lead with thousands of vertical holes in it, which only gamma rays travelling parallel to the holes can pass through.
- Scintillator — a sodium iodide crystal that emits a flash of light (scintillates) whenever a gamma ray hits it.
- Photomultiplier tubes — these detect the flashes of light from the scintillator and turn them into pulses of electricity.
- Electronic circuit — this collects the signals from the photomultiplier tubes and sends them to a computer for processing into an image.

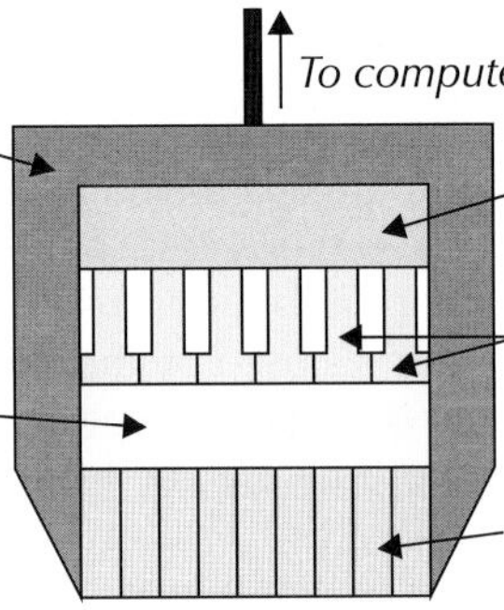

Figure 1: *The components of a gamma camera.*

The computer maps the spatial distribution and the frequency of gamma ray emissions detected by the gamma camera to make a two-dimensional image, effectively forming a snapshot of regions where the medical tracer has been taken up. The gamma camera is often rotated to take multiple images from different angles to provide different perspectives of the area.

Gamma cameras are useful in helping to diagnose patients without the need for surgery. They are cheaper than a PET scanner but are still fairly expensive, and they also require a patient to be exposed to ionising radiation, which is bad for you — see the next page.

PET scans

Positron Emission Tomography (PET) is another medical imaging technique that allows patients to be diagnosed without having to have surgery. PET scans involve injecting a patient with a substance used by the body, (e.g. glucose) that is bound to a positron-emitting radiotracer with a short half-life, (e.g. ^{13}N, ^{15}O or ^{18}F). The patient is left for a time to allow the radiotracer to move through the body to the organs that need to be examined. The positrons emitted by the radioisotope collide with electrons in the organs, and annihilate. This annihilation results in two high-energy gamma rays being emitted in opposite directions (see page 187).

Detectors all around the body detect and record these gamma rays (see Figure 2). The detectors then send the information to a computer, which builds up a map of the radioactivity in the body — this corresponds to the concentration of radiotracers in different parts of the body.

Tip: Because annihilation produces two gamma photons in opposite directions, the PET scanner can locate exactly where the annihilation took place, by measuring how soon one photon arrives at one side of the scanner compared to the other photon arriving at the opposite side.

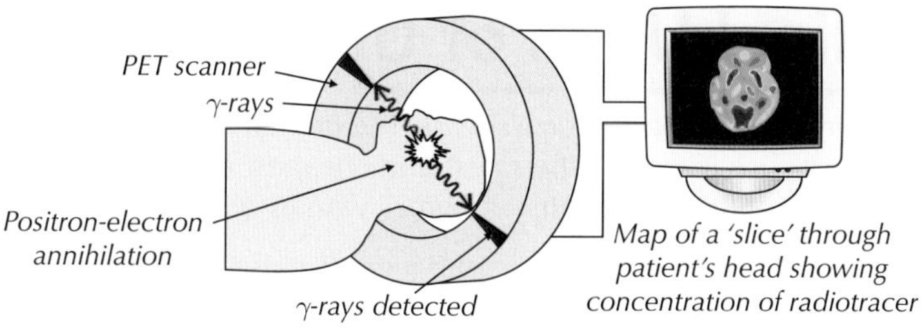

Figure 2: *Diagram showing how a PET scanner works.*

If radioactive glucose is used as a radiotracer, the distribution of radioactivity will correspond to metabolic activity in different parts of the body. This is because more of the radioactive glucose is taken up and used by cells that are doing more work, i.e. cells with a higher metabolism.

By looking at which cells are doing more work, doctors can help diagnose illnesses in patients — for example, cancer cells have a higher level of activity than healthy cells. Another advantage of PET scans is that brain activity can be investigated, which can be difficult with other non-invasive methods.

PET scan radiotracers have a short half-life, so the patient is exposed to radiation for only a short time. However, this short time period means there is only a limited time when a patient can be scanned — unlike with gamma cameras, where the tracer used takes much longer to decay. PET scanners are also incredibly expensive, meaning not many hospitals have one. This means doctors may have to balance the benefits of their patients having PET scans against the costs, inconvenience and time needed to send ill patients to a hospital with a PET scanner.

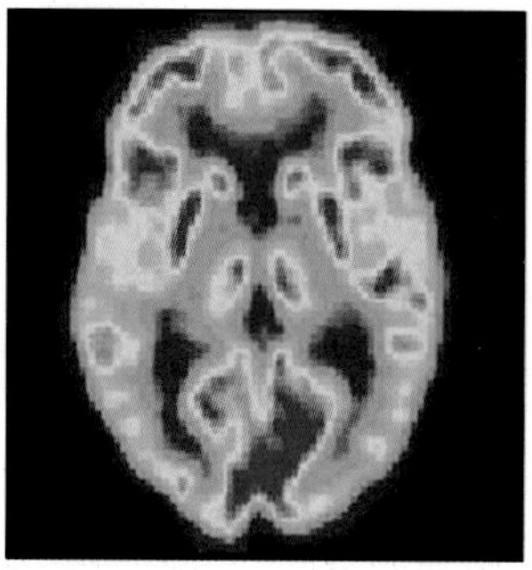

Figure 3: *Image of a healthy brain from a PET scanner.*

Tip: Other substances can also be labelled with medical tracers to investigate functions such as blood flow and oxygen use.

The risks of using ionising radiation

X-rays, γ-rays, and α and β particles are all types of ionising radiation. They interact with atoms or molecules to form ions — usually by removing an electron. This can damage cells, and can lead to:

HOW SCIENCE WORKS

- Cell mutations and cancerous tumours, if the cell's DNA is damaged or altered.
- Cell sterility — i.e. the cell may no longer be able to reproduce.
- Cell death — i.e. the cell could be destroyed completely.

The macroscopic effects of ionising radiation (i.e. the effects on the patient as a whole) can include tumours, skin burns, sterility, radiation sickness, hair loss and death. The result is that radiation is only used in the diagnosis and treatment of disease when absolutely necessary — i.e. when the benefits to the patient outweigh the risks. Even then, radiation doses are limited to the minimum possible.

Tip: See page 192 for more about γ-rays, α particles and β particles.

Practice Questions — Fact Recall

Q1 Describe how medical tracers are used to investigate body tissues and organ function.

Q2 Name a gamma-emitting isotope often used as a medical tracer. Describe the properties that make it suitable for this use.

Q3 Which medical imaging technique is fluorine-18 generally used in?

Q4 Name the main components of a gamma camera and describe what each of them does.

Q5 A tissue's rate of metabolic activity is linked to its rate of uptake of glucose. Describe how a PET scanner produces an image of metabolic activity in different parts of the body.

Learning Objectives:

- Know that ultrasound is a longitudinal wave with a frequency greater than 20 kHz.
- Be able to calculate the acoustic impedance of a medium using $Z = \rho c$.
- Know that ultrasound is reflected at boundaries, and how to calculate the fraction of wave intensity that is reflected using $\frac{I_r}{I_0} = \frac{(Z_2 - Z_1)^2}{(Z_2 + Z_1)^2}$.
- Know what the piezoelectric effect is.
- Know that an ultrasound transducer is a device that emits and receives ultrasound.
- Understand the importance of impedance (acoustic) matching, and how it can be achieved using a special gel in ultrasound scanning.
- Know what is meant by an ultrasound A-scan and B-scan.
- Understand how the Doppler effect can be used in ultrasound to find the speed of blood in a patient.
- Be able to use $\frac{\Delta f}{f} = \frac{2v\cos\theta}{c}$ to determine the speed, v, of blood in a patient.

Specification Reference 6.5.3

3. Medical Uses of Ultrasound

You'll have heard of ultrasound from its wide use in prenatal scans. The main reason for its use is that it's the safest and cheapest form of non-invasive imaging we have — but unfortunately it comes with its downsides.

What is ultrasound?

Ultrasound waves are longitudinal waves with higher frequencies than humans can hear (> 20 000 Hz, or 20 kHz). Ultrasound waves travel at the speed of sound.

When an ultrasound wave meets a boundary between two different materials, it is partially reflected and partially transmitted (undergoing refraction if the angle of incidence is not 0°). The reflected waves can be detected by an ultrasound scanner and used to generate an image.

Ultrasound scanning is a relatively safe medical imaging technique, and is particularly useful for viewing soft tissue. The most common frequencies used for medical purposes are usually from 1 to 15 MHz.

Acoustic impedance

The amount of reflection an ultrasound wave experiences at a boundary between different materials depends on the difference in **acoustic impedance**, Z, between them. The acoustic impedance of a material is defined as:

$$Z = \rho c$$

Z = acoustic impedance in $kgm^{-2}s^{-1}$

ρ = density of material in kgm^{-3}

c = speed of sound in the medium in ms^{-1}

Example — Maths Skills

At 20 °C the density of air is approximately 1.2 kgm^{-3} and the speed of sound in air is 340 ms^{-1}. Find the acoustic impedance of air at 20°C.

$Z = \rho c = 1.2 \times 340 = 408 = 410\ kgm^{-2}s^{-1}$ (to 2 s.f.)

Suppose an ultrasound wave with an intensity of I_0 travels through a material with acoustic impedance Z_1. It hits the boundary between this material and another material with acoustic impedance Z_2. If the two materials have a large difference in acoustic impedance, then most of the energy is reflected, and the intensity of the reflected wave I_r will be high — see Figure 1. If the acoustic impedance of the two materials is the same then there is no reflection.

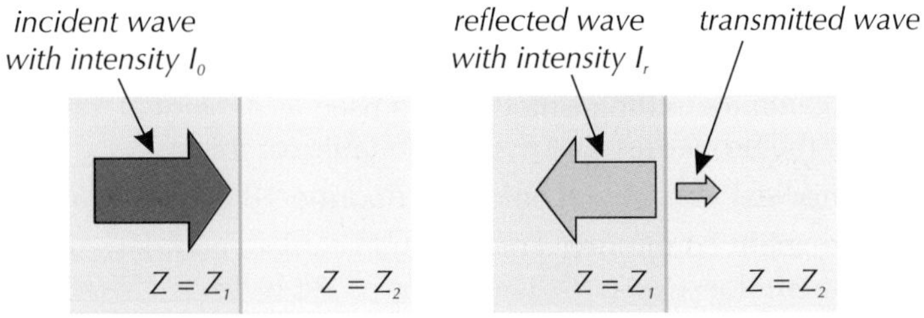

Figure 1: *When a wave travelling through a material of acoustic impedance Z_1 (left) hits a boundary with a material of a very different acoustic impedance Z_2, most of the energy is reflected (right).*

The equation for the fraction of wave intensity that is reflected is:

$$\frac{I_r}{I_0} = \frac{(Z_2 - Z_1)^2}{(Z_2 + Z_1)^2}$$

I_r = intensity of reflected wave in Wm^{-2}

I_0 = intensity of incident wave in Wm^{-2}

Z_1 = acoustic impedance of first material in $kgm^{-2}s^{-1}$

Z_2 = acoustic impedance of second material in $kgm^{-2}s^{-1}$

The fraction of wave intensity that is reflected has no units — it's just a ratio. To calculate the fraction of wave intensity that is transmitted, subtract this reflected fraction from 1.

Exam Tip
You don't need to understand where this equation comes from. And it'll be given in your data and formulae booklet, so you don't need to remember it — just make sure you know how to use it.

Example — Maths Skills

An ultrasound wave meets a boundary between two materials with acoustic impedances of Z_1 = 1.34 × 10^6 $kgm^{-2}s^{-1}$ and Z_2 = 2.57 × 10^6 $kgm^{-2}s^{-1}$. Find the fraction of the wave's intensity that is reflected.

$$\frac{I_r}{I_0} = \frac{(Z_2 - Z_1)^2}{(Z_2 + Z_1)^2} = \frac{(2.57 \times 10^6 - 1.34 \times 10^6)^2}{(2.57 \times 10^6 + 1.34 \times 10^6)^2}$$

$= 0.09895...$

$= 0.0990$ (to 3 s.f.)

The piezoelectric effect

Ultrasound images are produced using the **piezoelectric effect**. This describes the ability of certain materials, such as piezoelectric crystals, to produce a potential difference across them when they are deformed (squashed or stretched).

The particles inside piezoelectric crystals have one end that is more positively-charged, and one end that is more negatively-charged. Normally these particles are randomly oriented, but when stress is applied to the crystal, it shifts the orientation of their electric charges. Some particles move so that their charged ends line up in one direction — this generates a potential difference across the crystal (see Figure 2).

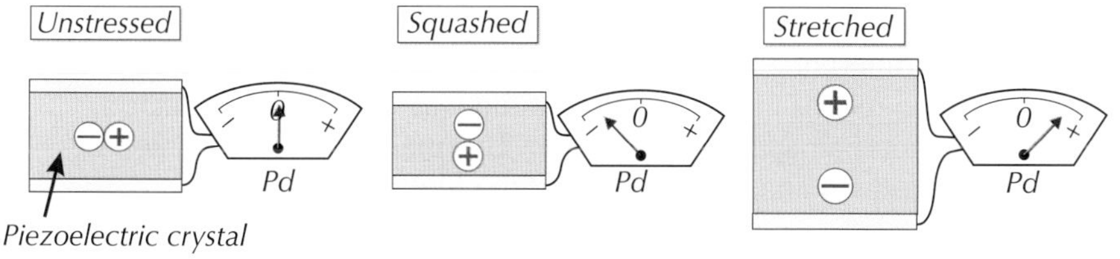

Figure 2: *When a piezoelectric crystal is deformed, one end becomes positively charged and one end becomes negatively charged, resulting in a p.d. across the crystal.*

Similarly, when you apply a p.d. across a piezoelectric crystal, the crystal deforms to rebalance the charge across it. If the p.d. is alternating, then the crystal vibrates at the same frequency.

Ultrasound waves are generated and detected by a device called an ultrasound transducer. Ultrasound transducers convert ultrasound into electrical signals, and vice versa. They contain lead zirconate titanate (PZT) crystals, which exhibit the piezoelectric effect. Inside the transducer, an alternating potential difference is applied to the PZT crystals, which vibrate to create an ultrasound wave. When the wave is reflected and returns, it causes the crystals to vibrate, which in turn creates an alternating potential difference that is detected in an adjoining circuit.

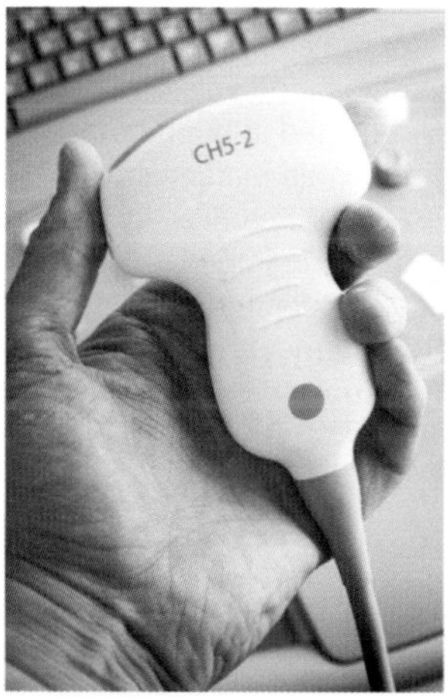

Figure 3: *An ultrasound probe containing an array of ultrasound transducers.*

Tip: The vibrations of the crystal are greatest when the frequency of the potential difference equals the natural frequency of the crystals (see page 55).

Tip: Resolution describes the ability of a device to distinguish two points that are close together.

The ultrasound waves produced by an ultrasound transducer have a wavelength equal to twice the thickness of the PZT crystals inside. Ultrasound with this wavelength will make the crystals resonate (see p.55) and produce a large signal. This means the ultrasound transducer only receives large signals from waves of the same wavelength that it produces, reducing unwanted interference from other sources.

The resolution of the ultrasound transducer increases as the length of each ultrasound pulse decreases, so the vibrations of the PZT crystal are reduced with damping material to produce short pulses.

Coupling media

Soft tissue has a very different acoustic impedance from air, so almost all the ultrasound energy is reflected from the surface of the body if there is air between the transducer and a patient's body. To avoid this, you need a **coupling medium** between the transducer and the body — this displaces the air and has an impedance much closer to that of body tissue. The coupling medium is usually an oil or gel that is smeared onto the skin. The use of coupling media is an example of impedance (acoustic) matching.

Using ultrasound in medical scans

HOW SCIENCE WORKS

Advantages

- There are no known hazards or side effects (unlike with X-rays, where the patient is exposed to ionising radiation, see page 227).
- Real-time images can be obtained.
- Ultrasound devices are relatively cheap and portable.
- The procedure is relatively comfortable for a patient, and they are allowed to move during the scan.

Disadvantages

- Ultrasound can't penetrate bone, since the difference in acoustic impedance between bone and the surrounding tissue is very large. So ultrasound can't be used to detect fractures or examine the brain.
- The large difference in acoustic impedance between air and body tissues also means most of an ultrasound wave's energy is reflected when it meets air spaces in the body. This means ultrasound can't be used to produce images from behind the lungs.
- Ultrasound images are low resolution (i.e. they're quite 'grainy'), so you can't see fine detail.
- Ultrasound can be used to identify solid masses, but it doesn't provide much information about their composition.

Tip: Remember, most of the energy from ultrasound waves is reflected at boundaries between materials with large differences in acoustic impedance — this includes where bone meets other tissues.

Types of ultrasound scan

There are two types of ultrasound scan you need to know about. A-scans are mostly used for measuring distances, while B-scans are used to form images.

The A-scan

An amplitude scan (A-scan) sends a short pulse of ultrasound into the body as an electron beam inside a cathode ray oscilloscope (CRO) starts to sweep across its screen. The receiver detects reflected ultrasound pulses, which are displayed as peaks on the CRO screen.

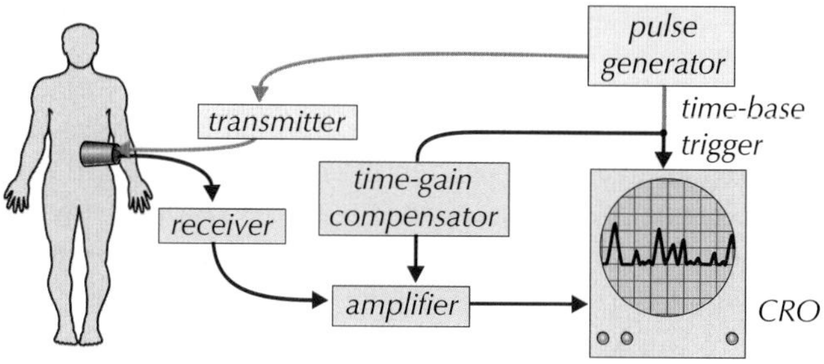

***Figure 4:** The set-up of an amplitude scan.*

The horizontal positions of the reflected pulses indicate the time the 'echo' took to return, and are used to work out distances between structures in the body (e.g. the distance between the front and back of an eyeball). Weaker pulses (that have travelled further in the body and arrive later) are amplified more than the stronger, earlier pulses to avoid the loss of valuable data — this process is called time-gain compensation (TGC). A stream of pulses can produce the appearance of steady peaks on the screen, although modern CROs can store a digital image of the peaks after just one pulse.

The B-scan

In a brightness scan (B-scan), the amplitude of the reflected pulses is displayed as the brightness of a spot on a screen. You can use a linear array of transducers (i.e. lots of transducers arranged in a line) to produce a two-dimensional image, such as in the prenatal scanning of a fetus.

Tip: Because the images in a B-scan are obtained in real time and the scan doesn't require anaesthesia, the patient can watch the images as they are formed.

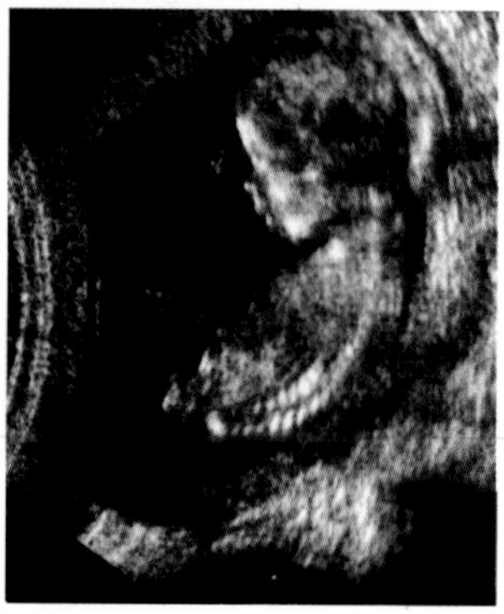

***Figure 5:** An ultrasound scan of a fetus at 13 weeks.*

The Doppler effect

Ultrasound waves reflected at an angle off moving cells undergo a change of frequency (or wavelength). This is caused by the **Doppler effect** (p.98). This change of frequency (known as the 'beat frequency') can allow doctors to find the speed at which those cells are moving (for example, blood cells in an artery). You can calculate the change in frequency as a fraction of the initial frequency using this equation:

$$\frac{\Delta f}{f} = \frac{2v\cos\theta}{c}$$

Δf = change in frequency of the ultrasound (beat frequency) in Hz

f = initial frequency of the ultrasound in Hz

v = speed of the moving cell in ms^{-1}

θ = angle between the ultrasound receiver and the line along which the cell is moving in °

c = speed of sound in the medium in ms^{-1}

Or:

$$v = \frac{c\Delta f}{2f\cos\theta}$$

Tip: The change in frequency will be positive if the blood is travelling towards the receiver, and negative if the blood is travelling away from the receiver.

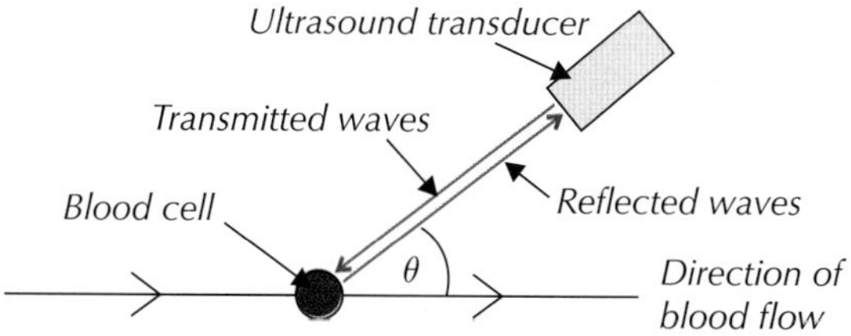

***Figure 6:** The ultrasound transducer detects ultrasound waves at an angle θ to the line along which the blood cells are moving.*

Tip: This technique is often used in conjunction with B-scan ultrasound. The B-scan provides an image of the area being studied, making it easier for the technician to target the Doppler beam correctly and adjust the angle of the receiver.

Example — **Maths Skills**

An ultrasound transducer directs longitudinal waves at a frequency of 8.5 MHz at an artery. The waves reflect off blood cells in the artery and are detected at an angle of 45° to the direction of blood flow. The change in frequency of the waves due to the Doppler effect is 3.1 kHz. If the speed of sound in the observed area is 1550 ms^{-1}, calculate the speed at which the blood cells in the artery are moving.

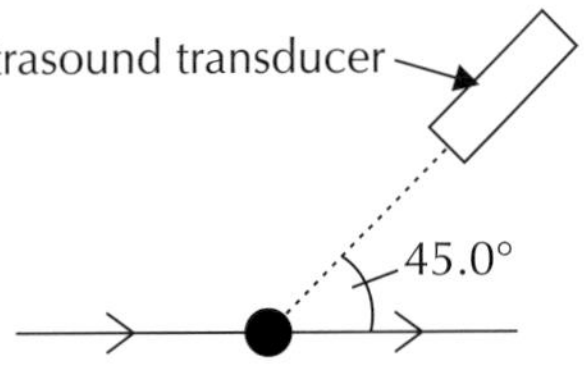

$f = 8.5$ MHz and $\Delta f = 3.1$ kHz, so:

$$v = \frac{c\Delta f}{2f\cos\theta} = \frac{1550 \times 3.1 \times 10^3}{2 \times 8.5 \times 10^6 \times \cos 45°} = 0.3997... = 0.40\ \text{ms}^{-1}\ \text{(to 2 s.f.)}$$

Practice Questions — Application

Q1 a) The density of water at 20 °C is 1000 kgm^{-3} (to 2 s.f.) and its acoustic impedance is 1.5×10^6 $kgm^{-2}s^{-1}$. What is the speed of sound, *c*, in water at this temperature?

b) A sound wave travels from bone ($Z = 8.0 \times 10^6$ $kgm^{-2}s^{-1}$) to water. What will the intensity of the reflected wave be as a percentage of the intensity of the incident wave?

Q2 An ultrasound transducer is used to direct longitudinal waves at a frequency of 12 MHz at an artery. The figure below shows the waves reflecting back to the transducer at an angle of 34° to the direction of blood flow.

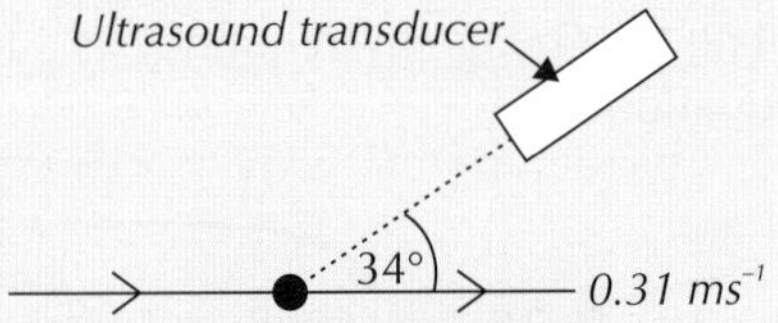

The blood cells are moving at a speed of 0.31 ms^{-1} in the artery, and the speed of sound in the area is 1550 ms^{-1}. Calculate the change in frequency of the wave you would expect to see due to the Doppler effect.

Practice Questions — Fact Recall

Q1 What is ultrasound?

Q2 Give the equation for calculating the acoustic impedance of a material. Define all the symbols used and give the units of each quantity.

Q3 What's the piezoelectric effect?

Q4 Explain how an ultrasound wave is generated and detected.

Q5 Why is a coupling medium needed to obtain a clear ultrasound image of body tissue?

Q6 a) Briefly describe what an A-scan is. Give one use of an A-scan.

b) Briefly describe what a B-scan is. Give one use of a B-scan.

Section Summary

Make sure you know...

- The structure of an X-ray tube and the function of its components, including the heater (cathode), anode (target metal) and high voltage supply.
- How X-ray photons can be produced using an X-ray tube.
- That the attenuation (absorption) coefficient, μ, of a material is a measure of how much the intensity of an X-ray beam decreases as it penetrates deeper into the material.
- That the intensity of an attenuated X-ray beam can be calculated using $I = I_0 e^{-\mu x}$.
- That X-ray attenuation mechanisms include simple scatter, the photoelectric effect, the Compton effect and pair production.
- That the high atomic numbers of barium and iodine cause them to show up very differently in X-ray images compared to body tissues, making them useful as contrast media in X-ray imaging.
- How a computerised axial tomography (CAT) scanner works, including the roles of the rotating X-ray tube, the ring of detectors, the computer software and the display.
- That CAT scans produce more detailed images than regular X-rays, especially for soft tissue.
- That the data from CAT scans can be manipulated to generate a 3D image.
- How radioactive substances such as technetium-99m and fluorine-18 can be used as medical tracers to show the structure and function of organs.
- How a gamma camera works, including the roles of the collimator, scintillator, photomultiplier tubes, computer and display, and are able to describe the way an image is formed.
- That a gamma camera can be used to help diagnose a patient without the need for surgery.
- That a positron emission tomography (PET) scanner relies on the annihilation of positron-electron pairs.
- That the positrons needed for a PET scanner are produced by a suitable radiotracer.
- That the detectors in a PET scanner detect and record the gamma rays produced by positron-electron annihilation and send the information to a computer, which builds up a map of the radioactivity in the body.
- That the distribution of radioactivity in a PET scan corresponds to the distribution of the medical tracer in the body, and knowledge of this can aid doctors when diagnosing disease.
- That ultrasound is a longitudinal wave with a frequency greater than 20 kHz.
- How to calculate the acoustic impedance of a medium using $Z = \rho c$.
- That ultrasound is partially reflected at boundaries, and how to calculate the fraction of wave intensity that is reflected using $\frac{I_r}{I_0} = \frac{(Z_2 - Z_1)^2}{(Z_2 + Z_1)^2}$.
- That the piezoelectric effect is when a potential difference is produced by piezoelectric crystals when they're deformed, or vice versa.
- That an ultrasound transducer is a device that emits and receives ultrasound.
- That soft tissue has a very different acoustic impedance from air, so almost all the ultrasound energy is reflected from the surface of the body if there is air between it and the transducer.
- The importance of impedance (acoustic) matching in ultrasound scanning, and how it can be achieved by using a special gel as a coupling medium.
- That an ultrasound amplitude scan (A-scan) is used to measure distances and diameters and an ultrasound brightness scan (B-scan) can be used to produce a two-dimensional image of a patient's body.
- That ultrasound waves reflected off moving cells at an angle to their line of movement undergo a change of frequency due to the Doppler effect, and how the speed of blood in a patient can be determined using $\frac{\Delta f}{f} = \frac{2v \cos\theta}{c}$.

Exam-style Questions

1 Calculate the fraction of wave intensity that is reflected when ultrasound waves are passing from soft tissue to muscle. The acoustic impedance of soft tissue is 1.58×10^6 kgm^{-2}s^{-1} and the acoustic impedance of muscle is 1.70×10^6 kgm^{-2}s^{-1}.

A 0.929

B 0.0731

C 0.0366

D 0.00134

(1 mark)

2 The intensity of an X-ray beam is reduced by 95% after passing through 4.5 cm of a material. Calculate the attenuation coefficient of the material.

A 0.011 cm^{-1}

B 0.36 cm^{-1}

C 0.57 cm^{-1}

D 0.67 cm^{-1}

(1 mark)

3 A doctor wants to use a non-invasive medical imaging technique to study the uptake of glucose in the lung cells of a patient. Which of the following techniques would be the most suitable?

A CAT scan

B PET scan

C Ultrasound scan

D X-ray imaging

(1 mark)

4 Gamma cameras can be used to help diagnose patients without the need for surgery.

(a) Describe how gamma rays incident on a gamma camera are converted into pulses of electricity.

(3 marks)

(b)* White blood cells are produced by the body to target and fight infections. Technetium-99m is one radioactive isotope that can be used to label white blood cells.

Use this information to discuss how a medical tracer could be used to identify areas of infection in the body. Explain why technetium-99m is suitable for use in medical tracers.

(6 marks)

* The quality of your response will be assessed in this question.

5 An X-ray tube can be used to generate X-rays for medical purposes.

(a) The power supply in an X-ray tube provides a potential difference of 31 kV across the circuit. Calculate the minimum wavelength of the X-rays produced by the X-ray tube.

(3 marks)

(b) As the X-rays pass through body tissues, they are attenuated. Besides simple scattering, suggest one attenuation mechanism that may cause this.

(1 mark)

(c) Iodine is often injected into the blood of a patient before X-rays of their blood vessels are taken. Explain why this is done.

(2 marks)

6 Ultrasound transducers can be used to image soft tissue.

(a) Explain how an A-scan can be used to measure the depth of an eyeball.

(3 marks)

(b) At 15 °C at sea level, the density of air is 1.23 kgm^{-3} and the speed of sound in air is 340 ms^{-1}. Calculate the acoustic impedance of air at this temperature and height.

(1 mark)

(c) Explain why it is difficult to use ultrasound to produce images from behind the lungs.

(2 marks)

An ultrasound operator wants to measure the speed of blood in an artery. He uses a transducer to direct pulses with a frequency of 8.1 MHz at the artery. The transducer is held at different angles to the direction of blood flow, and these angles are measured using B-scan ultrasound. The difference in frequency between the incident wave and the reflected wave is recorded for each angle. **Fig 6.1** shows the change in frequency plotted against the cosine of the angle between the receiver and the direction of blood flow. A line of best fit has been drawn.

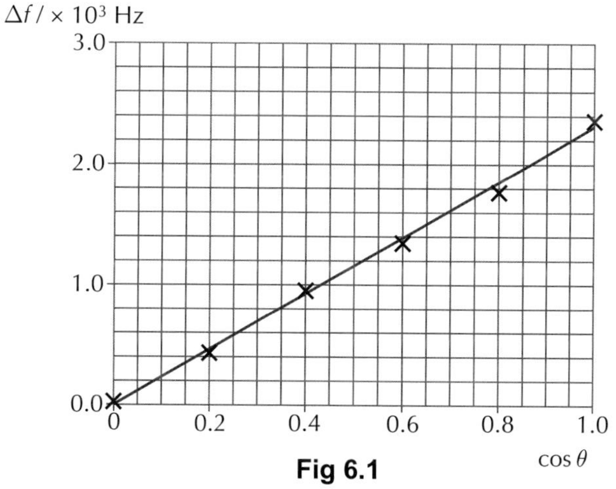

Fig 6.1

(d) Show that the gradient of the line of best fit in **Fig 6.1** is equal to $\frac{2vf}{c}$.

(1 mark)

(e) Assuming the speed of sound is 1550 ms^{-1}, calculate the speed of the blood cells.

(3 marks)

Maths Skills

1. Calculations

At least 40% of your exam marks will involve maths skills of some sort. You should be familiar with a lot of these skills from year 1 of A-level. This section covers the extra stuff you need to know on exponentials and logarithms, as well as briefly recapping some of the maths skills you'll have used before.

Tip: If you don't write a number in standard form, it's known as decimal form — e.g. 0.00012 or 34 500.

Standard form

You should be pretty familiar with using standard form by now, but don't forget that standard form must always look like this:

$$A \times 10^n$$

This number must always be between 1 and 10. (A)

'n' is the number of places the decimal point moves.

Make sure you know what the standard form button looks like on the calculator that you'll use in the exam. It'll probably say either 'Exp', 'EE' or '× 10ˣ' — see Figure 1.

Figure 1: *The 'Exp' or '×10ˣ' button is used to input standard form on calculators. Don't confuse it with the power buttons x^2 or $x^{■}$, which are used for raising numbers to a power.*

Orders of magnitude and logarithms

Orders of magnitude can be used to roughly compare the size of two (or more) numbers. If two numbers are roughly the same size, they will have the same order of magnitude. Usually, the order of magnitude of a number is the power that 10 would be raised to if the number were written in standard form. They are mostly used for approximate comparison.

Example — Maths Skills

- $12\,340 = 1.234 \times 10^4$, so its order of magnitude is 4.
- $0.000321 = 3.21 \times 10^{-4}$, so its order of magnitude is –4.
- So 12 340 is 8 orders of magnitude (10^8 times) larger than 0.000321.

Scales that use orders of magnitude are known as logarithmic. They allow data that ranges over many orders of magnitude to be displayed and interpreted easily, because instead of going up in equal amounts, each step on the scale is 'so many times bigger' than the last (e.g. 10 times bigger).

A logarithm base 10 ($\log_{10}$ or lg) of a number is defined as the power to which ten must be raised in order to get that number:

$$\log_{10}x = y \quad \text{means} \quad 10^y = x$$

You can have bases other than 10. The other common base in physics is the constant 'e'. e is equal to 2.71828... — it's stored in your calculator (see Figure 2). $\log_e$ is known as the natural logarithm, or 'ln'.

$$\ln x = y \quad \text{means} \quad e^y = x$$

Figure 2: *There are buttons on your calculator for log (log base 10) and ln. If you press shift or 'second function' first, you'll get the $10^{■}$ or $e^{■}$ functions, the inverse functions of log and ln. There's also a button for the constant e.*

There are a few log rules that work for both log and ln that you need to know:

$\log AB = \log A + \log B$ $\log\frac{A}{B} = \log A - \log B$ $\log x^n = n\log x$

Tip: These rules will be given in the data and formulae booklet in the exams.

Example — Maths Skills

Show that if $\ln y = \ln(ke^{-ax})$, the graph of $\ln y$ against x is a straight line.

If it is a straight line, it can be expressed in the form $\ln y = mx + c$.

$\ln y = \ln(ke^{-ax}) = \ln k + \ln e^{-ax} = \ln k - a\ln e^x = \ln k - ax$

$= -ax + \ln k$

So plotting $\ln y$ against x gives a straight line with gradient $-a$ and intercept $\ln k$.

Tip: ln and e are inverse functions, so $\ln e^x = e^{\ln x} = x$.

Tip: Remember, $e^{-ax} = (e^x)^{-a}$.

Significant figures

You should remember that you always give your answer to the lowest number of significant figures (s.f.) used in the calculation. Unfortunately it's not that simple when you're working with logarithms and raising to powers.

If you're taking a logarithm, the value of the result is accurate to the same number of decimal places as there are significant figures in the number you're taking a logarithm of.

If you're raising to a power, the value of the result is accurate to the same number of significant figures as there are decimal places in the power.

Tip: You should never round before you get to your final answer. It will introduce rounding errors and you may end up with the wrong answer.

Examples — Maths Skills

The number of undecayed nuclei in a sample of a radioactive isotope is given by $N = N_0e^{-\lambda t}$, where N_0 is the original number of undecayed nuclei, t is the time elapsed and λ is the decay constant.

Calculate N after exactly 2 minutes in a sample of an isotope with decay constant 3.40×10^{-2} s^{-1} that originally had 4.28×10^{24} undecayed nuclei.

$N = N_0e^{-\lambda t}$

$= 4.28 \times 10^{24} \times e^{-(3.40 \times 10^{-2}) \times (2 \times 60)}$

$= 4.28 \times 10^{24} \times e^{-4.08}$

$= 7.236... \times 10^{22}$

4.08 has 2 decimal places, so the $e^{-\lambda t}$ bit can be given to 2 s.f. and N_0 is given to 3 s.f.
The least number of s.f. is 2, so the answer is $N = 7.2 \times 10^{22}$ (to 2 s.f.)

Calculate the time taken for the number of undecayed nuclei to drop from 4.28×10^{24} to 2.14×10^{23}.

Rearrange the equation for t:

$$t = \frac{\ln(N_0) - \ln(N)}{\lambda} = \frac{\ln(4.28 \times 10^{24}) - \ln(2.14 \times 10^{23})}{3.40 \times 10^{-2}} = \frac{56.7159... - 53.7202...}{3.40 \times 10^{-2}}$$

$= 88.10977...$ s

4.28×10^{24} and 2.14×10^{23} are both given to 3 s.f. So the logarithms are accurate to 3 decimal places, which is equivalent to 5 s.f. for these numbers (i.e. 56.716 and 53.702). λ is given to 3 s.f., which is the least number of s.f., so the answer should be given to 3 s.f. too. So $t = 88.1$ s (to 3 s.f.)

Exam Tip
In your exam, you might lose a mark on an otherwise correct answer if it's not given to an appropriate number of significant figures. Make sure you always round your answers to the correct number of significant figures.

2. Algebra

Physics involves a lot of rearranging formulas and substituting values into equations. Easy stuff, but it's also easy to make simple mistakes.

Algebra symbols

Here's a quick recap of some of the symbols that you will come across:

Symbol(s)	Meaning
$=, \approx$	equal to, roughly equal to
$<, \leq, <<$	less than, less than or equal to, much less than
$>, \geq, >>$	greater than, greater than or equal to, much greater than
$\propto$	proportional to
Δ	change in (a quantity)
Σ	sum of

Tip: An example of using $\propto$ can be found on page 64.

Tip: Δ is the Greek capital letter 'delta'. An example of using Δ can be found on page 17.

Figure 1: *It can be easy to make a mistake rearranging equations when you're stressed in an exam. It's a good idea to double check rearrangements, especially if it's a tricky one where you've had to combine and rearrange equations.*

Rearranging equations

When rearranging equations, remember the golden rule — whatever you do to one side of the equation, you must do to the other side of the equation.

Example — Maths Skills

The equation for the charge on a discharging capacitor is $Q = Q_0 e^{-\frac{t}{CR}}$. Rearrange the equation to make t the subject.

$$Q = Q_0 e^{-\frac{t}{CR}}$$

Divide by Q_0

$$\frac{Q}{Q_0} = e^{-\frac{t}{CR}}$$

Take ln of both sides

$$\ln\left(\frac{Q}{Q_0}\right) = -\frac{t}{CR}$$

Multiply by $-CR$

$$t = -CR\ln\left(\frac{Q}{Q_0}\right)$$

Substituting into equations

Make sure you avoid common mistakes by putting values in the correct units for the equation and putting numbers in standard form before you substitute.

Tip: See page 187 for more on annihilation.

Tip: The values of h and c are given in your data and formulae booklet.

Tip: Converting all the values into the correct units <u>before</u> putting them into the equation helps you avoid silly mistakes.

Example — Maths Skills

Two protons collide and annihilate to produce two photons. The energy of one of the photons is 941.065 MeV.
Using the equation $E = \frac{hc}{\lambda}$, find the wavelength of one of the photons.

h = the Planck constant = 6.63×10^{-34} Js, c = speed of light = 3.00×10^{8} ms^{-1} and E = energy of photon = 941.065 MeV

h and c are in the right units but E is not. E must first be converted from MeV to eV and then from eV to J:

$$941.065 \times (1 \times 10^{6}) \times 1.60 \times 10^{-19} = 1.505... \times 10^{-10}\ \text{J}$$

To find the wavelength of the photon, first rearrange the equation to make λ the subject and then substitute in the correct values:

$$E = \frac{hc}{\lambda} \Rightarrow \lambda = \frac{hc}{E} = \frac{6.63 \times 10^{-34} \times 3.00 \times 10^{8}}{1.505... \times 10^{-10}} = 1.32 \times 10^{-15}\ \text{m (to 3 s.f.)}$$

3. Graphs

You can get a lot of information from a graph — you'll need to know what the area under a graph and the gradient represent, and be able to sketch and recognise simple graphs, given an equation.

Area under a graph

Many quantities in A-level physics can be found from the area between a curve or line and the horizontal axis of a graph. Here are a few examples from year 2:

- The work done to move an object from one point to another in a radial gravitational field is the area under a graph of gravitational force against radial distance (see p.70).
- Energy stored by a capacitor is the area under a graph of V against Q (p.113-114).

To find an area under a graph, you'll either need to work it out exactly or estimate the area — it depends on the graph's shape. You'll need to estimate the area if the graph is not made up of straight lines.

Tip: You'll need to remember all the examples you've seen in this book.

Example — Maths Skills

A capacitor is charged and the p.d.-charge graph is shown below. Use the graph to find the energy stored by the capacitor when the p.d. across it reaches 12 V.

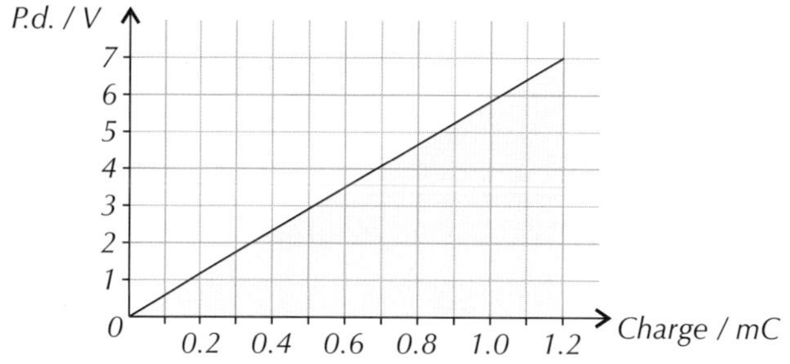

Energy stored (work done) = area under graph. The area is a triangle, so the energy stored $= \frac{1}{2} \times \text{base} \times \text{height} = \frac{1}{2} \times (1.2 \times 10^{-3}) \times 7$

$= 4.2 \times 10^{-3}$ J

Tip: Remember, you can end up with 'negative' areas (under the horizontal axis) if the quantity on the vertical axis is a vector. Just subtract their magnitude from the positive areas to work out the quantity represented by the total area.

Tip: The charge is given in mC, but you need it in C to be able to calculate the energy. So the change in charge is 0.7×10^{-3} C.

Example — Maths Skills

The graph shows how the gravitational force on an object changes with distance from a point mass. Find the work done to move an object from 1×10^5 m to 4×10^5 m away from the mass.

For a curved graph, you can estimate the area by counting the number of squares under the graph, which is approximately 8.

The area of one square $= 5 \text{ N} \times 0.5 \times 10^5 \text{ m} = 2.5 \times 10^5$ J.

So the work done $\approx 2.5 \times 10^5 \times 8$
$= 2 \times 10^6$ J

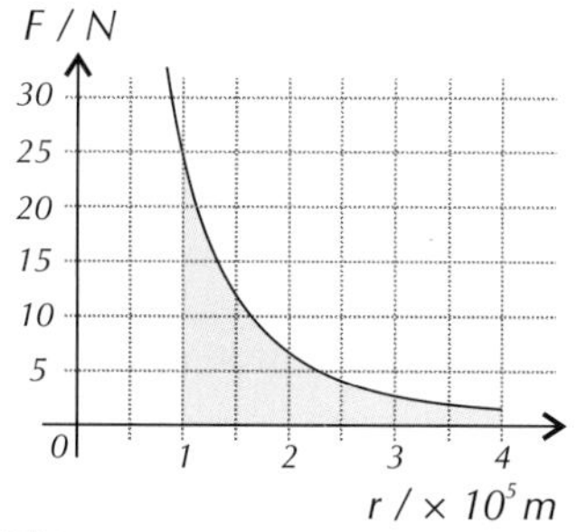

Tip: There's more on gravitational forces on pages 62-72.

Tip: Draw a dot or a cross inside every square you count, to help you keep track of them.

Rates of change

A graph is a plot of how one variable changes with another. The rate of change of the variable on the vertical axis with respect to the variable on the horizontal axis at any point is given by the gradient of the graph.

$$\text{Rate of change of } y \text{ with } x = \frac{\Delta y}{\Delta x} = \text{gradient of a } y\text{-}x \text{ graph.}$$

Often, the gradient represents a useful rate of change that you want to work out. For example, for a gravitational field created by a point mass, the gravitational field strength, $\mathbf{g}$, is given by minus the gradient of the graph of gravitational potential, V_g, against distance from the mass, r. That's because $\mathbf{g} = -\frac{\Delta V_g}{\Delta r}$ (see page 68).

If the graph is curved, you can find the instantaneous rate of change (at a point) by drawing a tangent to the curve at that point and finding the gradient of the tangent.

Tip: You can also find the average rate of change by calculating the total change in y over the total change in x.

Tip: When drawing a tangent, it helps to make it long — it will be easier to draw, and the tangent line will be more likely to intersect some grid lines, making the gradient easier to calculate. Ideally, your tangent should be longer than half the length of the line of the graph.

Example — Maths Skills

The graph below shows how the gravitational potential of the Moon's gravitational field varies with the distance from its surface.

Find the magnitude of the gravitational field strength of the Moon at a distance of 1500 km from its surface.

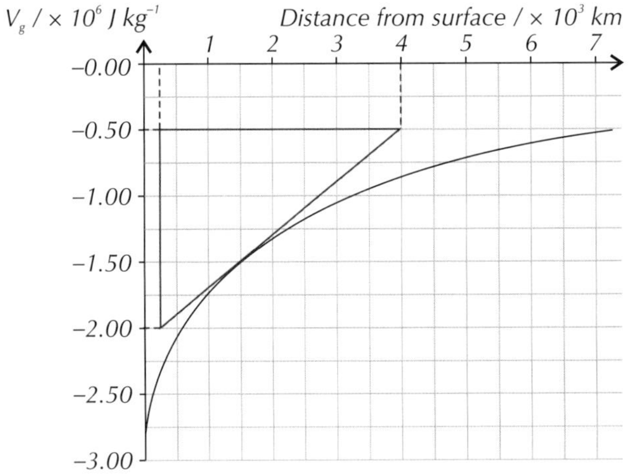

A tangent to the curve at 1.5×10^3 km is drawn on the graph. Its gradient is:

$$\frac{\Delta y}{\Delta x} = \frac{(-0.5 \times 10^6) - (-2.00 \times 10^6)}{(4 \times 10^3 \times 10^3) - (0.25 \times 10^3 \times 10^3)}$$

$$= \frac{1.50 \times 10^6}{3.75 \times 10^6}$$

$$= 0.4$$

So the gravitational field strength at 1500 km = -0.4 N kg^{-1}

Figure 1: Make sure you use a really sharp pencil and a ruler whenever you're drawing graphs and tangents.

Rate of change of a gradient

The gradient is already a 'rate of change of something', so the rate of change of a gradient is the 'rate of change of the rate of change'. Sometimes these represent useful quantities too.

A common example of this is on a displacement-time graph. The rate of change of its gradient is equal to the acceleration.

Example — **Maths Skills**

The graphs below show an object in simple harmonic motion (see pages 45-48). The rate of change of the displacement-time graph gives the velocity, and the rate of change of velocity gives the acceleration.

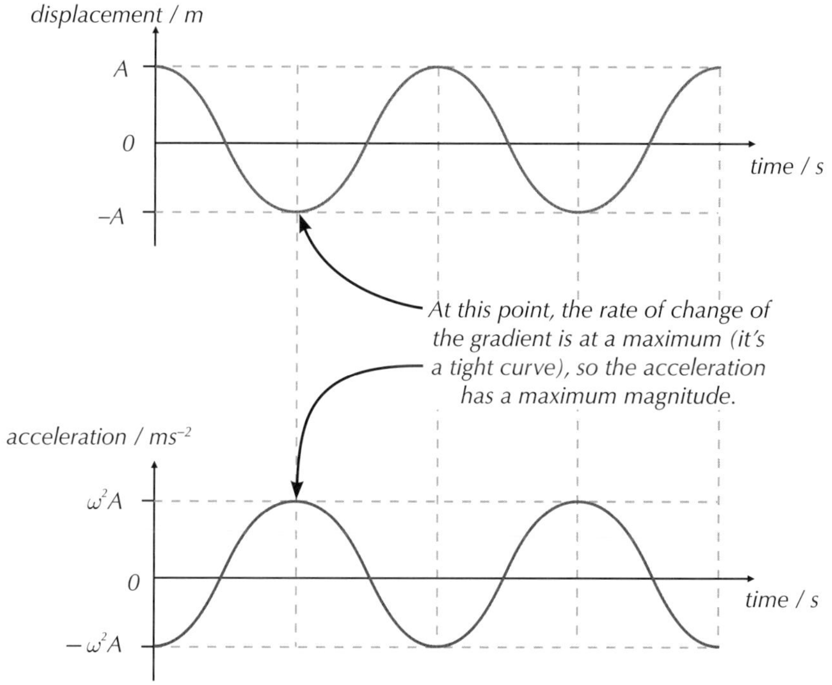

Tip: The maximum rate of change of gradient is where the curve is most tightly curved.

Tip: A is the maximum value of the displacement of an object oscillating with simple harmonic motion. ω is the angular frequency of the object. For more, see page 49.

Modelling rates of change

If you know the rate of change of something, you can use an iterative spreadsheet to model how the quantity itself changes over time.
For example:

- You can model how the charge on a discharging capacitor changes over time using the equation $\frac{\Delta Q}{\Delta t} = -\frac{Q}{CR}$ (p.123).
- You can model how the number of undecayed nuclei in a radioactive sample changes over time using the equation $\frac{\Delta N}{\Delta t} = -\lambda N$ (p.199).

Make sure you're familiar with your spreadsheet program before you start trying to model anything. You'll need to know how to reference and do calculations involving cells.

Sketching graphs

There are some graph shapes that crop up in physics all the time. The following graphs are examples you need to know how to recognise and sketch. k is constant in all cases.

Tip: There are loads of examples of $y = kx$ and $y = \frac{k}{x}$ in physics. For example, $Q = CV$ for a capacitor with fixed capacitance on page 113.

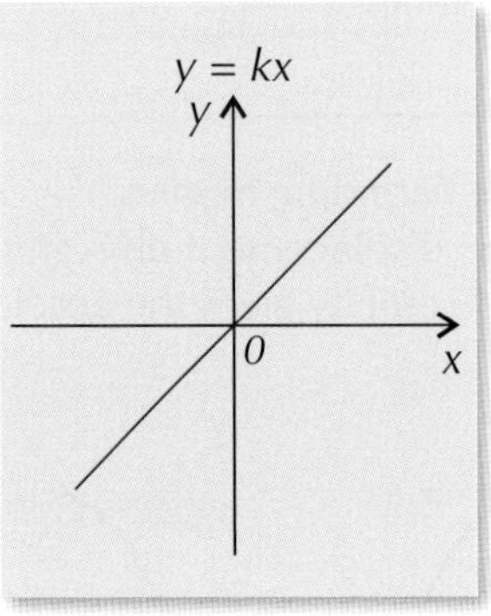

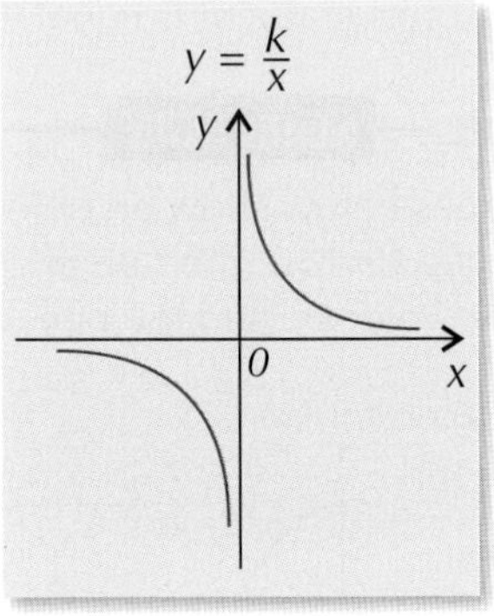

Tip: Make sure you can recognise a graph of e^{kx} too (see page 243).

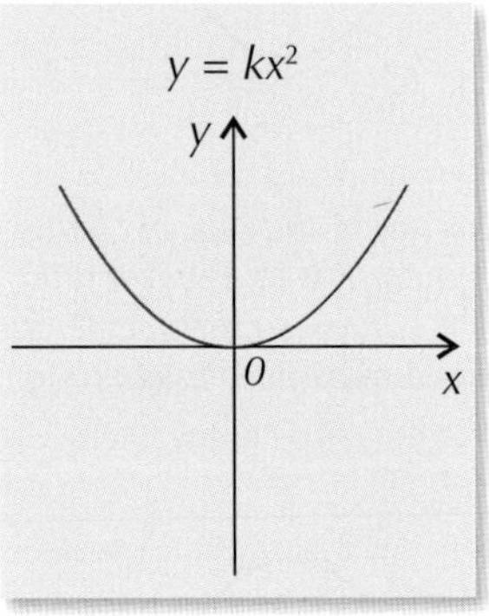

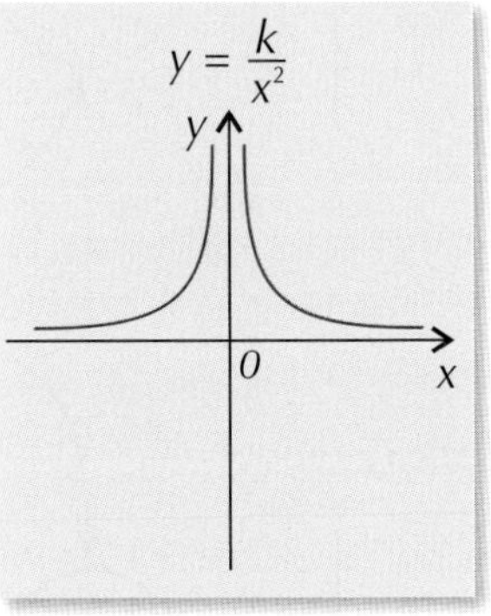

Tip: The *x-t*, *v-t* and *a-t* graphs for simple harmonic motion on page 46 are sin and cos graphs, e.g.: $x = A\cos(\omega t)$. The velocity and acceleration graphs can both be calculated from the displacement graph.

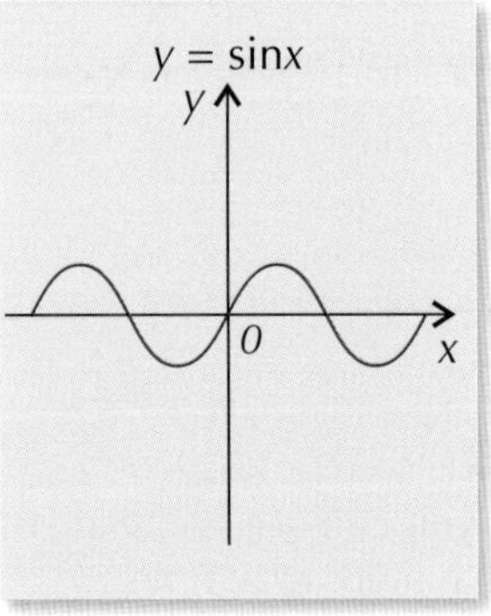

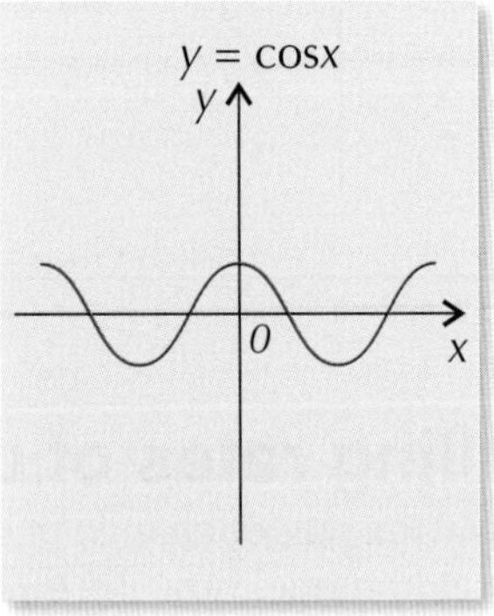

Tip: The graph for kinetic energy of an object in simple harmonic motion against time on p.48 has the same shape as the $y = \cos^2 x$ graph.

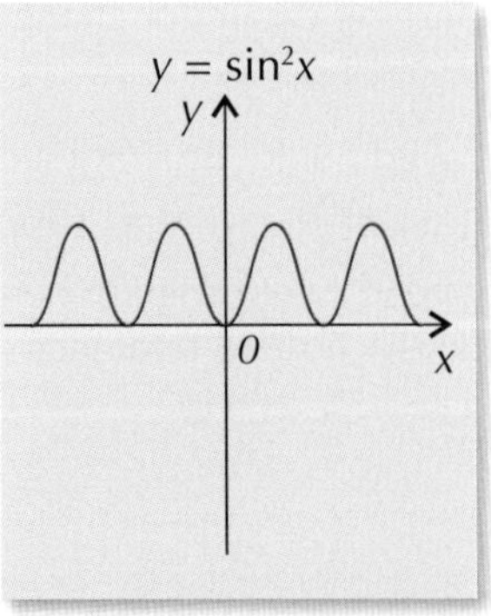

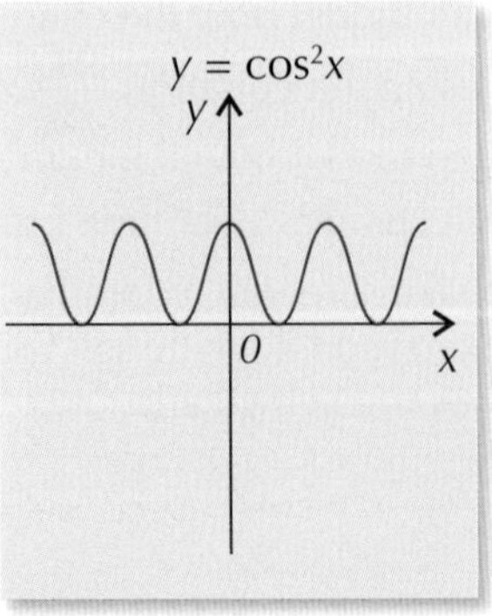

4. Exponential and Log Graphs

There are quite a few exponential relationships in year 2 of A-level, and you need to know what they'll look like as a graph. You also need to know how you can use logarithms to plot a seemingly complicated relationship as a nice straight line. Read on...

Exponential graphs

A fair few of the relationships you need to know about in year 2 of A-level physics are exponential — where the rate of change of a quantity is proportional to the amount of the quantity left. Here are a few that crop up in the A-level course (if they don't ring a bell, go have a quick read about them)...

- Charge on a capacitor (p.125) — the rate of decay of charge on a discharging capacitor is proportional to the amount of charge left on the capacitor:

$$Q = Q_0 e^{\frac{-t}{CR}}$$

There are also exponential relationships for I and V, and for charging capacitors (see pages 125-126).

- Radioactive decay (p.200) — the rate of decay of a radioactive sample is proportional to the number of undecayed nuclei in the sample:

$$N = N_0 e^{-\lambda t}$$

The activity of a radioactive sample behaves in the same way.

- X-ray attenuation (p.223) — the reduction of intensity of X-rays over distance is proportional to their intensity:

$$I = I_0 e^{-\mu x}$$

Because the rate of change is proportional to the amount, exponential growth gets faster and faster as the amount gets bigger. Exponential decay is just the opposite of exponential growth — it gets slower and slower as the amount gets smaller. Because of this, exponential graphs have a characteristic shape that you need to be able to recognise and sketch:

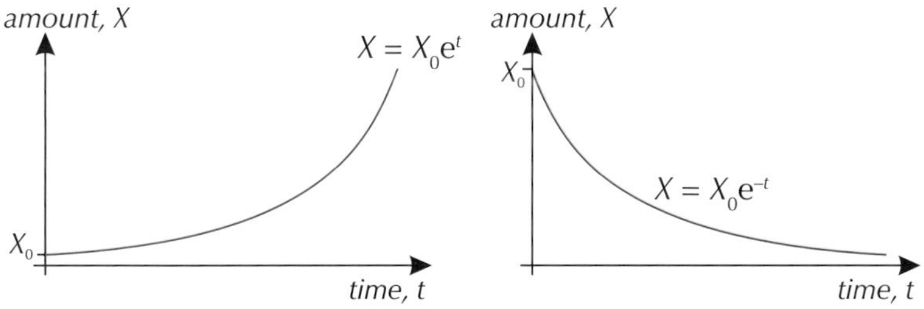

Figure 2: *(Left) The characteristic graph of exponential growth. The rate of growth gets faster and faster. (Right) The characteristic graph of exponential decay. The rate of decay gets slower and slower, never reaching zero.*

Tip: Don't forget, e is a constant equal to 2.718... — it'll be stored in your calculator. You can raise e to a power using the button e■ or e^x (see page 236).

Tip: These equations are all examples of exponential decay — see Figure 2.

Figure 1: *The growth of bacteria is a classic example of something that can be modelled with exponential growth. The more bacteria there are, the faster the rate of growth becomes.*

Tip: You should recognise the decay graph from the sections on capacitors (p.122) and nuclear decay (p.202).

> **Tip:** Remember the log rules from p.237 — you'll need them here.

Log-linear graphs

You can plot an exponential relationship as a straight line using the natural log, ln. For the relationship $y = ke^{-ax}$, if you take the natural log of both sides of the equation you get:

$$\ln y = \ln (ke^{-ax}) = \ln k + \ln (e^{-ax}) \quad \text{so} \quad \ln y = \ln k - ax$$

Then all you need to do is plot (ln y) against x. You get a straight-line graph with (ln k) as the vertical intercept, and $-a$ as the gradient.

> **Tip:** Graphs with the natural logarithm of a function on one of the axes are called log-linear graphs. There are examples of how they can be used on pages 128 and 200.

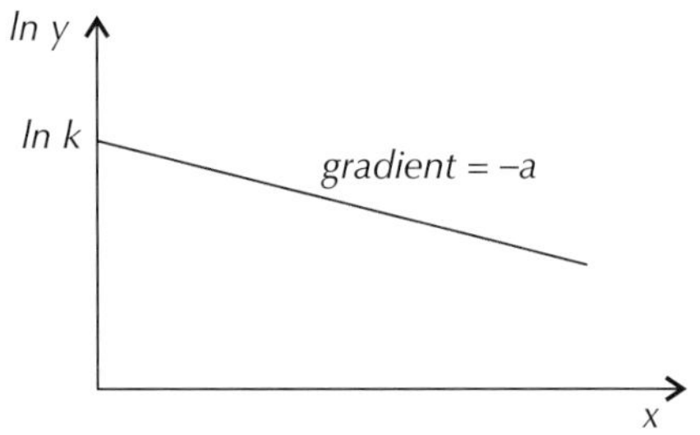

Figure 3: *The function $y = ke^{-ax}$ plotted on a log-linear graph of ln y against x.*

Log-log graphs

You can use logs to plot a straight-line graph for other relationships too. Say the relationship between two variables x and y is:

$$y = kx^n$$

> **Tip:** This relationship is called a power law relationship.

Take the log (base 10) of both sides to get:

$$\log y = \log k + n \log x$$

So, if you plot log y against log x, log k will be the y-intercept and n will be the gradient of the graph.

Example — Maths Skills

A physicist carries out an experiment to determine the nuclear radius, R (in m), of various elements with various nucleon numbers, A. She plots a line of best fit for her results on a graph of log R against log A. Part of the graph is shown. Using the equation $R = r_0 A^{1/3}$, find the value of the constant r_0 from the graph.

> **Tip:** Look back at pages 180-181 for more on nuclear radius.

First take logs of both sides:

$$\log R = \log (r_0 A^{1/3}) = \log r_0 + \log A^{1/3} = \log r_0 + \frac{1}{3}\log A$$

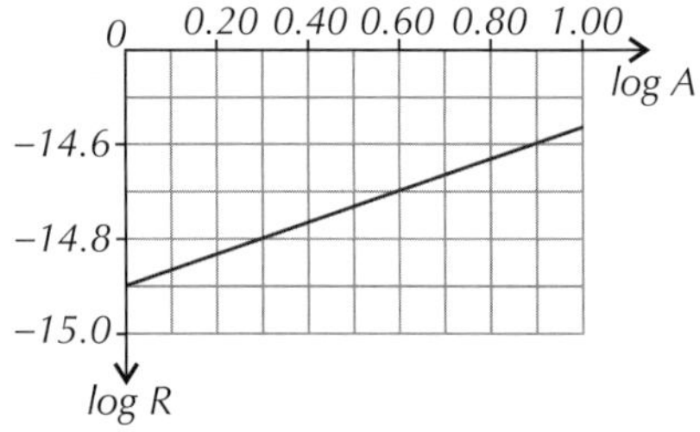

Comparing this to the equation of a straight line (in the form $y = mx + c$), you can see that the gradient of the graph is $\frac{1}{3}$ and the vertical intercept is log r_0.

So, reading from the graph, the vertical intercept is about –14.9.

$\log r_0 = -14.9$, so $r_0 = 10^{-14.9} = 1.258... \times 10^{-15} = 1.3$ fm (to 2 s.f.)

> **Tip:** If $a = \log_{10} b$, then $b = 10^a$.

5. Geometry and Trigonometry

You'll often find that you need to deal with different 2D and 3D shapes in physics. You should have used most of this in year 1, but here's a recap.

Geometry basics

Angle rules

These angle rules should be familiar from year 1 — make sure you know them.

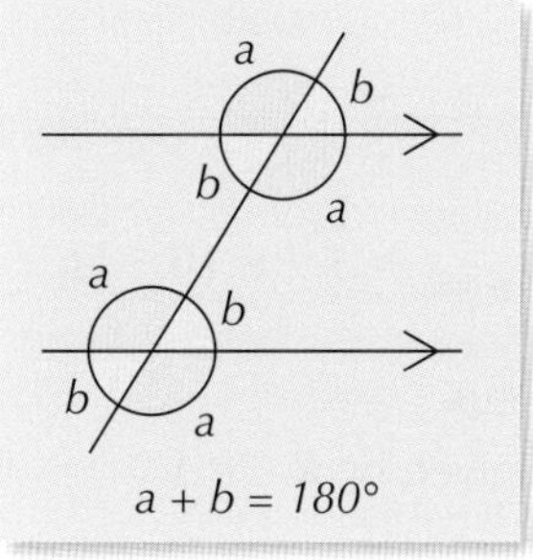

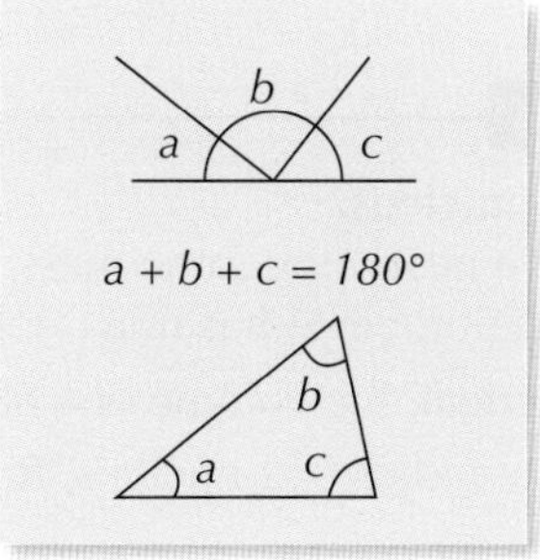

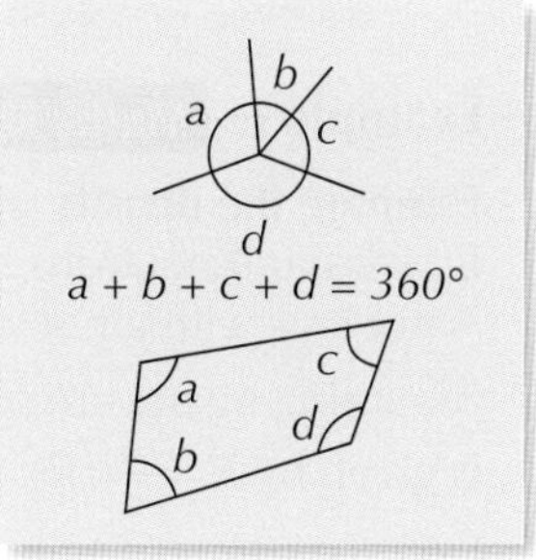

Tip: Remember, the arrows on the lines in the diagram mean that they're parallel.

Angles can be measured in degrees or radians — make sure you know how to convert between them:

- To convert from degrees to radians, multiply by $\frac{\pi}{180°}$.
- To convert from radians to degrees, multiply by $\frac{180°}{\pi}$.

Tip: You'll need to use radians when working with circular motion and simple harmonic motion. Don't forget to put your calculator into either degrees or radians.

Circumference and arc length

You will need to calculate the distance around the edge of a circle (or part of it).

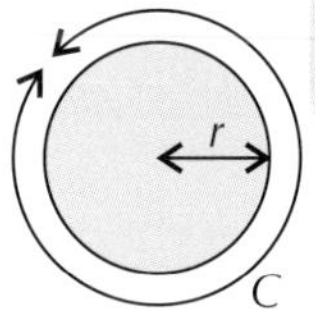

Circumference, $C = 2\pi r$

Arc length, $l = r\theta$ (θ in radians)

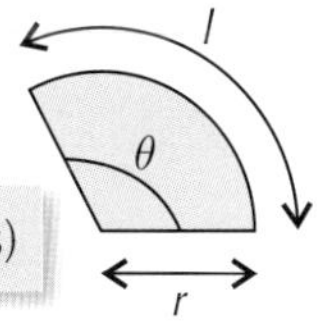

Exam Tip
You'll be given these in the data and formulae booklet. Just remember that θ is in radians.

Areas of shapes

Make sure you remember how to calculate the areas of these shapes:

Triangle

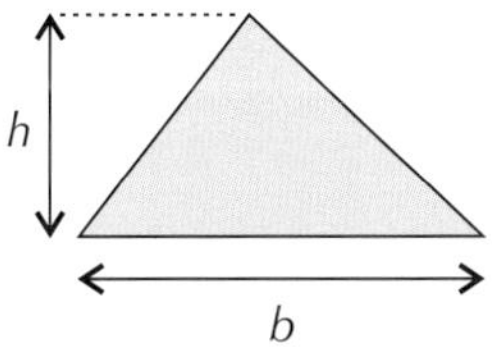

$A = \frac{1}{2} \times b \times h$

Circle

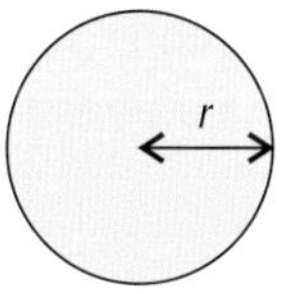

$A = \pi \times r^2$

Rectangle

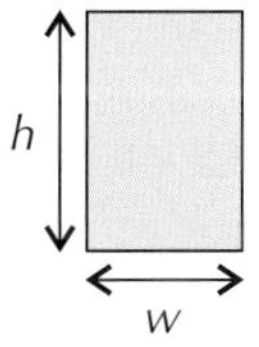

$A = h \times w$

Trapezium

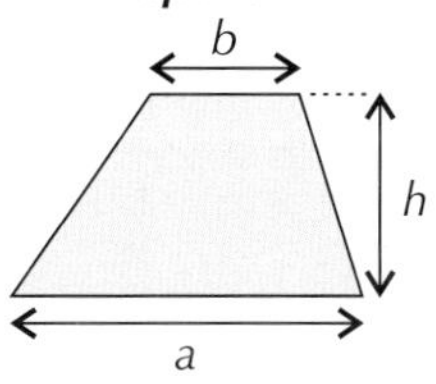

$A = \frac{1}{2} \times (a + b) \times h$

Exam Tip
You'll be given the formulae for the area of a circle and the area of a trapezium in the data and formulae booklet.

Surface areas

If you need to work out the surface area of a 3D shape, you just need to add up the areas of all the faces of the shape. A trickier shape to deal with is a sphere, where the surface area is given by $A = 4\pi r^2$ — this will be given to you in the data and formulae booklet.

Exam Tip
The curved surface area of a cylinder is given in the data and formulae booklet. This is the surface area of a cylinder with open ends. For a closed cylinder, you'll need to add the area of the circular ends ($2 \times \pi r^2$).

Volumes of shapes

Make sure you remember how to calculate the volumes of a cuboid, a sphere and a cylinder:

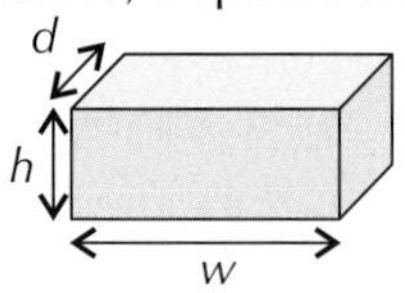

d
h
w

$V = w \times h \times d$

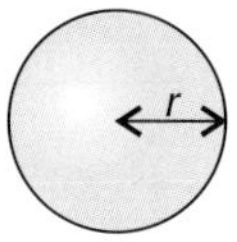

r

$V = \frac{4}{3}\pi r^3$

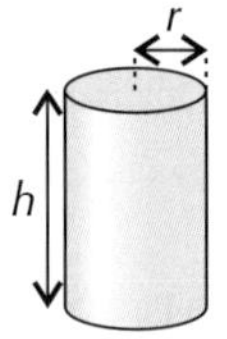

r
h

$V = \pi r^2 h$

> **Exam Tip**
> You'll be given the volume of a sphere and a cylinder in the data and formulae booklet — hoorah.

> **Tip:** r_0 is a constant, roughly equal to 1.4 fm — see page 181.

Example — Maths Skills

Estimate the density of a nucleus.

Protons and neutrons have nearly the same mass, $m_{nucleon} \approx 1.7 \times 10^{-27}$ kg.
A nucleus with a nucleon number A has mass $A \times m_{nucleon}$.

Assuming nuclei are spherical, the volume of a nucleus is given by $V = \frac{4}{3}\pi R^3$. The radius of a nucleus, R, is given by $R = r_0 A^{1/3}$

Density = mass ÷ volume, so:

$$\rho = \frac{\text{mass}}{\text{volume}} = \frac{A \times m_{nucleon}}{\frac{4}{3}\pi R^3} = \frac{A \times m_{nucleon}}{\frac{4}{3}\pi (r_0 A^{1/3})^3} = \frac{A \times m_{nucleon}}{\frac{4}{3}\pi r_0^3 A} = \frac{3m_{nucleon}}{4\pi r_0^3}$$

$$= \frac{3 \times 1.7 \times 10^{-27}}{4\pi \times (1.4 \times 10^{-15})^3} = 1.47... \times 10^{17} = 1.5 \times 10^{17}\ \text{kg m}^{-3}\ \text{(to 2 s.f.)}$$

Trigonometry basics

Right-angled triangles

For right-angled triangles, remember Pythagoras' theorem and SOH CAH TOA.

> **Tip:** You can find the buttons for sine (sin), cosine (cos) and tangent (tan) on your calculator, as well as their inverse functions ($\sin^{-1}$, $\cos^{-1}$ and $\tan^{-1}$).

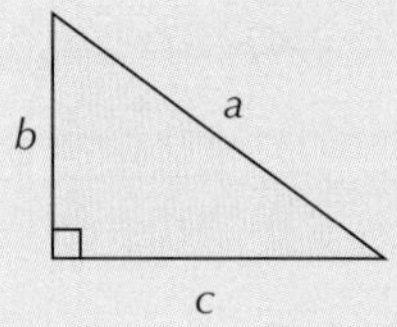

a
b
c

$a^2 = b^2 + c^2$

$$\sin\theta = \frac{\text{opposite}}{\text{hypotenuse}}$$

$$\cos\theta = \frac{\text{adjacent}}{\text{hypotenuse}}$$

$$\tan\theta = \frac{\text{opposite}}{\text{adjacent}}$$

Small-angle approximations

For really small angles in radians, you can make the following assumptions for the values of sin, cos and tan (see page 82 and 93):

$\sin\theta \approx \theta \qquad \tan\theta \approx \theta \qquad \cos\theta \approx 1$

> **Exam Tip**
> You'll be given these small-angle approximations in the data and formulae booklet.

Sine and cosine rules

If a triangle is not right angled, you need to use the sine and cosine rules to work out angles and side lengths.

The sine rule:

$$\frac{a}{\sin A} = \frac{b}{\sin B} = \frac{c}{\sin C}$$

The cosine rule:

$$a^2 = b^2 + c^2 - 2bc\cos A$$

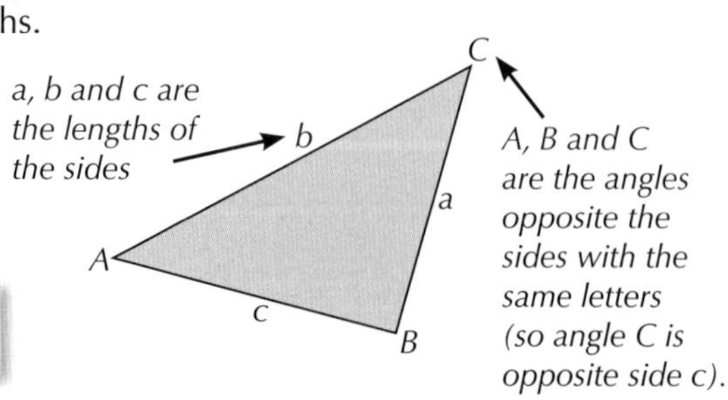

a, b and c are the lengths of the sides
A, B and C are the angles opposite the sides with the same letters (so angle C is opposite side c).
A
B
C
a
b
c

> **Tip:** You can use any two bits of the sine rule to make a normal equation with just one = sign. The sine rule also works if you flip all the fractions upside down: $\frac{\sin A}{a} = \frac{\sin B}{b} = \frac{\sin C}{c}$.

Exam Structure and Technique

Passing exams isn't all about revision — it really helps if you know how the exam is structured and have got your exam technique nailed so that you pick up every mark you can.

Figure 1: *The Room of Doom awaits you. But don't panic — prepare properly and there's no reason you can't ace the exam.*

Course structure

OCR A A-Level Physics is split into six modules:

Module 1 — Development of practical skills in physics

Module 2 — Foundations of physics

Module 3 — Forces and motion

Module 4 — Electrons, waves and photons

Module 5 — Newtonian world and astrophysics

Module 6 — Particles and medical physics

The first four Modules were covered in the Year 1 book. The final two are covered in this book.

Exam structure

For OCR A A-Level Physics you're gonna have to sit through three exams at the end of your course:

- Paper 1 — Modelling Physics — 37% of the A Level
- Paper 2 — Exploring Physics — 37% of the A Level
- Paper 3 — Unified Physics — 26% of the A Level

Papers 1 and 2 are both 2 hours and 15 minutes long. Paper 3 is 1 hour and 30 minutes long. The exams will cover material from both years of your A-Level course, so you need to learn everything from Year 1 as well as Year 2. Papers 1 and 2 are worth 100 marks each, and Paper 3 is worth 70 marks.

- Paper 1 covers material from Modules 1, 2, 3 and 5. It's split up into two sections. Section A has 15 marks' worth of multiple choice questions. Section B has 85 marks' worth of short and extended answer questions.
- Paper 2 covers material from Modules 1, 2, 4 and 6. It's split up into two sections. Section A has 15 marks' worth of multiple choice questions. Section B has 85 marks' worth of short and extended answer questions.
- Paper 3 covers content from all six modules. It has 70 marks' worth of short and extended answer questions.

Exam Tip
Make sure you have a good read through this exam structure. It might not seem important now but you don't want to get any nasty surprises at the start of an exam.

Exam Tip
It can be easy to lose track of time in long exams. Make sure to ration your time between questions (p.249) and check the clock as you go to make sure you're not surprised at the end of the exam.

Command words

It sounds obvious, but it's really important you read each question carefully, and give an answer that fits. Look for command words in the question — they'll give you an idea of the kind of answer you should write.

Exam Tip
Make sure you read the whole question carefully. Sometimes they specify the details (or the level of detail) they want you to include in your answer. For example, they may ask something like "Describe how the apparatus can be used to accurately and safely determine the *I-V* characteristic of a filament lamp." Make sure you talk about accuracy and safety to get all of the marks.

Some command words, like calculate, draw and complete are pretty self-explanatory. But command words for written questions can be a bit trickier. Common command words for these questions are:

- State — give a definition, example or fact.
- Identify — pick out information from data provided in the question, or say what something is.
- Describe — don't waste time explaining why a process happens — that's not what the question is after. It just wants to know what happens.
- Explain — give reasons for why something happens, not just a description.
- Suggest/Predict — use your scientific knowledge to work out what the answer might be.
- Compare — make sure you relate the things you're comparing to each other. It's no good just listing details about each one, you need to say how these things are similar or different.
- Discuss — you'll need to include more detail. Depending on the question you could need to cover what happens, what the effects are, and perhaps include a brief explanation of why it happens.
- Justify — show or prove that something is correct.
- Evaluate — Give the arguments both for and against an issue, or the advantages and disadvantages of something. You also need to give an overall judgement.

Exam Tip
Questions where you're marked on the quality of your extended response are marked with an asterisk (*) by the question.

Exam Tip
Don't worry too much about quality of extended response questions. Really, you should be looking to give full and coherent explanations for every question.

Exam Tip
Whilst these questions don't test your handwriting, spelling or grammar, if the examiner can't read what you've written or understand the point you're trying to make, it will lose you marks.

Quality of extended responses

For some extended answer questions, you'll be marked on the 'quality of your extended response'. These questions are designed to test how well you can put together a well structured and logical line of reasoning. They'll often require you to give a long answer in full written English, e.g. to explain, analyse or discuss something. To get top marks, you need to make sure that:

- you answer the question and all the information you give is relevant to the question you've been asked,
- you back up your points with clear evidence using the data given to you in the question,
- you organise your answer clearly, coherently and in a sensible order,
- you use specialist scientific vocabulary where it's appropriate.

These questions could also involve other tasks, like a calculation or having to draw an experimental set-up, like a circuit. Make sure any drawings are clear and use correct symbols where appropriate. When doing calculations, make sure your working is laid out logically and it's clear how you've reached your answer. That includes making sure any estimates and assumptions you've made in your working are clearly stated, e.g. assuming air resistance is negligible.

There's usually a lot to think about with this type of question, and it can be easy to write down a lot of great and relevant physics but forget to answer all parts of the question. It's always a good idea to double check you've done everything a question has asked you to do before moving on.

Time management

This is one of the most important exam skills to have. How long you spend on each question is really important in an exam — it could make all the difference to your grade.

Everyone has their own method of getting through the exam. Some people find it easier to go through the paper question by question and some people like to do the questions they find easiest first. The most important thing is to find out the way that suits you best before the exam — and that means doing all the practice exams you can before the big day.

Check out the exam timings given by OCR that can be found on page 247 and on the front of your exam paper. These timings give you just over 1 minute per mark.

However, some questions will require lots of work for only a few marks and other questions will be much quicker. So don't spend ages struggling with questions that are only worth a couple of marks — move on. You can come back to them later when you've bagged loads of other marks elsewhere.

It's worth keeping in mind that the multiple choice questions in Paper 1 and Paper 2 are all only worth 1 mark, even though some of them could be quite tricky and time-consuming. Don't make the mistake of spending too much time on these. If you're struggling with some of them, move on to the written answer questions where there are more marks available and then go back to the harder multiple choice questions later.

If you find that you're running out of time, go through the remaining questions and jot down what you can — for some of them you may get a mark for recalling the correct equation, or stating a definition. And if you've left any multiple choice questions, with only minutes to go, just make your best guess — you've got a 1 in 4 chance of getting an extra mark.

Exam Tip
Make sure you read the rest of the information given on the front of the exam paper before you start. It'll help make sure you're well prepared.

Exam Tip
Don't forget to go back and do any questions that you left the first time round — you don't want to miss out on marks because you forgot to do a question.

Exam Tip
When you're doing practice papers, set yourself a time limit so you get used to the exam timings.

Strange questions

You may get some weird questions that seem to have nothing to do with anything you've learnt. DON'T PANIC. Every question will be something you can answer using physics you know, it just may be in a new context.

Check the question for any key words that you recognise. For example, if a question talks about acceleration, think about the rules and equations you know, and whether any of them apply to the situation in the question. Sometimes you might have to pull together ideas from different parts of physics — read the question and try to think about what physics is being used. That way you can list any equations or facts you know to do with that topic and try to use them to answer the question.

Exam data and formulae booklet

When you sit your exams, you'll be given a data and formulae booklet as an insert within the exam paper. On it you'll find a lot of equations from the course, but not all of them. Make sure you know which equations you'll be given, and which you need to learn off by heart for the exam. There's also some useful data in the booklet to help you with your exam, including...

- the Planck constant, h
- the gravitational constant, G
- the length of a light year in metres, m

Constants that you'll be given in the exam are listed on page 284.

Exam Tip
If you're stuck on a question, have a quick look through the data and formulae booklet — you may find an equation that helps.

Answers

Module 5

Section 1 — Thermal Physics

1. Phases of Matter and Temperature

Page 16 — Application Questions

Q1 $T \approx \theta + 273 \Rightarrow \theta \approx T - 273 = 345 - 273 =$ **72 °C**

Q2 a) Net flow of thermal energy is from the filling to the pastry.

b) If the pastry and the filling are both in thermal equilibrium with the room, then the pastry and the filling must be in thermal equilibrium with each other. Since they are in thermal equilibrium with each other, they must be the same temperature.

Page 16 — Fact Recall Questions

Q1 a) Particles in solids vibrate about fixed positions in a lattice, and are close together.

b) Particles in liquids are free to move past one another and are constantly moving. They're fairly close together, in an irregular arrangement.

c) Particles in gases are far apart and free to move around with constant random motion. They are not in any particular order.

Q2 Brownian motion. It is caused by randomly moving water particles hitting the pollen particles unevenly.

Q3 The average kinetic energy stays the same because the energy supplied is altering the bonds, and therefore potential energy, of the particles instead of increasing their speeds.

Q4

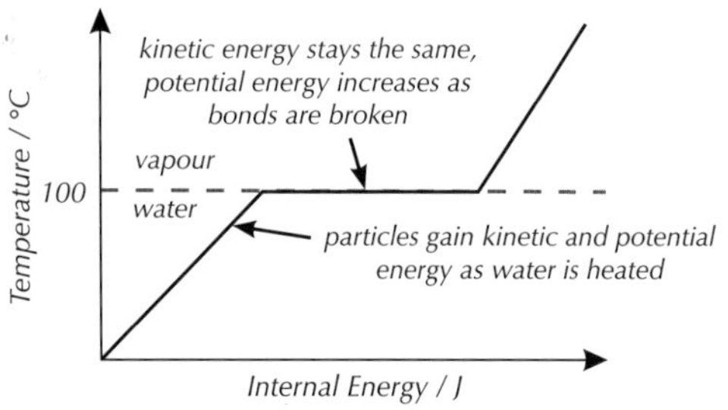

Q5 a) Absolute zero.

b) 0 K, –273 °C.

Q6 The internal energy of a body is the sum of the random distribution of kinetic and potential energies associated with the molecules of the body.

Q7 When the two objects are at the same temperature with no net flow of energy between them.

2. Thermal Properties of Materials

Page 21 — Application Questions

Q1 $E = mc\Delta\theta$, so $E = 0.45 \times 244 \times 3.0 = 329.4$
$= \mathbf{330\ J}$ **(to 2 s.f.)**

Q2 Convert masses to kg: 270 g ÷ 1000 = 0.27 kg
910 g ÷ 1000 = 0.91 kg

Expand and rearrange $m_l c_l (T_s - T_l) = m_b c_b (T_b - T_s)$ for T_s:

$\Rightarrow m_l c_l T_s - m_l c_l T_l = m_b c_b T_b - m_b c_b T_s$

$\Rightarrow m_l c_l T_s + m_b c_b T_s = m_b c_b T_b + m_l c_l T_l$

$\Rightarrow T_s(m_l c_l + m_b c_b) = m_b c_b T_b + m_l c_l T_l$

$$\Rightarrow T_s = \frac{m_b c_b T_b + m_l c_l T_l}{m_l c_l + m_b c_b}$$

$$= \frac{(0.27 \times 890 \times 82) + (0.91 \times 3600 \times 34)}{(0.91 \times 3600) + (0.27 \times 890)}$$

$= 37.28... =$ **37 °C (to 2 s.f.)**

You don't need to convert to kelvin here since changes in temperature are the same for both.

Q3 Convert mass to kg: 100.0 g ÷ 1000 = 0.1000 kg
Energy lost from 25.0 °C to 0 °C
$= mc_{water}\Delta\theta = 0.1000 \times 4180 \times 25.0 = 10\,450$ J
Energy lost in freezing $= mL_f = 0.1000 \times 334\,000$
$= 33\,400$ J
Energy lost from 0 °C to –5.00 °C
$= mc_{ice}\Delta\theta = 0.1000 \times 2110 \times 5.00 = 1055$ J
Total energy lost $= 10\,450 + 33\,400 + 1055$
$= 44\,905 =$ **44 900 J (to 2 s.f.)**

Page 21 — Fact Recall Questions

Q1 The amount of energy needed to raise the temperature of 1 kg of the substance by 1 K (or 1 °C).

Q2 E.g. Place an electric heater and digital thermometer in a block of the material and cover it with insulation. Turn the heater on and heat the material so that its temperature increases by about 10 K. Use a stopwatch to measure how long the heater was on for. Use an ammeter and voltmeter attached to the electric heater to measure the current and voltage supplied, and then calculate the energy supplied (E) using $E = W = VIt$. Then use $E = mc\Delta\theta$ to calculate c.

Q3 The quantity of thermal energy required to melt or freeze 1 kg of a substance.

Q4 E.g. Put equal masses of ice in two identical funnels above beakers. Put a heating coil in one of the funnels and turn it on for three minutes. Use an ammeter and voltmeter attached to the heating coil to measure the current and voltage supplied, and use these to calculate the energy transferred by the heating coil to the ice (E) using $E = W = VIt$. After switching off the heating coil, measure the mass of water collected in the beakers. Subtract the mass of water collected from the beaker without the heating coil from the mass of water from the beaker with the heating coil to get the mass of ice, m, that melted solely due to the presence of the heater. Use $E = mL$ to calculate L.

3. The Gas Laws

Page 26 — Application Questions

Q1 E.g.

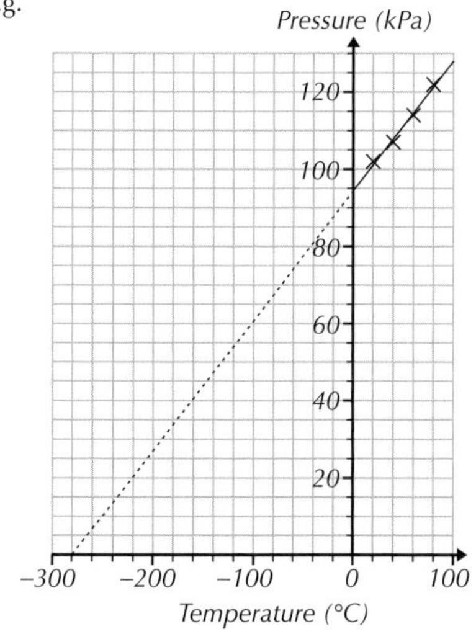

So absolute zero is approximately **−280°C.**

Depending on your line of best fit, accept any answer between −300°C and −260°C.

Q2 a) The volume has halved. As p and V are inversely proportional, the pressure will have doubled, so $p = 2 \times 1.4 \times 10^5 = \mathbf{2.8 \times 10^5}$ **Pa.**

b) Start temperature in K = 27 + 273 = 300 K.
End temperature in K = −173 + 273 = 100 K.
The temperature is divided by 3. As p and T are directly proportional, the pressure will also be divided by 3, so
$p = \frac{2.8 \times 10^5}{3} = 9.333... \times 10^4 = \mathbf{9.3 \times 10^4}$ **Pa (to 2 s.f.)**

Q3 a) 92 mm = 92×10^{-3} m
Volume of air bubble = $\pi r^2 l$
$= \pi \times (3.0 \times 10^{-3})^2 \times (92 \times 10^{-3})$
$= 2.6012... \times 10^{-6}$
$= \mathbf{2.6 \times 10^{-6}\ m^3}$ **(to 2 s.f.)**

b) pV = constant so $p_1V_1 = p_2V_2$
Rearrange to:
$V_2 = \frac{p_1V_1}{p_2} = \frac{350\,000 \times (2.6012... \times 10^{-6})}{420\,000}$
$= 2.167... \times 10^{-6} = \mathbf{2.2 \times 10^{-6}\ m^3}$ **(to 2 s.f.)**

c) So that the temperature has time to stabilise.

d) A straight line graph passing through the origin.

e) Boyle's law.

Page 26 — Fact Recall Questions

Q1 It must have a fixed mass.

Q2 The pressure law states that at constant volume, the pressure p of an ideal gas is directly proportional to its absolute temperature T, or $\frac{p}{T}$ = constant.

4. The Ideal Gas Equation

Page 28 — Application Questions

Q1 $pV = nRT \Rightarrow V = \frac{nRT}{p} = \frac{23 \times 8.31 \times (25 + 273)}{2.4 \times 10^5}$
$= 0.23731... = \mathbf{0.24\ m^3}$ **(to 2 s.f.)**

Q2 $pV = NkT \Rightarrow p = \frac{NkT}{V}$
$= \frac{(8.21 \times 10^{24}) \times (1.38 \times 10^{-23}) \times 500}{4.05}$
$= 1.3987... \times 10^4$
$= \mathbf{1.40 \times 10^4}$ **Pa (to 3 s.f.)**

Q3 $pV = NkT \Rightarrow T = \frac{pV}{Nk}$
$= \frac{(1.29 \times 10^5) \times 0.539}{(1.44 \times 10^{25}) \times (1.38 \times 10^{-23})}$
$= 349.894... =$ **350 K (to 3 s.f.)**

Q4 $pV = nRT \Rightarrow T = \frac{pV}{nR}$
$= \frac{(2.3 \times 10^5) \times 0.39}{20 \times 8.31} = 539.71...$
= **540 K (to 2 s.f.)**

Page 28 — Fact Recall Questions

Q1 The number of particles in one mole of a material. Its value is $6.02 \times 10^{23}\ mol^{-1}$.

Q2 $pV = nRT$
p = pressure (in Pa)
V = volume (in m^3)
n = number of moles of gas
R = molar gas constant (= 8.31 $J mol^{-1} K^{-1}$)
T = temperature (in K)

Q3 $k = \frac{R}{N_A}$
R = ideal gas constant (= 8.31 $JK^{-1}mol^{-1}$)
N_A = Avogadro's constant (= $6.02 \times 10^{23}\ mol^{-1}$)

Q4 $pV = NkT$
p = pressure (in Pa)
V = volume (in m^3)
N = number of molecules of gas
k = Boltzmann constant (= $1.38 \times 10^{-23}\ JK^{-1}$)
T = temperature (in K)

5. The Pressure of an Ideal Gas

Page 31 — Application Questions

Q1 $pV = \frac{1}{3}Nm\overline{c^2} \Rightarrow p = \frac{1}{3} \times \frac{Nm\overline{c^2}}{V}$ so:
$p = \frac{1}{3} \times \frac{(5 \times 6.02 \times 10^{23}) \times (5.31 \times 10^{-26}) \times (8.11 \times 10^6)}{1.44}$
$= 3.0005... \times 10^5 = \mathbf{3.00 \times 10^5}$ **Pa (to 3 s.f.)**

Q2 $pV = \frac{1}{3}Nm\overline{c^2} \Rightarrow \overline{c^2} = \frac{3pV}{Nm}$
905 cm^3 = $905 \times 10^{-6}\ m^3$
$\overline{c^2} = \frac{3 \times (7.40 \times 10^5) \times (905 \times 10^{-6})}{(0.310 \times 6.02 \times 10^{23}) \times (5.10 \times 10^{-26})}$
$= 2.110... \times 10^5\ m^2 s^{-2}$
$c_{rms} = \sqrt{\overline{c^2}} = \sqrt{2.110... \times 10^5} = 459.44... = \mathbf{459\ ms^{-1}}$ **(to 3 s.f.)**

Page 31 — Fact Recall Questions

Q1 Any four from: gas contains large number of particles / the particles move rapidly and randomly / the volume of the particles is negligible compared to the volume of the gas / collisions between particles themselves or between particles and the walls of the container are perfectly elastic / the duration of each collision is negligible when compared to the time between collisions / there are no forces between particles except for the moment when they are in a collision.

Q2 Each particle exerts a force on the wall of the container when it collides with it. The combined force from all the particles is spread all over the surface of each wall. This produces a steady, even force on all the walls of the box, which is pressure.

Q3 c_{rms} is the root mean square speed (of the particles in a gas). I.e. the square root of the mean of the squared speeds of the particles.

6. Internal Energy of an Ideal Gas

Page 34 — Application Questions

Q1 Curve B.

Q2 $E = \frac{3}{2}kT = \frac{3}{2} \times (1.38 \times 10^{-23}) \times 112 = 2.3184 \times 10^{-21}$

$= \mathbf{2.32 \times 10^{-21}}$ **J (to 3 s.f.)**

Page 34 — Fact Recall Questions

Q1 $pV = NkT$ and $pV = \frac{1}{3}Nm\overline{c^2}$.

Q2 That the potential energy is 0 J because there are no forces between the particles.

Exam-style Questions — pages 36-37

1 B ***(1 mark)***

pV = constant so $p_1V_1 = p_2V_2$. $V = \pi r^2 \times l$ so $p_1(\pi r^2 l_1) = p_2(\pi r^2 l_2)$.
Dividing by πr^2: $p_1 l_1 = p_2 l_2 \Rightarrow p_2 = (p_1 l_1) \div l_2$
$p_2 = (120 \times 30) \div 50 = 72$ kPa

2 D ***(1 mark)***

3 a) Increasing the temperature increases the kinetic energy of the gas molecules ***(1 mark)***, so the r.m.s. speed of the gas molecules increases ***(1 mark)***. So from $pV = \frac{1}{3}Nmc^2$ it can be seen that the pressure must also increase ***(1 mark)***.

You could also have answered this by talking about the change in momentum — at higher temperatures the particles collide with the walls of the container more often and on average there's a larger change in momentum during a collision. This means a greater force is exerted on the walls of the container in the same amount of time, and so the pressure is increased.

b)

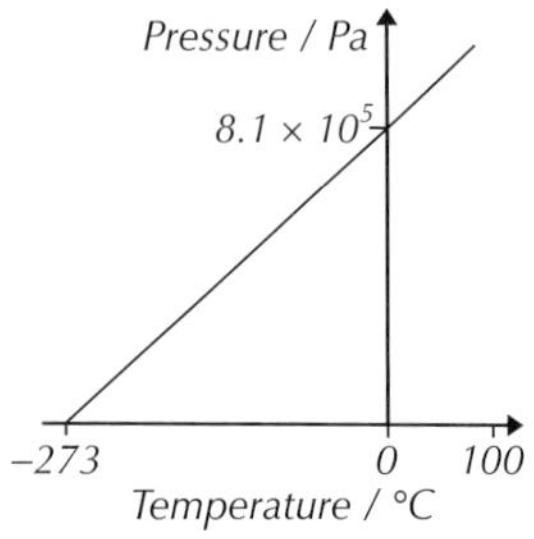

(1 mark for straight-line graph and 1 mark for crossing through (–273, 0) and (0, 8.1 × 10⁵).)

c) $pV = NkT \Rightarrow N = \frac{pV}{kT} = \frac{(8.1 \times 10^5) \times 0.51}{(1.38 \times 10^{-23}) \times 273}$

$= 1.096... \times 10^{26} = \mathbf{1.1 \times 10^{26}}$ **(to 2 s.f.)**

(2 marks for correct answer, otherwise 1 mark for correct working if answer incorrect.)

d) Total mass of gas $= (1.096... \times 10^{26}) \times (2.7 \times 10^{-26})$

$= 2.96...$ kg

$E = mc\Delta\theta = 2.96... \times (2.2 \times 10^3) \times 150$

$= 9.7699... \times 10^5 = \mathbf{9.8 \times 10^5}$ **J (to 2 s.f.)**

(2 marks for correct answer, otherwise 1 mark for finding the mass of gas if answer is incorrect.)

4 a) $E = mc\Delta\theta = (92 \times 10^{-3}) \times 2110 \times 25 = 4.853 \times 10^3$ J

$= \mathbf{4.9 \times 10^3}$ **J (to 2 s.f.)** ***(1 mark)***

b) $E = mL = (92 \times 10^{-3}) \times (3.3 \times 10^5)$

$= 3.036 \times 10^4$ J $= \mathbf{3.0 \times 10^4}$ **J (to 2 s.f.)** ***(1 mark)***

c) Total energy needed $= (4.853 \times 10^3) + (3.036 \times 10^4)$

$= 3.5213 \times 10^4$ J

Time taken $= \frac{\text{total energy needed}}{\text{rate of energy supplied}} = \frac{3.5213 \times 10^4}{50.0}$

$= 704.26 = \mathbf{700}$ **s (to 2 s.f.)**

(2 marks for correct answer, otherwise 1 mark for correct working if answer is incorrect.)

d) It would slow down the melting of the ice ***(1 mark)***. Heat is transferred from hotter substances to colder substances — as the ice is colder than 25 °C, insulating the beaker will stop heat transfer into the ice from the surrounding air ***(1 mark)***.

5 a) How to grade your answer (pick the description that best matches your answer):

0 marks: There is no relevant information.

1-2 marks: At least one *K* point for each of solid, liquid and gas have been covered. *B* points 1 and 3 have been covered. Answer is basic and has lack of structure, with information missing.

3-4 marks: At least two *K* points for each of solid, liquid and gas have been covered. All *B* points have been covered. Answer has some structure, and information is generally relevant.

5-6 marks: All *K* and *B* points have been covered. The answer is well structured, and information is relevant and presented clearly.

Here are some points your answer may include:

Kinetic Model of Matter (*K*)

1. Particles in solids are close together.
2. Particles in solids are arranged in a regular lattice structure.
3. Particles in solids vibrate around fixed positions.
4. Particles in liquids are fairly close together.
5. Particles in liquids have an irregular arrangement.
6. Particles in liquids are free to move past one another.
7. Particles in gases are far apart.
8. Particles in gases are free to move with constant random motion.
9. Particles in gases are not in any particular order.

Brownian Motion (*B*)

1. Brownian motion describes the random, zigzag motion of particles suspended in a fluid.
2. It is caused by randomly moving gas or liquid particles hitting the observed particles unevenly.
3. Brownian motion can be observed by putting smoke in an illuminated glass jar and viewing the motion of the smoke particles in the air with a microscope.

b) (i) $N = n \times N_A = 54.0 \times (6.02 \times 10^{23}) = 3.2508 \times 10^{25}$

$= \mathbf{3.25 \times 10^{25}}$ **(to 3 s.f.)** ***(1 mark)***

(ii) $pV = nRT \Rightarrow T = \frac{pV}{nR}$

$= \frac{1.00 \times 10^5 \times 4.18}{54.0 \times 8.31}$

$= 931.49...\text{K} = \mathbf{931\,K}$ **(to 3 s.f.)**

(2 marks for correct answer, otherwise 1 mark for correct working.)

(iii) Convert temperature to K:

$T \approx \theta + 273 = 855 + 273 = 1128$ K

$\frac{1}{2}m\overline{c^2} = \frac{3}{2}kT \Rightarrow \overline{c^2} = \frac{3kT}{m} = \frac{3 \times 1.38 \times 10^{-23} \times 1128}{3.40 \times 10^{-26}}$

$= 1.373... \times 10^6$ m²s⁻²

$c_{rms} = \sqrt{\overline{c^2}} = \sqrt{1.373... \times 10^6} = 1171.96...$

$= \mathbf{1170}$ **ms⁻¹ (to 3 s.f.)**

(3 marks for correct answer, otherwise 1 mark for converting temperature to kelvin and 1 mark for calculating the mean square speed.)

(iv) E.g.

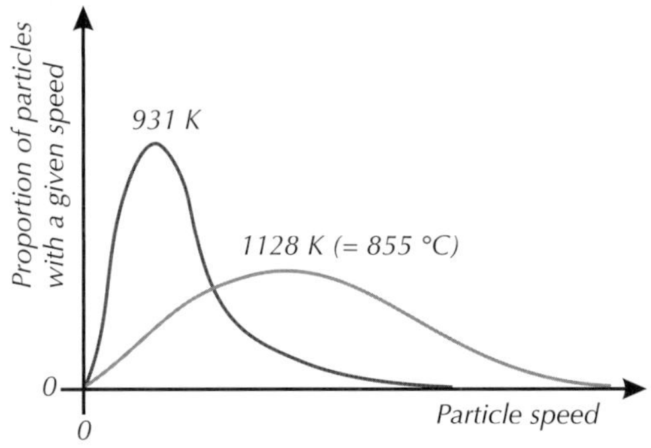

(1 mark for drawing Maxwell-Boltzmann curves, 1 mark for making high temperature curve lower and more spread out than low temperature curve.)

Section 2 — Circular Motion and Oscillations

1. Circular Motion

Page 40 — Application Questions

Q1 Period in seconds = 28 × 24 × 3600 = 2.419... × 10^6 s

Angular velocity = $\boldsymbol{\omega} = \frac{2\pi}{T} = \frac{2\pi}{2.419... \times 10^6}$

$= 2.5972... \times 10^{-6}$ rad s^{-1}

$=$ **2.6 × 10^{-6} rad s^{-1} (to 2 s.f.)**

$v = r\omega = (384\ 000 \times 1000) \times 2.5972... \times 10^{-6}$

$= 997.331... =$ **1000 ms^{-1} (to 2 s.f.)**

Q2 Frequency in rev s^{-1} = $f = \frac{\text{no. of revolutions}}{\text{time taken}} = \frac{4}{3600}$

$\omega = 2\pi f = 2 \times \pi \times \frac{4}{3600} = \frac{\pi}{450}$ rad s^{-1}

$\omega = \frac{\theta}{t} \Rightarrow \theta = \omega t = \frac{\pi}{450} \times 60 = \frac{2}{15}\pi = 0.41887...$

$=$ **0.419 rad (to 3 s.f.)**

$v = \omega r = \frac{\pi}{450} \times \frac{125}{2} = 0.43633...$ ms^{-1}

$=$ **0.436 ms^{-1} (to 3 s.f.)**

Q3 Frequency in rev s^{-1} = $f = 460 \div 60$.

2.0 cm from centre:

Angular speed = $\omega = 2\pi f = \frac{46}{3}\pi = 48.171...$

$=$ **48 s^{-1} (to 2 s.f.)**

Linear speed = $v = \omega r = \frac{46}{3}\pi \times 0.020$

$= 0.9634...$ ms^{-1} = **0.96 ms^{-1} (to 2 s.f.)**

4.0 cm from centre: Angular speed = **48 s^{-1} (to 2 s.f.)**

Angular speed is the same at any point on a solid rotating object.

Linear speed = $0.040 \times \frac{46}{3}\pi = 1.926... =$ **1.9 ms^{-1} (to 2 s.f.)**

Q4 Kinetic energy = $\frac{1}{2}mv^2$

$v = \omega r$ and $\omega = 2\pi f$

$\Rightarrow v = 2\pi fr$

$f = \frac{1}{T}$ and r is the length of the string l, so $v = \frac{2\pi l}{T}$

So kinetic energy $= \frac{1}{2}m(\frac{2\pi l}{T})^2$

$= \boldsymbol{\frac{2m\pi^2 l^2}{T^2}}$

Page 40 — Fact Recall Questions

Q1 Angle in radians = $\frac{\pi}{180}$ × angle in degrees

Q2 The angle that an object rotates through per second.

Q3 $v = \omega r$, where v is the linear speed, ω is the angular velocity, and r is the radius of the circle of rotation.

Q4 The period is the time taken for a complete revolution. The frequency is the number of complete revolutions per second.

Q5 $\omega = 2\pi f$, where ω is the angular velocity and f is the frequency.

2. Centripetal Force and Acceleration

Page 44 — Application Questions

Q1 Frequency in rev s^{-1} = f = 15 ÷ 60.0 = 0.25 rev s^{-1}

Angular velocity $\omega = 2\pi f = 2 \times \pi \times 0.25 = 0.5\pi$ rad s^{-1}

$\boldsymbol{F} = m\omega^2 r = 60.0 \times (0.5\pi)^2 \times \frac{8.5}{2} = 629.1...$

$=$ **630 N (to 2 s.f.)**

This might seem like a lot, but it's about the same as the force experienced by the rider due to gravity.

Q2 **C**

$\boldsymbol{a} = \omega^2 r$ so rearranging gives: $\omega = \sqrt{\frac{\boldsymbol{a}}{r}}$

$\omega = \frac{2\pi}{T}$ so rearranging gives: $T = \frac{2\pi}{\omega}$

So $T = \frac{2\pi}{\sqrt{\frac{\boldsymbol{a}}{r}}} \Rightarrow T = 2\pi\sqrt{\frac{r}{\boldsymbol{a}}}$

So time to complete 3 orbits = $3 \times T = 6\pi\sqrt{\frac{r}{\boldsymbol{a}}}$

Q3 Convert the washer's mass to kg:

$m_w = 310 \div 1000 = 0.31$ kg

So the weight of the washer is

$\boldsymbol{W} = m_w\boldsymbol{g} = 0.31 \times 9.81 = 3.0411$ N

This exerts a centripetal force $\boldsymbol{F}$ on the bung, i.e. $\boldsymbol{F} = m_b\omega^2 r$.

Convert the bung's mass to kg: $m_b = 22 \div 1000 = 0.022$ kg

So $3.0411 = 0.022 \times \omega^2 \times 0.60$

This gives $\omega = \sqrt{\frac{3.0411}{0.022 \times 0.60}} = 15.178...$ rad s^{-1}

Rearrange the formula $\omega = \frac{2\pi}{T}$ to give $T = \frac{2\pi}{\omega}$

So $T = \frac{2\pi}{15.178...} = 0.4139... =$ **0.41 s (to 2 s.f.)**

Q4 a) $\boldsymbol{a} = \frac{v^2}{r} = 31.1^2 \div 56.8$

$= 17.0283... =$ **17.0 ms^{-2} (to 3 s.f.)**

b) The formula for centripetal force is $\boldsymbol{F} = \frac{mv^2}{r}$, so if the linear speed is halved (i.e. v becomes $\frac{v}{2}$), then the centripetal force is quartered. So the new centripetal force is $\frac{F}{4}$.

Q5 If the biker's speed is the minimum possible speed for him to not fall, the centripetal acceleration towards the centre (i.e. down) at the top of the cylinder will be 9.81 ms^{-2}, due to gravity. The motorcycle will fall if its circular motion has a centripetal acceleration smaller than this.

So $\boldsymbol{a} = 9.81$ ms^{-2} $= \frac{v^2}{r}$

Rearranging this: $v = \sqrt{\boldsymbol{a}r} = \sqrt{9.81 \times 5.0}$

$= 7.003...$ ms^{-1} = **7.0 ms^{-1} (to 2 s.f.)**

This is only about 16 mph — not very fast at all.

Page 44 — Fact Recall Questions

Q1 If an object is moving in a circle, centripetal acceleration is the acceleration of the object directed towards the centre of the circle. Centripetal force is the force towards the centre of the circle responsible for the centripetal acceleration.

Q2 $\boldsymbol{a} = \omega^2 r$

$\boldsymbol{a}$ = centripetal acceleration in ms^{-2}

ω = angular velocity (in rad s^{-1})

r = radius of circular motion in m

Q3 $\boldsymbol{F} = \frac{mv^2}{r}$

$\boldsymbol{F}$ = centripetal force in N

m = mass of object in kg

v = magnitude of linear velocity in ms^{-1}

r = radius of circular motion in m

3. Simple Harmonic Motion

Page 48 — Application Questions

Q1 a) A and C

Maximum velocity occurs at the midpoint/equilibrium position.

b) B and E

Maximum acceleration occurs when displacement is maximum.

Q2

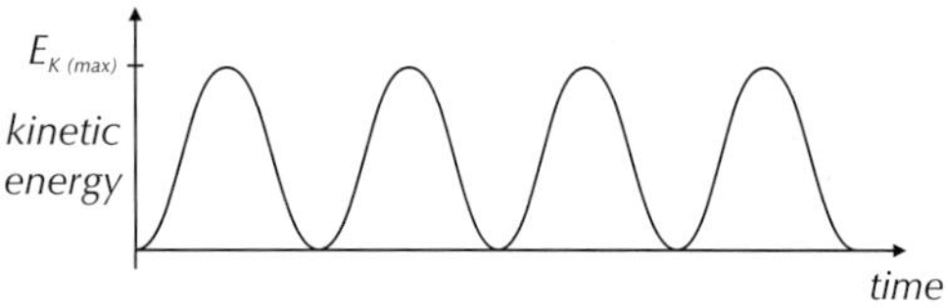

The kinetic energy starts at 0 as the girl starts from rest. It then varies sinusoidally (like a sine wave) between 0 and $E_{K\,(max)}$ as potential energy is converted to kinetic energy and back. The kinetic energy is at a maximum when the swing is at its lowest point. The swing passes through this point 4 times during two complete oscillations.

Q3 a) Maximum displacement = A = 0.60 m

Maximum speed = $\omega A = 2\pi f A$ = 0.90 ms^{-1}

Rearrange $v_{max} = 2\pi f A$ to give $f = \frac{v_{max}}{2\pi A}$

So $f = 0.90 \div (2 \times \pi \times 0.60) = 0.2387...$

$= $ **0.24 Hz (to 2 s.f.)**

b) $f = \frac{1}{T}$ so rearranging gives $T = \frac{1}{f}$

So $T = 1 \div 0.2387... = 4.188...$ = **4.2 s (to 2 s.f.)**

c) Maximum acceleration = $\omega^2 A$

$= (2\pi f)^2 A = (2 \times \pi \times 0.2387...)^2 \times 0.60 =$ **1.35 ms^{-2}**

Page 48 — Fact Recall Questions

Q1 $\mathbf{a} = -\omega^2 \mathbf{x}$

Q2 The frequency of oscillation is the number of complete cycles per second. The period of oscillation is the time taken for a complete cycle.

Q3 The velocity is $\frac{\pi}{2}$ radians ahead of the displacement.

Q4 a) Equilibrium

b) Maximum displacement

c) Maximum displacement

d) Equilibrium

Q5 E.g.

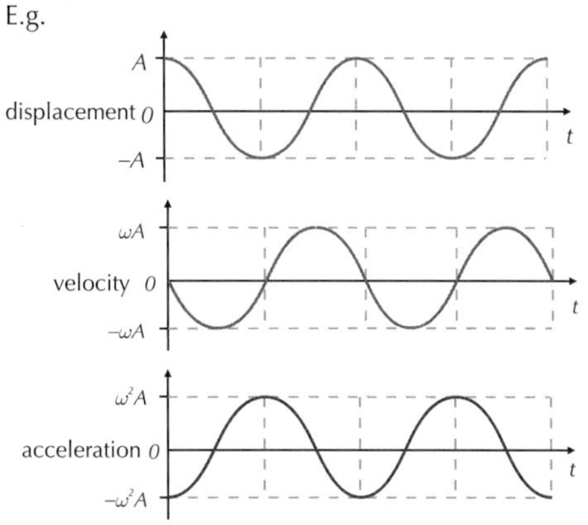

Q6 At its maximum displacement, the object's kinetic energy, E_K, is zero (so it has zero velocity). All of its energy is potential energy, E_P. As the object moves towards the equilibrium position, the restoring force does work on the object and transfers some E_P to E_K. At the equilibrium position, the object's E_P is said to be zero and its E_K is maximum — so its velocity is maximum. As the object moves away from the equilibrium, all that E_K is transferred back to E_P again.

4. Calculations with SHM

Page 51 — Application Questions

Q1 a) $\mathbf{x} = A$, so: $\mathbf{a} = -\omega^2 A = -1.5^2 \times 1.6 =$ **−3.6 ms^{-2}**

The question asks for the acceleration, not the magnitude of acceleration, which is why you can't use $a_{max} = \omega^2 A$. The displacement given in the question is a positive value, so the acceleration is negative.

b) Angular frequency = $\omega = 2\pi f$, rearranging gives $f = \frac{\omega}{2\pi}$

$= 1.5 \div (2 \times \pi) = 0.2387...$ Hz

Time to complete 1 oscillation = $T = \frac{1}{f} = \frac{1}{0.2387...}$

$= 4.188...$ s

Time to complete 15 oscillations

$= 15 \times 4.188... = 62.83...$ s = **63 s (to 2 s.f.)**

Q2 a) Frequency = $f = \frac{1}{T} = \frac{1}{0.75} = 1.333...$ s^{-1}

Maximum speed is given by: $v_{max} = \omega A = 2\pi f A$

Substituting this in and rearranging for amplitude gives:

$A = \frac{v_{max}}{2\pi f} = \frac{0.85}{2 \times \pi \times 1.3333...} = 0.101...$ m

$=$ **0.10 m (to 2 s.f.)**

b) Velocity at $\mathbf{x}$ = 0.080 m:

$\mathbf{v} = \pm\omega\sqrt{A^2 - \mathbf{x}^2} = \pm 2\pi f\sqrt{A^2 - \mathbf{x}^2}$

$= \pm 2 \times \pi \times 1.333... \times \sqrt{0.101...^2 - 0.080^2}$

$= \pm 0.5228...$ ms^{-1} = **± 0.52 ms^{-1} (to 2 s.f.)**

The answer has a ± sign at the front because you don't know the direction of the velocity.

Q3 Period = T = time to complete exactly 5 oscillations ÷ 5

$= \frac{15.5}{5} = 3.1$ s

Frequency = $f = \frac{1}{T} = \frac{1}{3.1} = 0.3225...$ s^{-1}

Amplitude = A = 0.45 m

Displacement at time t: $\mathbf{x} = A\cos(\omega t) = A\cos(2\pi f t)$

At time t = 10.0, $\mathbf{x} = 0.45 \times \cos(2\pi \times 0.3225... \times 10.0)$

$= 0.0681...$ m = **0.068 m (to 2 s.f.)**

Don't forget to put your calculator into radians here.

Q4 Pendulum passes through equilibrium twice every period, so if it's set to 120 ticks per minute:

Period = $T = 120 \div 2 \div 60 = 1$ s

Frequency = $f = \frac{1}{T} = \frac{1}{1} = 1$ Hz

Amplitude A = 6.2 cm = 0.062 m

Magnitude of max acceleration = $a_{max} = \omega^2 A = (2\pi f)^2 A$

$a_{max} = (2\pi \times 1)^2 \times 0.062 = 2.447...$ ms^{-2} = **2.4 ms^{-2} (to 2 s.f.)**

Page 51 — Fact Recall Questions

Q1 Displacement = $\mathbf{x} = A\cos(\omega t)$, where $\mathbf{x}$ is the object's displacement, ω its angular frequency, and A its maximum displacement.

Q2 Acceleration = $\mathbf{a} = -\omega^2 \mathbf{x}$ Velocity = $\mathbf{v} = \pm\omega\sqrt{A^2 - \mathbf{x}^2}$

Here, $\mathbf{a}$ is the acceleration of the object, $\mathbf{x}$ is its displacement, $\mathbf{v}$ its velocity, ω its angular frequency, and A its maximum displacement.

Q3 Max speed = $v_{max} = \omega A$, where v_{max} is the object's maximum velocity, ω its angular frequency, and A its maximum displacement.

Q4 Max acceleration = $a_{max} = \omega^2 A$, where a_{max} is the object's maximum acceleration, ω its angular frequency, and A its maximum displacement.

5. Investigating SHM

Page 54 — Application Question

Q1 a) It takes 0.50 s for each complete oscillation, T = **0.500 s**.

b) $f = \frac{1}{T}$, so $f = 1 \div 0.500$ = **2.00 Hz**

c) $\omega = 2\pi f = 2\pi \times 2 = 4\pi = 12.566...$ = **12.6 rad s^{-1} (to 3 s.f.)**

Page 54 — Fact Recall Questions

Q1 E.g. You could use a mass attached to the end of a pendulum, which is made to oscillate in front of a piece of card with a reference mark on it to show the midpoint of the oscillations. Attached at the same point that the pendulum is suspended from, there should be a protractor to measure the angle that the pendulum makes with a vertical reference line as it is released.

Q2 a) This means the time period you calculate for a single oscillation will be more accurate, since any human error introduced by starting or stopping the stopwatch too late (or too early) will be 'shared' over several oscillations.

b) Using a fiducial marker helps you start and stop your stopwatch at the same point in the oscillations, meaning the measurement of the time period for one complete oscillation will be more accurate.

6. Free and Forced Oscillations

Page 58 — Fact Recall Questions

Q1 A free vibration involves no transfer of energy between the oscillating object and its surroundings. The object will continue to oscillate at its natural frequency and with the same amplitude forever. A forced vibration occurs if there's a periodic external driving force acting on the object.

Q2 Resonance occurs when the driving frequency approaches the natural frequency of an object and the object begins to oscillate with a rapidly increasing amplitude.

Q3 E.g. any three from: a radio's electric circuit resonating when it's tuned to the same frequency as a radio station / a glass resonating when driven by a sound wave at its natural frequency / a column of air in an organ pipe resonating when driven by the motion of air at its base / a swing in a playground resonating when it's pushed by someone at its natural frequency.

Q4 A damping force is a force that acts on an oscillator and causes it to lose energy to its surroundings, reducing the amplitude of its oscillations.

Q5 Light damping — damping such that an oscillating system takes a long time to stop, and the amplitude of the system reduces by only a small amount each period.
Heavy damping — damping such that the system takes less time to stop oscillating than a lightly damped system, and the amplitude gets much smaller each period.
Critical damping — damping such that the amplitude of an oscillating system is reduced (and so the system returns to equilibrium) in the shortest possible time.
Overdamping — extremely heavy damping such that an oscillating system takes longer to return to equilibrium than a critically damped system.

Q6

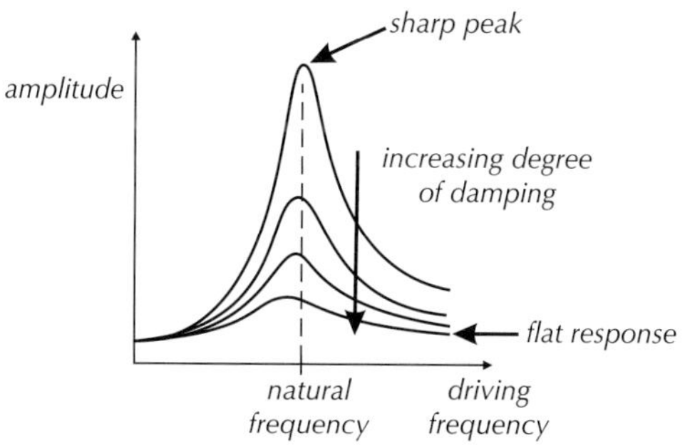

Exam-style Questions – pages 60-61

1 D *(1 mark)*

The equation for linear speed is $v = \omega r$, where $r = 3.0$ m.
Frequency = 6 revolutions per minute, so $f = 6 \div 60 = 0.1$ Hz.
$\omega = 2\pi f = 2 \times \pi \times 0.1 = 0.628...$ rads^{-1}.
This means $v = 0.628... \times 3.0 = 1.884... = 1.9$ ms^{-1} (to 2 s.f.).

2 C *(1 mark)*

A critically damped system takes the shortest possible time to return to equilibrium. An overdamped system takes longer to return to equilibrium.

3 C *(1 mark)*

The correct graph needs to show 1 complete cycle (i.e. 2 'peaks') in 3 seconds and zero potential energy at 0, 1.5 s and 3 s.

4 a) The maximum displacement of the block (A) is 0.20 m and the time period of the oscillations (T) is 1.147 s.
Velocity is given by:

$$v = \pm\omega\sqrt{A^2 - x^2}$$
$$= \pm\frac{2\pi}{T}\sqrt{A^2 - x^2}$$
$$= \pm\frac{2\times\pi}{1.147}\times\sqrt{0.20^2 - 0.15^2}$$
$$= \pm 0.7246...\ \text{ms}^{-1} = \pm\mathbf{0.72\ ms^{-1}}\ \textbf{(to 2 s.f.)}$$

Don't forget to convert 20 cm and 15 cm into metres.
(2 marks for the correct answer, otherwise 1 mark for correct working if answer incorrect.)

b) Maximum kinetic energy is: $E_{K\,(max)} = \frac{1}{2}mv_{max}^2$
Maximum velocity is:

$$v_{max} = \omega A = \frac{2\pi}{T}A = \frac{2\times\pi}{1.147}\times 0.20 = 1.095...\ \text{ms}^{-1}$$

Substitute this into the equation for maximum kinetic energy:

$$E_{K\,(max)} = \frac{1}{2}\times 0.60\times(1.095...)^2 = 0.3600...$$
$$= \mathbf{0.36\ J\ (to\ 2\ s.f.)}$$

(3 marks for correct answer, otherwise 1 mark for calculating maximum velocity, and 1 mark for substituting this into the formula for maximum kinetic energy.)

c) If t is the number of seconds shown on the stopwatch, then at $t = 0$, the block is at its equilibrium position (i.e. when $t = 0$, the displacement (x) = 0). This means x is given by the formula $x = A\sin(\omega t)$. So at $t = 3.0$:

$$x = A\sin(\omega t)$$
$$= 0.20\times\sin\left(\frac{2\pi}{T}\times 3.0\right)$$
$$= 0.20\times\sin\left(\frac{2\times\pi}{1.147}\times 3.0\right)$$
$$= -0.1327...\ \text{m}$$

So the block is **0.13 m** (to 2 s.f.) from its equilibrium position.
(2 marks for correct answer, otherwise 1 mark for correct working if answer incorrect.)

5 a) The centripetal force acting on the motorcyclist is given by:

$$F = \frac{mv^2}{r} = \frac{210\times v^2}{5.0}$$

To avoid sliding off the track, the motorcyclist needs to travel with a minimum speed of v_{min}, where:

$$1500 = \frac{210\times v_{min}^2}{5.0}$$

$$\text{So } v_{min}^2 = \frac{1500\times 5.0}{210} = 35.714...$$

This means $v_{min} = \sqrt{35.714...} = 5.976...$

$= \mathbf{6.0\ ms^{-1}}$ **(to 2 s.f.)**
(2 marks for the correct answer, otherwise 1 mark for correct working if answer incorrect.)

b) $v = \omega r$, so $\omega = \frac{v_{min}}{r} = \frac{5.976...}{5.0} = 1.195...$

$= \mathbf{1.2\ rad\ s^{-1}}$ **(to 2 s.f.)**

(2 marks for the correct answer, otherwise 1 mark for correct working if answer incorrect.)

c) $\omega = \frac{2\pi}{T}$ So $T = \frac{2\pi}{\omega} = \frac{2\pi}{1.195...} = 5.256...$

$= \mathbf{5.3\ s}$ **(to 2 s.f.)**

(2 marks for the correct answer, otherwise 1 mark for correct working if answer incorrect.)

6 a) The amplitude of the pendulum's oscillation will increase rapidly around its natural frequency ***(1 mark)***. This is called resonance ***(1 mark)***.

b) E.g. friction with the water acts to dampen the oscillation ***(1 mark)***. The resonance will occur at a frequency slightly less than the pendulum's natural frequency ***(1 mark)***. The pendulum's response will be flatter (i.e. the resonance peak will be less sharp) ***(1 mark)***.

Section 3 — Gravitational Fields

1. Gravitational Fields

Page 64 — Application Questions

Q1 $F = \frac{GMm}{r^2}$

$= \frac{(6.67 \times 10^{-11}) \times (2.15 \times 10^{30}) \times (2.91 \times 10^{30})}{(1 \times 10^{11})^2}$

$= 4.17308... \times 10^{28}$

$\approx \mathbf{4.2 \times 10^{28}\ N}$

We can ignore the minus sign in Newton's law of gravitation here because we're not interested in the direction of the force.

Q2 $\frac{2.5}{0.5} = 5$, so the force will be $5^2 = 25$ times larger:

$25 \times 25 = \mathbf{625\ N}$

Q3 a) 6 370 000 m + 10 000 m = 6 380 000 m = **6380 km**

b) The upwards force must balance the downwards force due to gravity:

$F = -\frac{GMm}{r^2}$

$= -\frac{(6.67 \times 10^{-11}) \times (2500) \times (5.97 \times 10^{24})}{(6380 \times 10^3)^2}$

$= -24\,456.7...$ N

So upwards force = **24 500 N (to 3 s.f.)**

Page 64 — Fact Recall Questions

Q1 A gravitational field is a force field generated by any object with mass which causes any other object with mass to experience an attractive force.

Q2 a)

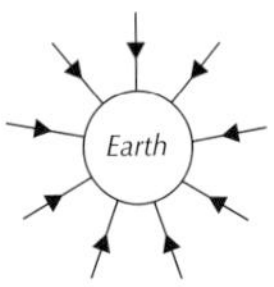

b)

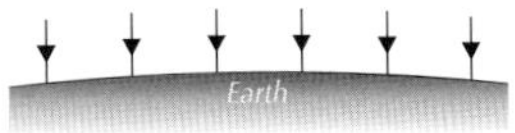

Q3 $F = -\frac{GMm}{r^2}$,

where F is the force on mass m due to mass M,
G is the gravitational constant, 6.67×10^{-11} Nm^{2}kg^{-2},
and r is the distance between M and m.

2. Gravitational Field Strength

Page 67 — Application Questions

Q1 $g = \frac{F}{m} = \frac{581}{105} = 5.5333... = \mathbf{5.53\ Nkg^{-1}}$ **(to 3 s.f.)**

Q2 $g = \frac{F}{m}$, so $F = gm$. The magnitude of the value of g is lower on the Moon, which means an object with the same mass will experience less force due to gravity (and so be easier to lift).

Q3 $g = -\frac{GM}{r^2} = -\frac{6.67 \times 10^{-11} \times 4.87 \times 10^{24}}{(6050 \times 10^3)^2}$

$= -8.874... = \mathbf{-8.87\ Nkg^{-1}}$ **(to 3 s.f.)**

Q4 a) $F = -\frac{GMm}{r^2}$

You just want to find the radius, so you're not really interested in direction, and you can just use $F = \frac{GMm}{r^2}$

Rearrange for r:

$r = \sqrt{\frac{GMm}{F}} = \sqrt{\frac{6.67 \times 10^{-11} \times 7.34 \times 10^{22} \times 65}{105}}$

$= 1.7408... \times 10^6 = \mathbf{1.7 \times 10^6\ m}$ **(to 2 s.f.)**

If you'd used the minus sign in the equation, you should have used $F = -105$ too, so the minus signs cancel out.

b) $g = -\frac{GM}{r^2} = -\frac{6.67 \times 10^{-11} \times 7.34 \times 10^{22}}{((1.7408... \times 10^6) + (640 \times 10^3))^2}$

$= -0.8636... = \mathbf{-0.86\ Nkg^{-1}}$ **(to 2 s.f.)**

Page 67 — Fact Recall Questions

Q1 g is the gravitational field strength (or the force per unit mass due to gravity), measured in Nkg^{-1}.

Q2 M is the mass of the object creating the gravitational field.

Q3 The correct graph is c).

3. Gravitational Potential and Energy

Page 72 — Application Questions

Q1 a) No change (G is a constant).

b) V_g will double (twice as negative) as it's related to $\frac{1}{r}$ ($V_g = -\frac{GM}{r}$).

c) No change (mass is the same everywhere).

d) g will be four times bigger as it's related to $\frac{1}{r^2}$ ($g = -\frac{GM}{r^2}$).

Q2 $\Delta W = m\Delta V_g = 1.72 \times 531$

$= 913.32$

$= \mathbf{913\ J}$ **(to 3 s.f.)**

Q3 Number of squares underneath the graph between 2×10^6 m and 5×10^6 m is approximately 2 squares. Each square on the graph is worth $20 \times (1 \times 10^6) = 2 \times 10^7$ Nm. Therefore work done $\approx 2 \times 2 \times 10^7 = \mathbf{4 \times 10^7\ J}$.

Q4 a) $E = -\frac{GMm}{r} = -\frac{6.67 \times 10^{-11} \times 7.34 \times 10^{22} \times 95.0}{-1.74 \times 10^6}$

$= 2.6729... \times 10^8 = \mathbf{2.67 \times 10^8\ J}$ **(to 3 s.f.)**

b) Using the equation for escape velocity:

$v = \sqrt{\frac{2GM}{r}} = \sqrt{\frac{2 \times (6.67 \times 10^{-11}) \times (7.34 \times 10^{22})}{(1.74 \times 10^6)}}$

$= 2.3722... \times 10^3$

$= \mathbf{2.37 \times 10^3\ ms^{-1}}$ **(to 3 s.f.) (= 2.37 kms^{-1})**

Page 72 — Fact Recall Questions

Q1 The gravitational potential at a point is the work done in moving a unit mass from infinity to that point.
It's measured in Jkg^{-1}.

Q2 The negative energy sign means that the gravitational potential is negative and increases towards zero as you move further away from the source of the gravitational field. At infinity, the gravitational potential is zero. You can think of this negative potential as being caused by you having to do work against the gravitational field to move an object out of it.

Q3

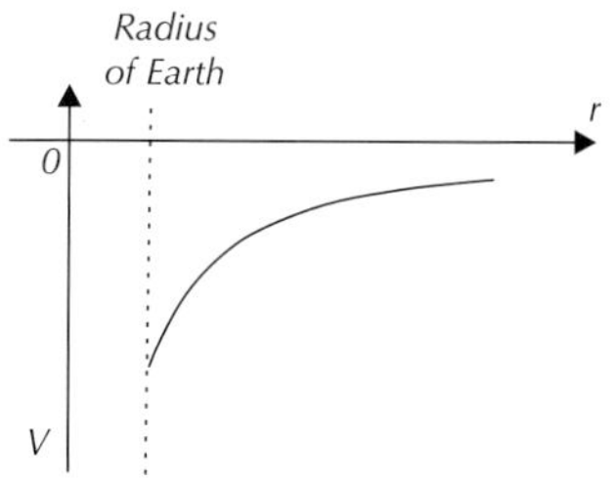

Q4 Escape velocity is the velocity needed so an object has just enough kinetic energy to escape a gravitational field.

4. Motion of Masses in Gravitational Fields

Page 76 — Application Questions

Q1 $\frac{mv^2}{r} = \frac{GMm}{r^2} \Rightarrow v^2 = \frac{GMmr}{r^2 m}$

Cancelling m's and r's: $v^2 = \frac{GM}{r}$

So $v = \sqrt{\frac{GM}{r}}$

Q2 $T^2 = \left(\frac{4\pi^2}{GM}\right)r^3 \Rightarrow M = \left(\frac{4\pi^2}{G}\right)\frac{r^3}{T^2}$

Convert T to s:

42 hours = 42 × 3600 = 151 200 s

$M = \left(\frac{4\pi^2}{6.67 \times 10^{-11}}\right)\frac{(3.95 \times 10^8)^3}{(151200)^2}$

$= 1.5955... \times 10^{27}$ kg = **1.6×10^{27} kg (to 2 s.f.)**

Q3 $T^2 \propto r^3$

So $\frac{T^2}{r^3} = \frac{T_{new}^2}{\left(\frac{r}{2}\right)^3}$

$\frac{T^2}{r^3} = \frac{8T_{new}^2}{r^3}$

$T^2 = 8T_{new}^2$

$T_{new}^2 = \frac{T^2}{8}$

$T_{new} = \frac{\mathbf{T}}{\sqrt{8}} = \frac{\mathbf{T}}{2\sqrt{2}}$

Page 76 — Fact Recall Questions

Q1 A satellite is a small mass that orbits a larger mass.

Q2 The orbital speed of a satellite is inversely proportional to the square root of the radius of its orbit ($v \propto \frac{1}{\sqrt{r}}$), so as the radius increases the speed decreases.

Q3 The orbital period of a satellite is proportional to the square root of the radius cubed ($T \propto \sqrt{r^3}$), so as the radius increases the orbital period increases.

Q4 A geostationary satellite is a satellite that orbits directly over the equator and is always above the same point on Earth. Their orbit takes exactly 1 day. Geostationary satellites are always above the same point of the Earth, so receivers don't need to be repositioned to keep up with them.

Q5 Kepler's first law: Each planet moves in an ellipse around the Sun, with the Sun at one focus.
Kepler's second law: A line joining the Sun to a planet will sweep out equal areas in equal times.
Kepler's third law: The period of the orbit and the mean distance between the Sun and the planet are related by $T^2 \propto r^3$.

Exam-style Questions — pages 78-79

1 B ***(1 mark)***

Gravitational field strength decreases as you move away from the mass by the inverse square law.

2 A ***(1 mark)***

3 a) Time taken for one orbit = T and distance for circular orbit = $2\pi r$, so speed = $\frac{\text{distance}}{\text{time}}$ becomes:

$v = \frac{2\pi r}{T} \Rightarrow T = \frac{2\pi r}{v}$ ***(1 mark)***

Substitute in expression for v:

$T = \frac{2\pi r}{v} = \frac{2\pi r}{\left(\sqrt{\frac{GM}{r}}\right)} = \frac{2\pi r\sqrt{r}}{\sqrt{GM}}$

So $T^2 = \left(\frac{4\pi^2}{GM}\right)r^3$ ***(1 mark)***

b) Kepler's third law states that $T^2 \propto r^3$, so:

$\frac{T_I^2}{r_I^3} = \frac{T_E^2}{r_E^3} \Rightarrow r_E^3 = T_E^2\frac{r_I^3}{T_I^2}$

So $r_E = \sqrt[3]{T_E^2\frac{r_I^3}{T_I^2}} = \sqrt[3]{3.6^2 \times \frac{420\,000^3}{1.8^2}}$

$= 666\,708.44... =$ **670 000 km (to 2 s.f.)**

(2 marks for correct answer, otherwise 1 mark for correct working.)

c) $\mathbf{g} = \frac{\mathbf{F}}{m} \Rightarrow \mathbf{F} = m\mathbf{g}$

$\mathbf{F}$ = 65 × 1.8 = 117 = **120 N (to 2 s.f.)** ***(1 mark)***

d) A line joining Jupiter to a Carpo will sweep out equal areas in equal times. ***(1 mark)***

4 a) A force field is a region in which a body experiences a non-contact force ***(1 mark)***.

b) E.g.

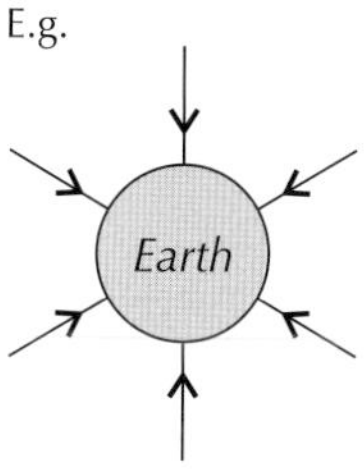

(1 mark for field lines pointing to the centre of Earth.)

c) No — the value of $\mathbf{g}$ close to the surface of the Earth is uniform, so the height difference in a classroom will have a negligible effect as the distance to the centre of the Earth is about the same for both students. ***(1 mark for correct answer and correct reasoning.)***

d) $\mathbf{g} = -\frac{GM}{r^2}$

$\Rightarrow r = \sqrt{-\frac{GM}{\mathbf{g}}} = \sqrt{-\frac{(6.67 \times 10^{-11}) \times (5.97 \times 10^{24})}{(-6.31)}}$

$= 7.94392... \times 10^6$ m $= 7943.92...$ km

So altitude = 7943.92... – 6370 = 1573.92...

= **1570 km (to 3 s.f.)**

(2 marks for correct answer, otherwise 1 mark for correct working.)

e) First find the distance r:

$V_g = -\frac{GM}{r}$

$\Rightarrow r = -\frac{GM}{V_g} = -\frac{(6.67 \times 10^{-11}) \times (5.97 \times 10^{24})}{-20.6 \times 10^6}$

$= 1.933... \times 10^7$ m

Then find $\mathbf{g}$:

$\mathbf{g} = -\frac{GM}{r^2} = -\frac{(6.67 \times 10^{-11}) \times (5.97 \times 10^{24})}{(1.933... \times 10^7)^2}$

$= -1.0656... = -$ **1.07 Nkg^{-1} (to 3 s.f.)**

(3 marks for correct answer, otherwise 1 mark for rearranging to make r the subject and 1 mark for correct r.)

f) $v = \sqrt{\frac{2GM}{r}} = \sqrt{\frac{2 \times (6.67 \times 10^{-11}) \times (5.97 \times 10^{24})}{1.933... \times 10^{7}}}$

$= 6.4187... \times 10^3 =$ **6.42×10^3 ms^{-1} (to 3 s.f.)**

(2 marks for correct answer, otherwise 1 mark for use of correct formula)

The value of r in this equation is the distance from the centre of the mass M (in this case Earth), so use the distance you found in e).

g) A geostationary orbit ***(1 mark)***.

5 a) You can use the graph to find the magnitude of the force, F, on the asteroid at a given distance r from the centre of the planet. So you can use Newton's law of gravitation $F = \frac{GMm}{r^2}$.

You don't need the minus sign since you're using the magnitude of the force, F.

Pick a set of values from the graph e.g. $F = 10 \times 10^5$ N and $r = 1 \times 10^9$ m

So $m = \frac{Fr^2}{GM} = \frac{10 \times 10^5 \times (1 \times 10^9)^2}{6.67 \times 10^{-11} \times 2.14 \times 10^{24}}$

$= 7.005... \times 10^9 = 7 \times 10^9$ kg (to 1 s.f.)

(2 marks for correct answer, otherwise 1 mark for correct working.)

b) The gravitational potential energy of the asteroid is the gravitational potential of the planet's gravitational field at a distance, r, from the planet multiplied by the asteroid's mass ***(1 mark)***. The gravitational potential is the work done in moving a unit mass from infinity to that point. Work is done moving a mass away from the planet since there is a force of attraction between them, so to move a mass towards a planet, the work done is negative. So the gravitational potential (and so the gravitational potential energy) is negative ***(1 mark)***.

c)

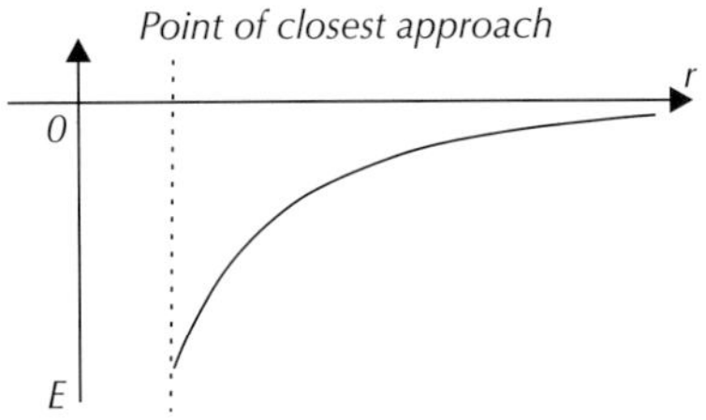

(1 mark for a graph below the r axis showing a 1/r relationship with E tending to zero as r tends to infinity)

Section 4 — Astrophysics and Cosmology

1. The Solar System

Page 81 — Fact Recall Questions

Q1 A cluster of stars and planets held together by gravity.

Q2 A star and all the objects which orbit it (and the objects which orbit those objects).

Q3 The Oort cloud.

Q4 The unit of distance that is equal to the mean distance between the Earth and the Sun.

2. Astronomical Distances

Page 84 — Application Questions

Q1 $20 \text{ pc} = 20 \times 3.1 \times 10^{16} \text{ m} = 62 \times 10^{16}$ m

$1 \text{ ly} = 9.5 \times 10^{15}$ m

so, $20 \text{ pc} = (62 \times 10^{16}) \div (9.5 \times 10^{15}) = 65.263...$ ly

$=$ **70 ly (to 1 s.f.)**

Q2 a) First convert 0.37 arcseconds into degrees and then radians:

$0.37 \times \frac{1}{3600} = 1.0277... \times 10^{-4\,\circ}$

$1.0277... \times 10^{-4\,\circ} \times \frac{\pi}{180} = 1.79... \times 10^{-6}$ rad

Then $d = \frac{r}{\theta} = \frac{1.50 \times 10^{11}}{1.79... \times 10^{-6}}$

$= 8.36... \times 10^{16}$ m $=$ **8.4×10^{16} m (to 2 s.f.)**

b) The distance to Sirius is $8.36... \times 10^{16}$ m.

1 light year $= 9.5 \times 10^{15}$ m,

so Sirius is $(8.36... \times 10^{16}) \div (9.5 \times 10^{15}) = 8.80...$

$= 8.8$ light years away (to 2 s.f.).

So light from Sirius will take **8 years and 10 months** to reach Earth.

Page 84 — Fact Recall Questions

Q1 Half the angle by which a nearby star appears to move in relation to the background stars in 6 months as the Earth moves between opposite points of its orbit.

Q2 A parsec is a unit of distance equal to 3.1×10^{16} m. A star is exactly one parsec (pc) away from Earth if the angle of parallax, $\theta = 1$ arcsecond $= \left(\frac{1}{3600}\right)^{\circ}$.

Q3 The distance that electromagnetic waves travel in a vacuum in one year.

3. Stellar Evolution

Page 88 — Fact Recall Questions

Q1 gravity

Q2 the main sequence

Q3 The pressure produced from the hydrogen fusion in the core of the star balances the gravitational force compressing the star, so it is stable.

Q4 When the core of a star runs out of hydrogen, hydrogen fusion, and the pressure caused by it, stops and the core starts to contract and heat up. The heat from the core of the star heats up the surrounding hydrogen layer until it is hot enough for the hydrogen to fuse into helium.

Q5 Core hydrogen burning stops in the star and the core of the star begins to contract, causing the outer layers to expand and cool, and the star becomes a red giant.

Q6 Once fusion stops in the core of a low-mass star, the pressure created by fusion is lost, so the core begins to contract under its own weight and heat up. Once the core gets to about the size of the Earth, electron degeneracy pressure stops it contracting any further. The outer layers of the star are ejected, and the hot dense core that is left over is a white dwarf.

Q7 The remnants of the outer layers of a red giant that are ejected as the star becomes a white dwarf.

Q8 The electron degeneracy pressure cannot withstand the gravitational forces at this mass, so when the fusion reactions have stopped, the core will continue to contract beyond the point at which a white dwarf would form.

Q9 The core contracts and the outer layers fall onto the core and rebound, causing a huge shockwave that propels the outer layers into space.

Q10 A neutron star. They are made mostly of neutrons.

Q11 A black hole is an object whose escape velocity is greater than the speed of light. They are formed when a star of core mass more than 3 solar masses contracts and collapses into an infinitely dense point.

4. Stellar Radiation and Luminosity

Page 92 — Application Questions

Q1 a) A is bigger, because A has the greater luminosity. Luminosity is proportional to surface area and the fourth power of temperature, and since A and B are both at the same temperature, A must be bigger.

b) Main sequence

c) White dwarf

d) Star B

Q2 a) Rigel's peak wavelength will be shorter than the Sun's, because Rigel has a higher temperature than the Sun.

b) $L = 4\pi r^2 \sigma T^4$

$$\text{so } r = \sqrt{\frac{L}{4\pi\sigma T^4}} = \sqrt{\frac{4.5\times10^{31}}{4\pi\times5.67\times10^{-8}\times12\,000^4}}$$

$= 5.518... \times 10^{10}$ m $=$ **5.5×10^{10} m (to 2 s.f.)**

Page 92 — Fact Recall Questions

Q1 A white star

Q2 $L = 4\pi r^2 \sigma T^4$, where L is the luminosity of the star, r is the radius of the star, σ is Stefan's constant and T is the surface temperature of the star.

Q3

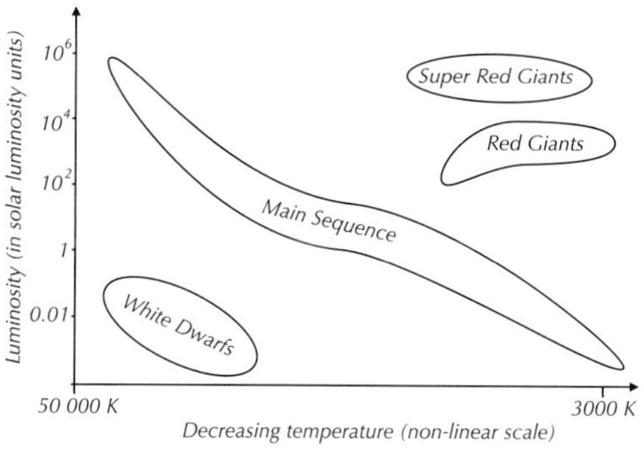

5. Stellar Spectra

Page 97 — Application Questions

Q1 B

Q2 $\Delta E = -3.4 - (-13.6) = 10.2 \text{ eV} = 10.2 \times 1.6 \times 10^{-19}$
$= 1.632 \times 10^{-18}$ J

$$\Delta E = hf, \text{ so } f = \frac{\Delta E}{h} = \frac{1.632\times10^{-18}}{6.63\times10^{-34}} = 2.461...\times10^{15}$$

= **2.5×10^{15} Hz (to 2 s.f.)**

Q3 $$\Delta E = \frac{hc}{\lambda} = \frac{6.63\times10^{-34}\times3.00\times10^{8}}{3.60\times10^{-7}} = 5.525\times10^{-19}\text{ J}$$

Convert to eV, $\Delta E = 5.525 \times 10^{-19} \div (1.60 \times 10^{-19})$
$= 3.453...$ eV

So original energy level $= -2.00 - 3.453... = -5.453...$
= **-5.45 eV (to 3 s.f.)**

Q4 $d\sin\theta = n\lambda$

$$\text{so, } d = \frac{n\lambda}{\sin\theta} = \frac{3\times4.7\times10^{-7}}{\sin(0.020)} = 0.000070504...\text{ m}$$

= **7.1×10^{-5} m (to 2 s.f.)**

Page 97 — Fact Recall Questions

Q1 The zero order maximum is a line of maximum brightness at the centre of a diffraction pattern. It's in line with the incident beam.

Q2 White light is made up of a range of different wavelengths. These diffract by different amounts when they pass through the diffraction grating, forming a spectrum.

Q3 A photon. The photon's energy is equal to the difference between the energy levels (the change in energy of the electron).

Q4 Each atom in the star absorbs particular wavelengths of radiation that correspond with the differences between its electron energy levels. So there will only be absorption lines in the spectra at the particular wavelengths corresponding to the elements found in the star.

6. The Big Bang Theory

Page 102 — Application Questions

Q1 1.4×10^7 pc = 14 Mpc
$v = H_0 d = 70 \times 14 = 980 \text{ kms}^{-1}$ = **9.8×10^5 ms^{-1} (to 2 s.f.)**

Q2 Convert v to ms^{-1}
$v = 463 \times 1000 = 463\,000$ ms^{-1}

$$\frac{\Delta\lambda}{\lambda} = \frac{v}{c}$$

$$\text{so } \Delta\lambda = \frac{v\lambda}{c} = \frac{463\,000\times0.211}{3\times10^{8}} = 0.00032564...\text{m}$$

= **3.26×10^{-4} m (to 3 s.f.)**

Page 102 — Fact Recall Questions

Q1 On a large scale the universe is homogeneous and isotropic and the laws of physics are the same everywhere.

Q2 The sound waves travelling in the same direction as the police car are 'bunched up' in front of the police car. The bunching up causes the frequency and therefore the pitch of the sound waves to be higher.

Q3 Red shift is where waves emitted from a source that is moving away from the observer are detected by the observer with a longer wavelength and lower frequency than they were emitted at. Blue shift is where waves emitted from a source that is moving towards the observer are detected by the observer with a shorter wavelength and higher frequency than they were emitted at.

Q4 $v \approx H_0 d$, where v is the recessional velocity of an object in kms^{-1}, H_0 is the Hubble constant in kms^{-1}Mpc^{-1} and d is the distance of the object from Earth in Mpc.
You could also have given the units as ms^{-1} for v, s^{-1} for H_0 and m for d.

Q5 The theory that the universe started off very hot and very dense and has been expanding ever since.

Q6 Cosmic microwave background radiation is electromagnetic radiation in the microwave region that is found everywhere in the universe, and is largely homogeneous and isotropic. It is a continuous spectrum of radiation that equates to a temperature of 2.7 K.

7. The Evolution of the Universe

Page 105 — Application Question

Q1 a) 13.7 billion years $= 13.7 \times 10^9 \times 365 \times 24 \times 60 \times 60$
$= 4.3204... \times 10^{17}$ s

$$t \approx H_0^{-1}, \text{ so } H_0 \approx t^{-1} \approx \frac{1}{4.3204...\times10^{17}}$$

$= 2.314... \times 10^{-18}$ s^{-1}
$= (2.314... \times 10^{-18} \times 3.1 \times 10^{22})$ ms^{-1}Mpc^{-1}
$= 71752.08...$ ms^{-1}Mpc^{-1}
$= (71752.08... \div 1000)$ kms^{-1}Mpc^{-1}
= **71.8 kms^{-1}Mpc^{-1} (to 3 s.f.)**

b) That H_0 has been constant since the universe began.

Page 106 — Fact Recall Questions

Q1 If you assume the expansion of the universe has been constant, then the age of the universe is H_0^{-1}, where H_0 is Hubble's constant.

Q2 We can only measure the size of the observable universe, because only the light from within that distance has had enough time to reach us.

Q3 a) The temperature is about 10^{12} K. E.g. quarks join together to form particles like protons and neutrons.

b) The temperature is about 3000 K. E.g. electrons combine with hydrogen and helium nuclei to form atoms.

c) The temperature is about 2.7 K. E.g. stars, galaxies and galactic clusters form, due to gravitational attraction from density fluctuations in the universe.

Q4 Dark matter could explain why galactic clusters have a greater mass than is calculated from their luminosity, and why stars at the edge of a galaxy move faster than they should.

Q5 MACHOs are Massive Compact Halo Objects, e.g. black holes and brown dwarfs.

Q6 WIMPs are Weakly Interacting Massive Particles.

Q7 The expansion of the universe is thought to be accelerating, not decelerating as astronomers expected — dark energy might explain this.

Exam-style Questions — pages 108-110

1 C ***(1 mark)***

What the star becomes depends on its core's mass compared to the Chandrasekhar limit, which is 1.4 solar masses. The ratio of the core mass to the Sun's mass is 1.09..., so the mass is below the Chandrasekhar limit, and the star becomes a white dwarf.

2 D ***(1 mark)***

3 B ***(1 mark)***

Calculate the energy of the photon using $E = hf$, and convert it to eV by dividing it by 1.6×10^{-19}. Then just add this number to −13.6 eV.

4 a)

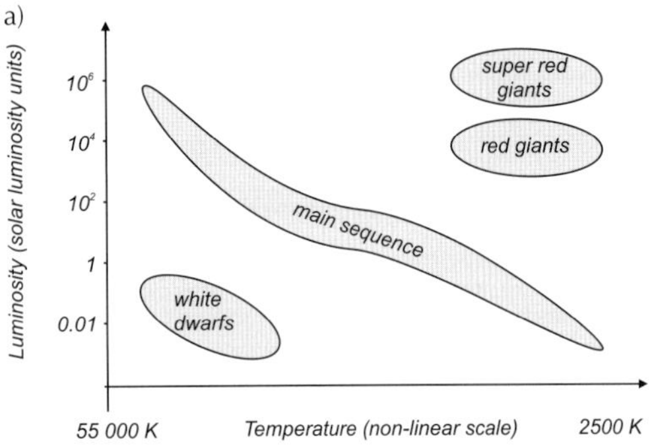

(1 mark for correct shape of main sequence, 1 mark for drawing the red giants and super red giants correctly, and 1 mark for drawing white dwarfs correctly.)

b)

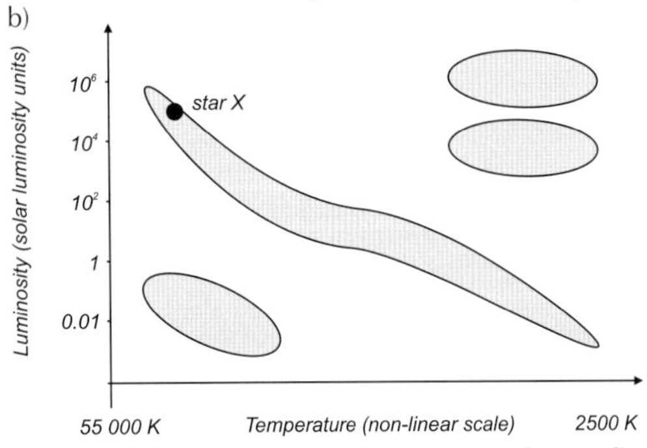

(1 mark for correctly plotting Star X on the HR diagram at a roughly correct position for its luminosity and temperature.)

c) When fusion in the core of the star stops, the core will begin to contract ***(1 mark)*** and the outer layers will be ejected in a supernova, leaving the core behind ***(1 mark)***. The gravitational forces will be so large that the core will not stop at a white dwarf or a neutron star, but will collapse to an infinitely dense point (a black hole) ***(1 mark)***.

5 a) Wien's displacement law states that:

$$\lambda_{max} \propto \frac{1}{T}, \text{ so } \lambda_{max}T = \text{constant}$$

$$\lambda_{maxMu}T_{Mu} = \lambda_{maxSun}T_{Sun}$$

$$T_{Mu} = \frac{\lambda_{maxSun}T_{Sun}}{\lambda_{maxMu}} = \frac{500.0 \times 10^{-9} \times 5800}{828.6 \times 10^{-9}} = 3499.87...\text{ K}$$

$$= \mathbf{3500\ K\ (to\ 2\ s.f.)}$$

(2 marks for correct answer, otherwise 1 mark for correct method)

b) $L = 4\pi r^2 \sigma T^4$, so

$$r = \sqrt{\frac{L}{4\pi\sigma T^4}}$$

$$= \sqrt{\frac{1.083 \times 10^{32}}{4\pi \times 5.67 \times 10^{-8} \times 3499.87...^4}}$$

$= 1.0064... \times 10^{12}$ m

$= \mathbf{1.0 \times 10^{12}}$ **m (to 2 s.f.)**

(2 marks for correct answer, otherwise 1 mark for correct method)

c) For a parallax angle in arcseconds, $p = \frac{1}{d}$

So, $d = \frac{1}{p}$

$$d = \frac{1}{5.5 \times 10^{-4}} = 1818.18...\text{ pc}$$

1 pc = 3.1×10^{16} m

1 ly = 9.5×10^{15} m

Convert d into light years:

$d = (1818.18... \times 3.1 \times 10^{16}) \div 9.5 \times 10^{15}$

$= 5933.014...$ ly

= 5900 ly (to 2 s.f.)

(3 marks for correct answer, otherwise 1 mark for correct method for calculation of distance in parsecs, and 1 mark for correct calculation to convert parsecs to light years)

d) The outer layers of the star collapse towards the centre and rebound off the core ***(1 mark)***. This triggers powerful shockwaves which cause a very bright explosion — this is a supernova ***(1 mark)***.

6 a) The universe started off very hot and very dense (perhaps as an infinitely hot, infinitely dense singularity) and expanded to form the present universe ***(1 mark)***.

b) Hubble's law says that the further away a galaxy is, the faster it is moving away from us. It suggests that the universe is expanding so at one point it must have been denser and hotter ***(1 mark)***.

c) $H_0 = 65 \text{ kms}^{-1}\text{Mpc}^{-1} = \frac{65 \times 10^3}{3.1 \times 10^{22}} = 2.09... \times 10^{-18}\text{ s}^{-1}$

$$t = \frac{1}{H_0} = \frac{1}{2.09... \times 10^{-18}} = 4.769... \times 10^{17}\text{ s}$$

= 15 billion years (to 2 s.f.)

(3 marks for correct answer, otherwise 1 mark for correctly converting H_0 to s^{-1} and 1 mark for substituting into correct formula)

d) E.g. Cosmic microwave background radiation: The Big Bang theory predicts that lots of gamma radiation was produced in the early universe ***(1 mark)***. The cosmic microwave background radiation is consistent with this prediction, as the predicted radiation would be Doppler shifted by the expansion of the universe from the gamma range of the spectrum to the microwave range ***(1 mark)***.

7 a) $d\sin\theta = n\lambda$, so:
$\lambda = d\sin\theta \div n = 5.4 \times 10^{-6} \times \sin(0.18) \div 2$
$= 4.833... \times 10^{-7}$ m
$=$ **4.8×10^{-7} m (to 2 s.f.)**
(2 marks for correct answer, otherwise 1 mark for correct method)

b) $\Delta E = \frac{hc}{\lambda} = \frac{6.63\times10^{-34}\times3\times10^{8}}{4.833...\times10^{-7}} = 4.114... \times 10^{-19}$ J
$= 4.114... \times 10^{-19} \div 1.60 \times 10^{-19}$
$= 2.57173...$ eV $=$ **2.6 eV (to 2 s.f.)**
(3 marks for correct answer, otherwise 1 mark for correct method and 1 mark for correct conversion from J to eV)

c) Electrons in hydrogen atoms absorb and emit photons by moving between energy levels ***(1 mark)***. Since the electron can only move between these discrete energy levels, they absorb the same energies (and so frequencies/wavelengths) of photon that they emit, and these energies would be different for different atoms ***(1 mark)***. The dark lines are caused as the light emitted from inside the Sun passes through hydrogen in the Sun, which absorbs these photons, and prevents them from reaching the observer ***(1 mark)***

d) $\frac{\Delta\lambda}{\lambda} = \frac{v}{c}$, so:
$v = \frac{\Delta\lambda c}{\lambda} = \frac{0.019\times10^{-9}\times3\times10^{8}}{4.833...\times10^{-7}}$
$= 11791.96...\ \text{ms}^{-1}$
$=$ **1.2×10^{4} ms^{-1} (to 2 s.f.)**
(2 marks for correct answer, otherwise 1 mark for correct method)

Module 6

Section 1 — Capacitors

1. Capacitors

Page 115 — Application Questions

Q1 $C = \frac{Q}{V}$ rearranged gives $Q = CV = 0.10 \times 230 =$ **23 C**

Q2 $W = \frac{1}{2}V^2C = \frac{1}{2} \times 230^2 \times 40 \times 10^{-3} = 1058$ J
$=$ **1060 J (to 3 s.f.)**

Q3 E.g. it would have to be very large to provide enough power, which might hinder the portability of the media player / it could only power the device for a short time, so it would need charging very often / the voltage would decrease as the capacitor discharged, so it would be difficult to produce a constant output.

Q4 $W = \frac{1}{2}\frac{Q^2}{C}$, so
$C = \frac{Q^2}{2W} = \frac{(2.25\times10^{-3})^2}{2\times1.30}$
$= 1.9471... \times 10^{-6}$ F
$=$ **1.95×10^{-6} F (to 3 s.f.)**

Page 115 — Fact Recall Questions

Q1 The capacitance of an object is the amount of charge it is able to store per unit potential difference (p.d.) across it.

Q2 Electrons flow from the power supply to one of the plates of the capacitor. The electrons cannot move across the gap between the plates, as they are separated by an electrical insulator, so a negative charge builds up on the plate. The negative charge on this plate repels electrons in the opposite plate. This causes electrons to flow away from the opposite plate, and a positive charge builds up on this plate.

Q3 Find the area under the V-Q graph.

Q4 $W = \frac{1}{2}QV$, $W = \frac{1}{2}V^2C$, $W = \frac{1}{2}\frac{Q^2}{C}$
Where W is energy stored by the capacitor, Q is the charge on the capacitor, V is the potential difference across the capacitor and C is the capacitance of the capacitor.

2. Capacitors in Circuits

Page 120 — Application Questions

Q1 In series, $\frac{1}{C_{total}} = \frac{1}{C_1} + \frac{1}{C_2}$.
If $C_1 = C_2 = C$, then $\frac{1}{C_{total}} = \frac{2}{C}$, so $C_{total} = \frac{C}{2}$
So the total capacitance of the circuit is half the capacitance of one capacitor. Potential difference, V, is fixed, so the total charge stored by the capacitors, Q, is also half the total charge stored by a single capacitor if it were connected on its own in the same circuit.
So, $Q = 1.2 \times 10^{-3} \div 2 = 6.0 \times 10^{-4}$ C
$Q = It$, so $I = Q \div t$
$t = 2 \times 60 = 120$ s
So, $I = 6.0 \times 10^{-4} \div 120$
$=$ **5.0×10^{-6} A**

Q2 a) Total capacitance is given by the gradient of the graph.
gradient $= (3.6 \times 10^{-3} - 0) \div (0.8 - 0)$
$= 4.5 \times 10^{-3}$ F
So the total capacitance is 4.5×10^{-3} F.

b) For capacitors connected in parallel, $C_{total} = C_1 + C_2 + ...$
Since all the capacitors have the same capacitance,
$C_{total} =$ number of capacitors $\times C$
So, number of capacitors $= C_{total} \div C$
$= 4.5 \times 10^{-3} \div 1.5 \times 10^{-3}$
$=$ **3**

Page 120 — Fact Recall Questions

Q1 $\frac{1}{C_{total}} = \frac{1}{C_1} + \frac{1}{C_2} + ...$

Q2 The total capacitance of the capacitors.

3. Investigating Charging and Discharging Capacitors

Page 124 — Application Question

Q1 $\frac{\Delta Q}{\Delta t} = -\frac{Q}{CR}$, so $Q = -\frac{CR\,\Delta Q}{\Delta t}$
Capacitor is discharging, so $\Delta Q = -85$ nC $= -8.5 \times 10^{-8}$ C
$Q = -\frac{4.2\times10^{-3}\times5.5\times10^{3}\times(-8.5\times10^{-8})}{4.5\times10^{-6}}$
$= 0.4363...$ C
$=$ **0.44 C (to 2 s.f.)**

Page 124 — Fact Recall Questions

Q1

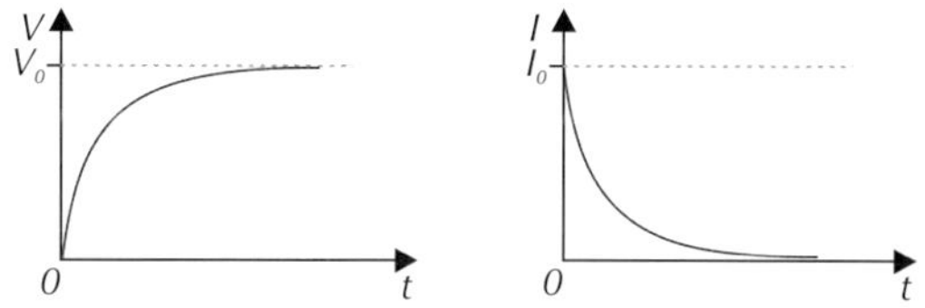

Q2 E.g. remove the power source and reconnect the circuit.

Q3 E.g. Attach a fully charged capacitor in a circuit containing an ammeter and a fixed resistor, and with a voltmeter/voltage sensor connected across the capacitor. Complete the circuit to allow the capacitor to discharge. Use a data logger connected to the ammeter and voltmeter/voltage sensor to measure the current through the resistor and the potential difference across the capacitor at different times. Once the capacitor has fully discharged, connect the data logger to a computer and use the results obtained by the data logger to plot graphs of current through the circuit and potential difference across the capacitor against time.

4. Charging and Discharging Calculations

Page 129 — Application Questions

Q1 $V = V_0 e^{-\frac{t}{CR}} = 15.0 \times e^{-\frac{1.5\times10^{-3}}{(0.60\times10^{-6})\times(2.6\times10^{3})}} = 5.7345...$
$= \mathbf{5.7\ V\ (to\ 2\ s.f.)}$

Don't forget to convert all the values you've been given into SI units.

Q2 When the capacitor is fully charged, the charge is $Q = CV = 15 \times 10^{-9} \times 230 = 3.45 \times 10^{-6}$ C
Then after 0.01 seconds,

$Q = Q_0 e^{-\frac{t}{CR}} = 3.45 \times 10^{-6} \times e^{-\frac{0.01}{(15\times10^{-9})\times(50\times10^{3})}}$
$= 5.587... \times 10^{-12} = \mathbf{6 \times 10^{-12}\ C\ (to\ 1\ s.f.)}$

Q3 a) The capacitor losing 63% of its charge is the same as the capacitor discharging to 37% of its original charge, which takes a time equal to the time constant, $\tau = CR$.
$\tau = CR = 15 \times 10^{-6} \times 400 \times 10^{3} = \mathbf{6\ s}$

b) $C = \frac{\tau}{R} = \frac{60}{400 \times 10^{3}} = \mathbf{1.5 \times 10^{-4}\ F}$

c) The capacitor would have to be a lot bigger than the original one. The resistance of the resistor could be increased instead.

Q4 $\frac{Q}{Q_0} = 0.3$

$\frac{Q}{Q_0} = e^{-\frac{t}{CR}}$ so $0.3 = e^{-\frac{20.0}{CR}} \Rightarrow \ln 0.3 = \ln(e^{-\frac{20.0}{CR}})$

$\Rightarrow \ln 0.3 = -\frac{20.0}{CR} \Rightarrow CR = -\frac{20.0}{\ln 0.3} = \mathbf{16.6\ s\ (to\ 3\ s.f.)}$

Q5 a) $\ln(V_0)$, where V_0 is the initial potential difference across the capacitor.

b) Vertical axis intercept = $\ln V_0 = 2.708$
Rearranging to find V_0:
$V_0 = e^{2.708} = 14.99... = \mathbf{15.0\ V\ (to\ 3\ s.f.)}$

c) The gradient of the graph is:
$-\frac{2.708 - 1.001}{2.00\times10^{-6} - 0} = -853\ 500$
The gradient $= -\frac{1}{CR}$ so $CR = -1 \div -853\ 500$
$= 1.171... \times 10^{-6}$
$= 1.17$ µs (to 3 s.f.)
So the time constant τ = **1.17 µs (to 3 s.f.)**

Page 129 — Fact Recall Questions

Q1 $Q = Q_0(1 - e^{-\frac{t}{CR}})$

Q2 $V = V_0 e^{-\frac{t}{CR}}$

Q3 The resistance of the circuit and the capacitance of the capacitor.

Q4 The time taken for the capacitor to discharge to $\frac{1}{e}$ (about 37%) of its original charge. It is given by $\tau = CR$.

Exam-style Questions — pages 131-132

1 A ***(1 mark)***.
The charge increases quickly and then levels off as it charges, and decreases quickly and then levels off as it discharges.

2 C ***(1 mark)***.
$Q = Q_0 e^{-\frac{t}{CR}}$ so $\frac{Q}{Q_0} = e^{-\frac{t}{CR}}$ where $\frac{Q}{Q_0}$ is 0.65.
So $\ln(0.65) = -\frac{1}{100\times10^{3}\times C}$
so $C = -\frac{1}{100\times10^{3}\times\ln(0.65)} = 2.3\times10^{-5}$ F (to 2 s.f.)

3 B ***(1 mark)***.
First find the total capacitance of the three capacitors in series, $\frac{1}{C_{series}} = \frac{1}{20} + \frac{1}{20} + \frac{1}{20} = \frac{3}{20}$, so $C_{series} = 6.66...$ µF.
Then find the total capacitance of the circuit by adding this value to the capacitance of the capacitor in parallel, $C_{total} = 12.0 + 6.66... = 18.66... = 18.7$ µF.

4 a) i)

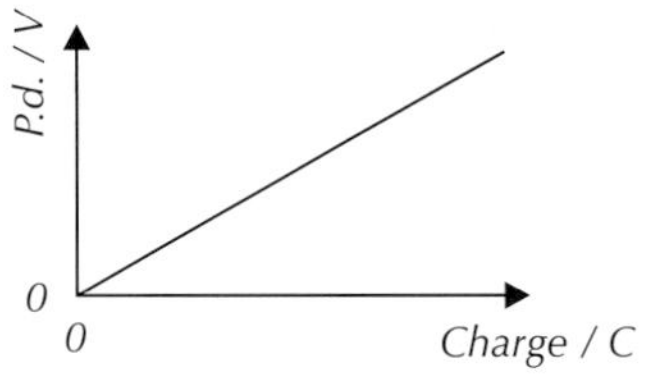

(1 mark for straight line through the origin)

ii) Energy stored by the capacitor ***(1 mark)***.

b) $C = \frac{Q}{V} = \frac{18\times10^{-6}}{12} = \mathbf{1.5 \times 10^{-6}\ F}$
(2 marks for correct answer, 1 mark for correct method if answer incorrect)
Remember, when a capacitor is fully charged, the potential difference across it is equal to the p.d. of the source used to charge it.

c) $V = V_0 e^{-\frac{t}{CR}} = 12 \times e^{-\frac{0.030}{1.5\times10^{-6}\times29\times10^{3}}}$
$= 6.020... = \mathbf{6.0\ V\ (to\ 2\ s.f.)}$
(2 marks for correct answer, 1 mark for correct method if answer incorrect)

d)

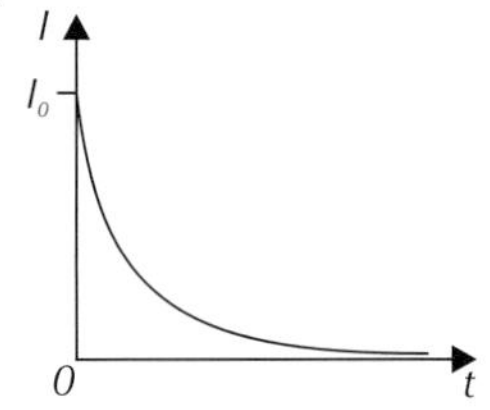

(1 mark for correct shape)

e) $I = I_0 e^{-\frac{t}{CR}}$, so $t = -CR\ln\left(\frac{I}{I_0}\right)$
Looking for when the current has fallen to 10% of its initial value, so $\frac{I}{I_0} = 0.1$
$t = -(1.5 \times 10^{-6} \times 29 \times 10^{3}) \times \ln(0.1)$
$= 0.1001...$ s $= \mathbf{0.10\ s\ (to\ 2\ s.f.)}$
(2 marks for correct answer, 1 mark for correct method if answer incorrect)

5 a) $W = \frac{1}{2}V^2C = \frac{1}{2} \times 50.0^2 \times 3.0 \times 10^{-3}$
$= 3.75$ J = **3.8 J (to 2 s.f.)**
(2 marks for correct answer, 1 mark for correct method if answer incorrect)

b) $\tau = CR = 2.0 \times 10^{3} \times 3.0 \times 10^{-3} = \mathbf{6\ s}$ ***(1 mark)***

c) After 14 s, the charge on the capacitor is:
$Q = Q_0(1 - e^{-\frac{t}{CR}}) = CV(1 - e^{-\frac{t}{CR}})$
$Q = 3.0\times10^{-3} \times 50.0(1 - e^{-\frac{14}{6}})$
$= 0.135... = \mathbf{0.14\ C\ (to\ 2\ s.f.)}$
(2 marks for correct answer, 1 mark for correct method if answer incorrect)

d)

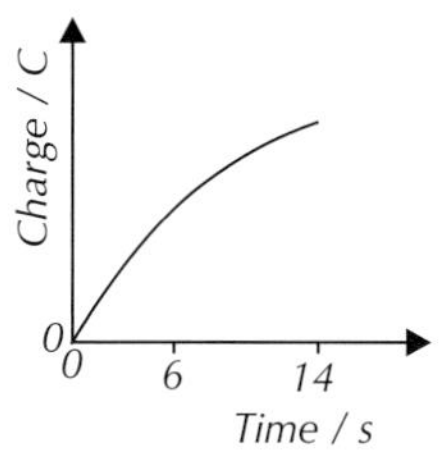

(1 mark for correct shape)

The charge should increase to around 63% at 6 s and 90% at 14 s.

Section 2 — Electric Fields

1. Electric Fields

Page 136 — Application Questions

Q1 $F = \frac{Qq}{4\pi\varepsilon_0 r^2}$

$= \frac{(-1.60 \times 10^{-19})^2}{4\pi \times (8.85 \times 10^{-12}) \times (5.22 \times 10^{-13})^2}$

$= 8.44784... \times 10^{-4}$

$= \mathbf{8.45 \times 10^{-4}}$ **N (to 3 s.f.)**

Q2 Start by rearranging the formula for the magnitude of $\boldsymbol{E}$ to make r the subject:

$\boldsymbol{E} = \frac{Q}{4\pi\varepsilon_0 r^2} \Rightarrow r^2 = \frac{Q}{4\pi\varepsilon_0 \boldsymbol{E}}$

$\Rightarrow r = \sqrt{\frac{Q}{4\pi\varepsilon_0 \boldsymbol{E}}}$

Then put the numbers in:

$r = \sqrt{\frac{4.15 \times 10^{-6}}{4\pi \times (8.85 \times 10^{-12}) \times 15\,000}}$

$= 1.57725...$

$= \mathbf{1.6}$ **m (to 2 s.f.)**

Q3 $\boldsymbol{E} = \frac{F}{Q} = \frac{0.080}{5.0 \times 10^{-5}} = \mathbf{1600\ NC^{-1}}$

Page 136 — Fact Recall Questions

Q1 $\boldsymbol{E}$ is a measure of the force per unit positive charge (in an electric field).

Q2

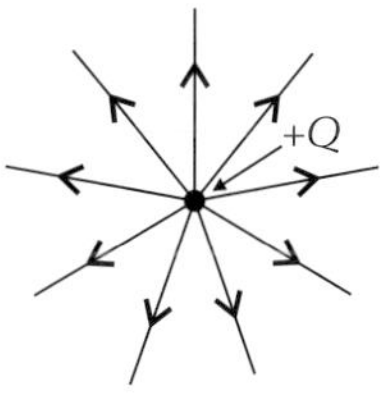

Q3 $\boldsymbol{F} = \frac{Qq}{4\pi\varepsilon_0 r^2}$

Q4

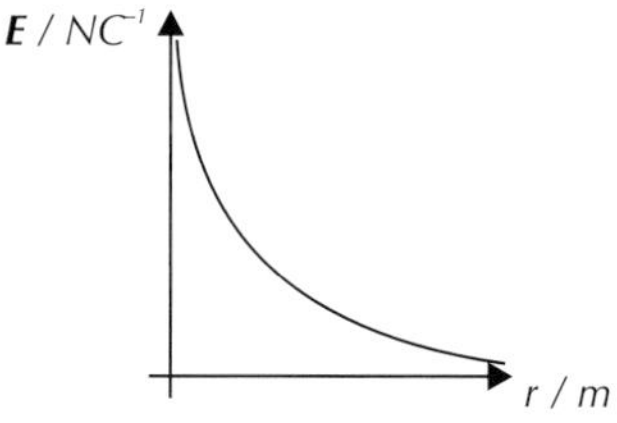

2. Uniform Electric Fields

Page 138 — Application Questions

Q1

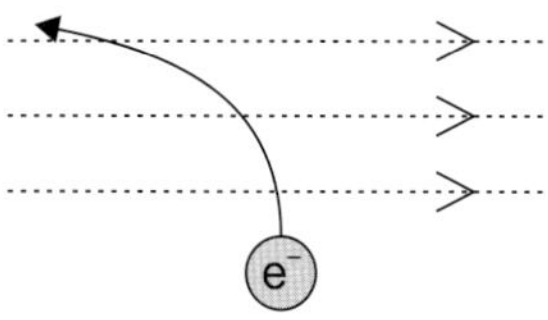

Q2 If the particle isn't moving, the upwards force from the electric field must balance the particle's weight, i.e.

$mg = \boldsymbol{E}Q = \frac{VQ}{d}$

This comes from weight $= mg$, $\boldsymbol{E} = \frac{F}{Q}$ and $\boldsymbol{E} = \frac{V}{d}$.

Rearrange this to make d the subject, then put in the numbers:

$mg = \frac{VQ}{d} \Rightarrow d = \frac{VQ}{mg}$

$= \frac{(5.00 \times 10^{-9}) \times 2(1.60 \times 10^{-19})}{(6.64 \times 10^{-27}) \times 9.81}$

$= 0.02456...\text{ m} = \mathbf{2.46}$ **cm (to 3 s.f.)**

Q3 $C = \frac{\varepsilon_0 A}{d} = \frac{8.85 \times 10^{-12} \times 7.0 \times 10^{-4}}{8.5 \times 10^{-6}} = 7.28823... \times 10^{-10}$

$= \mathbf{7.3 \times 10^{-10}}$ **F (to 2 s.f.)**

$(= 0.73\text{ nF})$

Page 138 — Fact Recall Question

Q1 $\varepsilon = \varepsilon_r \varepsilon_0$ where ε is the permittivity of the material, ε_r is the relative permittivity and ε_0 is the permittivity of free space.

3. Electric Potential

Page 141 — Application Question

Q1 a) $V = \frac{Q}{4\pi\varepsilon_0 r} = \frac{12.6 \times 10^{-6}}{4\pi \times 8.85 \times 10^{-12} \times 5.19 \times 10^{-2}}$

$= 2.18298... \times 10^6$

$= \mathbf{2.18 \times 10^6}$ **V (to 3 s.f.)**

b) First find the new potential at the centre of the smaller sphere:

$V_2 = \frac{Q}{4\pi\varepsilon_0 r}$

$= \frac{12.6 \times 10^{-6}}{4\pi \times 8.85 \times 10^{-12} \times ((5.19 \times 10^{-2}) + (12.9 \times 10^{-2}))}$

$= 6.26... \times 10^5\text{ V}$

The electric potential energy at a point $= Vq$, and work done is the difference in electric potential energy between two points, so:

Work done

$= V_2 q - V_1 q = (V_2 - V_1)q$

$= (2.18298... \times 10^6 - 6.26... \times 10^5) \times 0.152 \times 10^{-6}$

$= 0.2366163...$

$= \mathbf{0.237}$ **J (to 3 s.f.)**

You could also have worked out the electric potential energy at each point using $E = (Qq) \div (4\pi\varepsilon_0 r)$ and then found the difference between the two values.

c) $C = 4\pi\varepsilon_0 R = 4\pi \times 8.85 \times 10^{-12} \times 2.11 \times 10^{-2}$

$= 2.346... \times 10^{-12}$

$= \mathbf{2.35 \times 10^{-12}}$ **F (to 3 s.f.)**

Page 141 — Fact Recall Questions

Q1 The electric potential at a point is the work done to move a unit positive charge from infinity to that point in an electric field.

Q2 a)

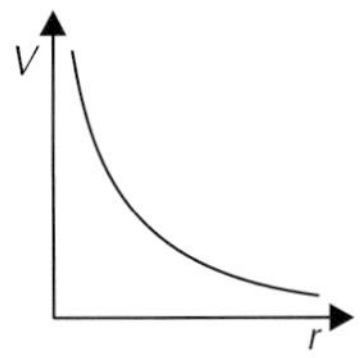

b)

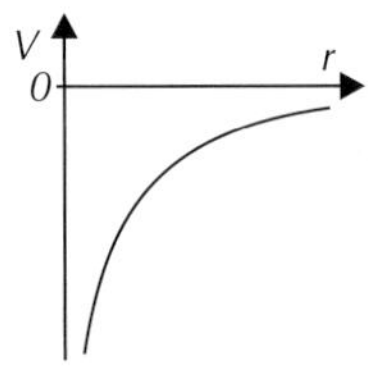

Q3 The equation for electric potential, $V = \frac{Q}{4\pi\varepsilon_0 r}$, and the charge on a capacitor $Q = CV$.

4. Comparing Electric and Gravitational Fields

Page 142 — Fact Recall Questions

Q1 Any three from: E.g.

1. Gravitational field strength $\boldsymbol{g}$ is the force a unit mass would experience in a gravitational field. Electric field strength $\boldsymbol{E}$ is the force a unit positive charge would experience in an electric field.
2. Newton's law and Coulomb's law are the same but with G switched for $\frac{1}{4\pi\varepsilon_0}$ and M and m switched for Q and q.
3. Field lines for a radial gravitational field and a radial electric field have the same shape.
4. Gravitational potential V_g and absolute electric potential V give the energy a unit mass or charge would have at a point.

Q2 Any two from: E.g.

1. Gravitational forces are always attractive, but electric forces can be attractive or repulsive.
2. Objects can be shielded from electric fields, but not from gravitational fields.
3. The size of an electric force is dependent on the medium between the charges, but for gravitational fields it makes no difference.

Exam-style Questions — pages 144-145

1 A ***(1 mark)***

There will be a uniform electric field pointing from the 100 V plate to the 0 V plate. The alpha particle is positively charged, so there will be a force on it in this direction due to the electric field.

2 B ***(1 mark)***

Capacitance of an isolated charged sphere is $C = 4\pi\varepsilon_0 R$, where R is the radius of the sphere.

3 C ***(1 mark)***

The magnitude of the force decreases as you move away from the charged sphere by the inverse square law.

4 A ***(1 mark)***

First find an equation for the electric field strength between the lithium nucleus and electron: $\mathbf{E}_1$ = electric field strength due to lithium nucleus electric field – electric field strength due to electron electric field. Remember, electric field strength is a vector, so you <u>take away</u> the electric field strength due to the electron. Charge of proton = $+Q$ and charge of electron = $-Q$, so $\mathbf{E}_1 = \frac{3(+Q)}{4\pi\varepsilon_0 r^2} - \frac{(-Q)}{4\pi\varepsilon_0 r^2} = \frac{4Q}{4\pi\varepsilon_0 r^2}$. Then find the electric field strength between the helium nucleus and the electron:

$\mathbf{E}_2 = \frac{2(+Q)}{4\pi\varepsilon_0 r^2} - \frac{(-Q)}{4\pi\varepsilon_0 r^2} = \frac{3Q}{4\pi\varepsilon_0 r^2}$. From the two equations you can see that $\frac{\mathbf{E}_1}{4} = \frac{\mathbf{E}_2}{3}$, so $\mathbf{E}_2 = \frac{3\mathbf{E}_1}{4}$.

5 a) Sphere A is stationary, so forces acting on it must be equal, so force due to gravity = force due to electric field.

$$\boldsymbol{F} = \frac{Qq}{4\pi\varepsilon_0 r^2} \Rightarrow r = \sqrt{\frac{Qq}{4\pi\varepsilon_0 \boldsymbol{F}}}$$
$$= \sqrt{\frac{(-92.5 \times 10^{-6})(-34.7 \times 10^{-6})}{4\pi \times 8.85 \times 10^{-12} \times 1.99}}$$
$$= 3.8083... = \mathbf{3.81\ m\ (to\ 3\ s.f.)}$$

(2 marks for correct answer, otherwise 1 mark for correct working.)

b) A uniform electric field is produced ***(1 mark)***, pointing from the plate with the more positive potential (the top plate) to the plate with the less positive potential (the bottom plate) ***(1 mark)***.

c) First calculate the electric field strength:

$$\boldsymbol{E} = \frac{\boldsymbol{F}}{Q} = \frac{0.18}{92.5 \times 10^{-6}} = 1945.9...\ \text{NC}^{-1}$$

Then calculate the electric potential:

$$\boldsymbol{E} = \frac{V}{d} \text{ so } V = \boldsymbol{E}d = 1945.9... \times 0.104$$
$$= 202.378...$$
$$= 200\ \text{V (to 2 s.f.)}$$

This is the potential difference between the plates, so the electric potential of the top plate is **200 V (to 2 s.f.)**.

(3 marks for the correct answer, otherwise 1 mark for calculating the electric field strength and 1 mark for correct working for calculating the electric potential.)

d) Rearrange $C = \frac{\varepsilon A}{d}$ for ε:

$$\varepsilon = \frac{Cd}{A} = \frac{34.5 \times 10^{-12} \times 0.104}{0.31}$$
$$= 1.157... \times 10^{-11}\ \text{Fm}^{-1}$$ ***(1 mark)***

This is larger than ε_0, so the space between the plates is not a vacuum ***(1 mark)***.

Section 3 — Electromagnetism

1. Magnetic Fields

Page 148 — Application Questions

Q1

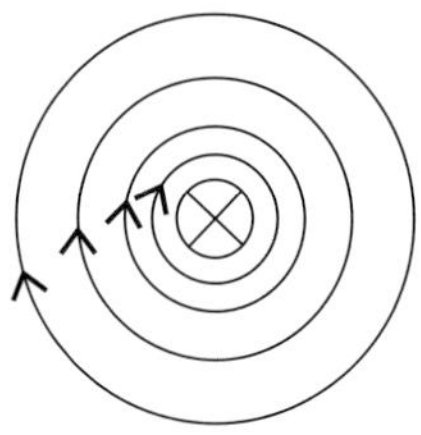

Q2 The force will act upwards.

Page 148 — Fact Recall Questions

Q1 Magnetic fields are found around permanent magnets and moving charges.

Q2 From north to south.

Q3 The strength of the magnetic field.

Q4 You can use your left hand. The first (index) finger represents the direction of the magnetic field, the second (middle) finger represents the direction of the current, and the thumb represents the direction of the force (or motion).

2. Magnetic Flux Density

Page 153 — Application Questions

Q1 a) $F = BIL\sin\theta = 0.20 \times 4.0 \times (25 \times 10^{-2}) \times \sin 53°$
$= 0.1597... =$ **0.16 N (to 2 s.f.)**

b) There is no force acting on the wire.

Q2 a) E.g.

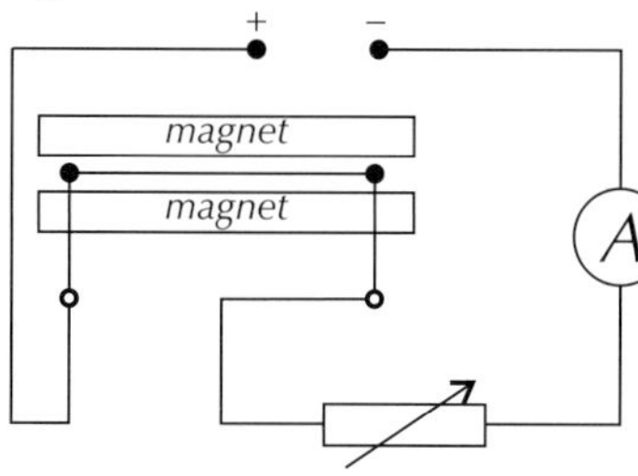

b) A variable resistor should be used so that the resistance, and therefore the current, can be changed (since $V = IR$) in order to determine a value for $\mathbf{B}$.

c) Only that length of wire will experience a vertical force, which will contribute to the reading on the mass balance.

d) Convert the mass readings into force using $\mathbf{F} = m\mathbf{g}$, then plot your data on a graph of force $\mathbf{F}$ against current I. Because $\mathbf{F} = \mathbf{B}IL$, the gradient of the line of best fit is equal to $\mathbf{B}L$. You know the length, L, so just divide the gradient by L to get the magnetic flux density, $\mathbf{B}$.

Page 153 — Fact Recall Questions

Q1 The current-carrying wire must be perpendicular to the magnetic field.

Q2 The force on a wire is $\mathbf{F} = \mathbf{B}IL \sin\theta$. At 90°, $\sin\theta = 1$, so the force is at a maximum.

3. Forces on Charged Particles

Page 158 — Application Questions

Q1 Only charged particles experience a force from magnetic fields (neutrons have no charge).

Q2 The force will act upwards (using the left-hand rule).
Because the particle is negatively charged, you need to point your second finger in the opposite direction to its actual motion.

Q3 $F = BQv$
$= (640 \times 10^{-3}) \times (3.2 \times 10^{-19}) \times (5.5 \times 10^{3})$
$= 1.1264 \times 10^{-15} =$ **1.1×10^{-15} N (to 2 s.f.)**

Q4 $v = \frac{E}{B} \Rightarrow E = Bv = (20.0 \times 10^{-3}) \times 5.2 \times 10^{6}$
$= 1.04 \times 10^{5} =$ **1.0×10^{5} NC^{-1} (to 2 s.f.)**

Q5 a) $F = \frac{mv^2}{r}$, $F = BQv$

$\Rightarrow \frac{mv^2}{r} = BQv$

$\Rightarrow B = \frac{mv^2}{Qvr}$

Cancelling the 'v's:
$B = \frac{mv}{Qr}$

b) $B = \frac{mv}{Qr} = \frac{(1.673 \times 10^{-27}) \times (1.99 \times 10^{7})}{(1.60 \times 10^{-19}) \times 5.49}$
$= 0.03790... =$ **0.0379 T (to 3 s.f.)**

Page 158 — Fact Recall Question

Q1 It will follow a circular path.

4. Magnetic Flux and Flux Linkage

Page 162 — Application Question

Q1 a) Flux linkage is at its greatest when the area is normal to the magnetic field. If the angle of the field with the normal to the area is increased, the flux linkage decreases.
This is because the number of field lines passing through the area decreases.

b) At highest value of flux linkage $\cos\theta = \cos 0 = 1$, so:

$N\phi = \mathbf{B}AN \Rightarrow \mathbf{B} = \frac{N\phi}{AN} = \frac{1.3 \times 10^{-6}}{(25 \times 10^{-4}) \times 10}$
$= 5.2 \times 10^{-5}\ \text{T} =$ **52 μT**

Don't forget to convert the area to m^2.

Q2 $N\phi = \mathbf{B}AN\cos\theta \Rightarrow \theta = \cos^{-1}\left(\frac{N\phi}{\mathbf{B}AN}\right)$

So $\theta = \cos^{-1}\left(\frac{(10.0 \times 10^{-3})}{(2.06 \times 10^{-3}) \times (145 \times 10^{-4}) \times (355)}\right)$
$= 19.429... =$ **19.4° (to 3 s.f.)**

Page 162 — Fact Recall Questions

Q1 Magnetic flux is measured in webers (Wb).

Q2 An e.m.f. (and a current if the conductor is part of a circuit).

Q3 The number of turns on the coil cutting the flux (N).

5. Faraday's Law and Lenz's Law

Page 167 — Application Questions

Q1 The plane's wings act as a conductor, and the Earth has a magnetic field that the plane cuts through, so electromagnetic induction occurs.

Q2 $\varepsilon = -\frac{\Delta(N\phi)}{\Delta t} = -\frac{N\Delta(\mathbf{B}A)}{\Delta t} = -\frac{50 \times (-1.5) \times 0.24}{0.32} = 56.25$

So magnitude of e.m.f. = **56 V (to 2 s.f.)**
Here the magnetic flux density decreases, so ΔB is negative.

Q3 a) The current is going away from the reader.
By Lenz's law, the e.m.f. produces a force that opposes the motion that caused it, so the force produced acts downwards. The direction of the field is to the right, so using Fleming's left-hand rule, the current induced must be in the direction going away from the reader.

b) Distance travelled by wire, $\mathbf{s} = \mathbf{v}\Delta t$.
So area of flux cut $A = L\mathbf{v}\Delta t$ where L = length of wire.
So total magnetic flux cut $\phi = \mathbf{B}A = \mathbf{B}L\mathbf{v}\Delta t$

$\varepsilon = -\frac{\Delta(N\phi)}{\Delta t} = -\frac{\Delta\phi}{\Delta t}$ (since $N = 1$), so

$\varepsilon = -\frac{\mathbf{B}L\mathbf{v}\Delta t}{\Delta t} = -\mathbf{B}L\mathbf{v}$

$= -(5.4 \times 10^{-3}) \times 1.2 \times 0.50 = -0.00324$ V
So magnitude of e.m.f. = **0.0032 V (to 2 s.f.)**

Page 167 — Fact Recall Questions

Q1 Faraday's law states that the induced e.m.f. is directly proportional to the rate of change of flux linkage (induced e.m.f., $\varepsilon = -\frac{\Delta N\phi}{\Delta t}$).

Q2 The negative of the induced e.m.f.

Q3 Calculate the negative of the area under the graph to get the total flux linkage change and divide this by the number of turns on the coil.

Q4 Lenz's law states that the induced e.m.f. is always in such a direction as to oppose the change that caused it.

6. Uses of Electromagnetic Induction

Page 172 — Application Questions

Q1 a) $\frac{n_s}{n_p} = \frac{V_s}{V_p}$, so $V_s = V_p\frac{n_s}{n_p} = 190 \times \frac{420}{250} = 319.2$

$= \textbf{320 V (to 2 s.f.)}$

b) $\frac{I_p}{I_s} = \frac{V_s}{V_p}$, so $I_p = \frac{I_s V_s}{V_p} = \frac{13 \times 75}{120} = 8.125$

$= \textbf{8.1 A (to 2 s.f.)}$

Q2 a) E.g. By using a laminated core.

b) (i) A step-down transformer.

(ii) $\frac{n_s}{n_p} = \frac{V_s}{V_p} \Rightarrow n_s = n_p\frac{V_s}{V_p}$

$n_s = 110 \times \frac{19}{230} = 9.086...$

= **9 turns (to the nearest whole number)**

Page 172 — Fact Recall Questions

Q1 Generators convert kinetic energy into electrical energy. They induce an electric current by rotating a coil in a magnetic field. The output voltage and current change direction with every half rotation of the coil, producing an alternating current.

Q2 An alternating current flowing in the primary (or input) coil produces a changing magnetic field in the iron core. The changing magnetic field is passed through the iron core to the secondary (or output) coil, where it induces an alternating voltage (e.m.f.) of the same frequency as the input voltage. The more turns there are in the secondary coil compared to the primary coil, the bigger the induced e.m.f. is compared to the input e.m.f..

Remember, e.m.f. is just another way of saying voltage.

Q3 Any one from: the current through the primary coil / the voltage across the primary coil.

Exam-style Questions — pages 174-175

1 C ***(1 mark)***

Magnetic flux cut through, $\Delta\phi = BA = BLv\Delta t$ where L is the length of the rod and v is the velocity of the rod.

Faraday's law: $\varepsilon = -\Delta(N\phi) \div \Delta t = -\Delta\phi \div \Delta t$ (as $N = 1$)

So $\varepsilon = -BLv\Delta t \div \Delta t = -BLv$

Rearranging: $v = -\varepsilon \div (BL)$

$= -(4.5 \times 10^{-3}) \div ((21 \times 10^{-3}) \times (33 \times 10^{-2}))$

$= -0.649...$

So the velocity of the bar = $0.65\ ms^{-1}$ (to 2 s.f.)

2 A ***(1 mark)***

Convert v to ms^{-1}: $42\ kms^{-1} = 42 \times 10^3\ ms^{-1}$

$v = E \div B \Rightarrow B = E \div v = (37 \times 10^3) \div (42 \times 10^3)$

$= 0.8809... = 0.88$ T (to 2 s.f.)

3 a) The particle will follow a circular path ***(1 mark)***.

b) $F = BQv$

$= 0.93 \times (1.60 \times 10^{-19}) \times (8.1 \times 10^7) = 1.20528 \times 10^{-11}$

$= \mathbf{1.2 \times 10^{-11}}$ **N (to 2 s.f.)**

The force would act downwards.

(1 mark for correct force and 1 mark for direction.)

The magnitude of the charge on an electron is given in the exam data and formulae booklet. Find the direction of the force with Fleming's left-hand rule — because electrons have a negative charge, point your second finger in the opposite direction to its velocity.

c) $F_{+2e} = 2 \times F_{-e} = 2 \times 1.20528 \times 10^{-11} = 2.41056 \times 10^{-11}$

The force will be $\mathbf{2.4 \times 10^{-11}}$ **N (to 2 s.f.) upwards**.

(1 mark for correct force and 1 mark for direction.)

If you didn't get the correct answer for b), you'll get the marks for c) as long as the force is double the force for the electron and in the opposite direction.

4 a) Downwards ***(1 mark)***

Fleming's left-hand rule can be used to find the direction of the force on the wire. The force on the cradle is in the opposite direction.

b) E.g. The magnetic flux density can be found by varying the current using the variable d.c. power supply and noting the mass shown on the top pan balance for a range of currents ***(1 mark)***. After converting the masses to weights using $\boldsymbol{W = mg}$ ***(1 mark)***, a graph of force against current can be plotted ***(1 mark)***. Drawing a line of best fit and finding the gradient gives you BL, so dividing the gradient by L gives the magnetic flux density $\boldsymbol{B}$ ***(1 mark)***.

c) $F = BIL \Rightarrow I = \frac{F}{BL}$

$I = \frac{68 \times 10^{-3}}{44 \times 10^{-3} \times 52 \times 10^{-2}} = 2.972... = \mathbf{3.0\ A}$ **(to 2 s.f.)**

(2 marks for correct answer, otherwise 1 mark for correct working)

5 a) The magnetic field produced by a primary coil carrying direct current is constant after being switched on, so will not produce a changing magnetic field in the core ***(1 mark)***. Faraday's law dictates that a changing magnetic field is required in the core for an e.m.f. to be generated in the secondary coil ***(1 mark)***.

b) $\frac{V_s}{V_p} = \frac{n_s}{n_p} \Rightarrow n_s = n_p \times \frac{V_s}{V_p}$

$n_s + n_p = 250$ so $n_p = 250 - n_s$

So $n_s = (250 - n_s) \times \frac{V_s}{V_p} = \left(250 \times \frac{V_s}{V_p}\right) - \left(n_s \times \frac{V_s}{V_p}\right)$

Bringing the n_s's to one side:

$$n_s \times \left(1 + \frac{V_s}{V_p}\right) = \left(250 \times \frac{V_s}{V_p}\right) \Rightarrow n_s = \frac{\left(250 \times \frac{V_s}{V_p}\right)}{\left(1 + \frac{V_s}{V_p}\right)}$$

$$n_s = \frac{\left(250 \times \frac{7}{230}\right)}{\left(1 + \frac{7}{230}\right)} = 7.383... = \textbf{7 turns (to 1 s.f.)}$$

(2 marks for correct answer, otherwise 1 mark for correct working.)

c) How to grade your answer (pick the description that best matches your answer):

0 marks: There is no relevant information.
1-2 marks: A diagram has been drawn, and method points 1, 3 and 4 have been covered. One accuracy point has been mentioned. Answer is basic and has lack of structure and information is missing.
3-4 marks: A labelled diagram has been drawn and method points 1, 3, 4, 5 and 6 have been covered. Two accuracy points have been mentioned. Answer has some structure, and information is presented fairly clearly.
5-6 marks: A labelled diagram has been drawn, all of the method points have been covered. All accuracy points have been mentioned. The answer is well structured and information is clearly presented.
Here are some points your answer may include:
Diagram

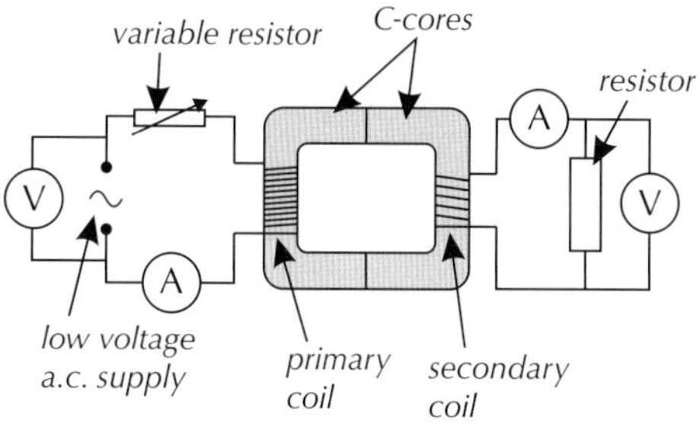

Method
1. Put two C-cores together and wrap wire around each to make the coils.
2. Make sure the two coils have different numbers of turns.
3. Turn on the a.c. supply to the primary coil.
4. Adjust the variable resistor to change the input current.
5. Record the current through and voltage across each coil for a range of input currents.
6. Keep numbers of turns on the coils constant throughout.
7. Use the results to see how the ratio of the currents through the coils affects the ratio of the voltages across the coils.
8. Results won't quite give $\frac{V_s}{V_p} = \frac{I_p}{I_s}$ due to energy losses in the transformer.

Accuracy
1. Use low currents.
2. Use a laminated core.
3. Use low-resistance wire.

Section 4 — Nuclear and Particle Physics

1. Atomic Structure

Page 179 — Application Questions

Q1 a) 8
b) 8 + 8 = **16**
c) $^{16}_{8}O$

Q2 a) $^{45}_{21}X$ (the nucleon number is 21 + 24 = 45)
b) E.g. $^{46}_{21}X$
Isotopes have the same number of protons but a different number of neutrons. So the proton number must be the same as in part a), but the nucleon number must be different.

Q3 a) 2
b) 2
Isotopes have the same proton number.

Page 179 — Fact Recall Questions

Q1 The fact that most alpha particles passed straight through the foil showed that most of the atom is empty space.
Some positively-charged alpha particles were deflected by the atom by a large angle, showing that there must be a nucleus with a large positive charge.
Very few alpha particles bounced back, showing that the nucleus is very small.
High momentum alpha particles were deflected backwards by the nucleus, showing that most of the mass of the atom must be in the nucleus.

Q2 Inside an atom there is a nucleus which contains neutrons and protons. Electrons orbit the nucleus. Most of the atom is empty space, as the electrons orbit at relatively large distances.

Q3 The number of protons in the nucleus of an atom.

Q4 The total number of protons and neutrons in the nucleus of an atom.

Q5 Atoms with the same number of protons but different numbers of neutrons.

2. The Nucleus

Page 183 — Application Questions

Q1 Rearrange $R = r_0A^{1/3}$ for A,
$A = \left(\frac{R}{r_0}\right)^3 = \left(\frac{7}{1.4}\right)^3 = 5^3 = 125$
number of protons = A – number of neutrons
= 125 – 73
= **52**

Q2 a) For separations less than around 0.5 fm the strong nuclear force is repulsive. So at 0.4 fm it would be repulsive.
b) For separations between about 0.5 fm and 3 fm the strong nuclear force is attractive. So at 1.5 fm it would be attractive.
c) For separations bigger than about 3 fm the strong nuclear force has little effect. So at 4.2 fm, the strong interaction is small enough that it can be ignored.

Q3 a) $\rho = \frac{m}{V}$,
Model the atom as a sphere, so $V = \frac{4}{3}\pi r^3$
So, $\rho = \frac{3m}{4\pi r^3}$
Rearrange for r:
$r = \left(\frac{3m}{4\pi\rho}\right)^{\frac{1}{3}}$
$= \left(\frac{3 \times 8.5 \times 10^{-26}}{4\pi \times 8.4 \times 10^4}\right)^{\frac{1}{3}}$
$= 6.22802... \times 10^{-11}$ m
= **6.2×10^{-11} m (to 2 s.f.)**

b) $\rho = \frac{3m}{4\pi r^3}$
$= \frac{3 \times 8.5 \times 10^{-26}}{4\pi \times (5.2 \times 10^{-15})^3}$
$= 1.44317... \times 10^{17}$ kg m^{-3}
= **1.4×10^{17} kg m^{-3} (to 2 s.f.)**

Page 183 — Fact Recall Questions

Q1 Approximate diameter of an atom = 0.1 nm (1×10^{-10} m)
Approximate diameter of a small nucleus = a few fm

Q2 Nuclear radius increases proportionally with the cube root of the nucleon number ($R = r_0A^{1/3}$).

Q3 The nucleus is assumed to be a sphere and its volume is calculated using $V = \frac{4}{3}\pi r^3$

Q4 a)

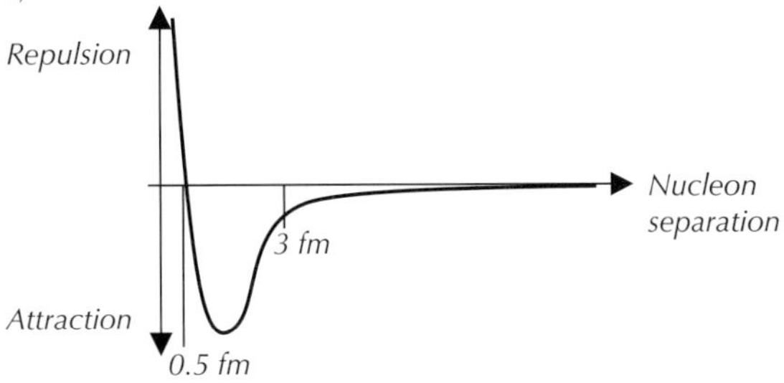

b) Between 0 and about 0.5 fm.
c) Between about 0.5 and 3 fm.

3. Particles and Antiparticles

Page 188 — Application Questions

Q1 $E_\gamma = 2mc^2$
$= 2 \times 1.673 \times 10^{-27} \times (3.00 \times 10^8)^2$
$= 3.0114 \times 10^{-10}$ J
$=$ **3.01×10^{-10} J (to 3 s.f.)**

Q2 a) $E_\gamma = mc^2$
$= 1.983 \times 10^{-27} \times (3.00 \times 10^8)^2$
$= 1.7847 \times 10^{-10}$ J
$=$ **1.78×10^{-10} J (to 3 s.f.)**

b) $E_\gamma = hf$, so:
$$f = \frac{E_\gamma}{h} = \frac{1.7847 \times 10^{-10}}{6.63 \times 10^{-34}}$$
$= 2.6918... \times 10^{23}$ Hz $=$ **2.69×10^{23} Hz (to 3 s.f.)**
$E_\gamma = \frac{hc}{\lambda}$, so
$$\lambda = \frac{hc}{E} = \frac{6.63 \times 10^{-34} \times 3.00 \times 10^8}{1.7847 \times 10^{-10}}$$
$= 1.1144... \times 10^{-15}$ m $=$ **1.11×10^{-15} m (to 3 s.f.)**

Q3 $E_\gamma = 2mc^2$, and $E_\gamma = hf$, so
$hf = 2mc^2$, so:
$$f = \frac{2mc^2}{h} = \frac{2 \times 1.88 \times 10^{-28} \times (3.00 \times 10^8)^2}{6.63 \times 10^{-34}}$$
$= 5.10407... \times 10^{22}$ Hz $=$ **5.10×10^{22} Hz (to 3 s.f.)**

Page 188 — Fact Recall Questions

Q1 Hadrons
Q2 Any one from: the weak nuclear force / gravity / the electrostatic force.
Q3 The neutrino has zero charge and (almost) zero mass. It is a lepton.
Q4 a) An antiparticle has the opposite charge to its corresponding particle.
b) An antiparticle has the same mass as its corresponding particle.
Q5 The positron
Q6 a) –1
b) 0
c) 0
Q7 $\Delta E = \Delta mc^2$, where ΔE is the change in energy, Δm is the change in mass, and c is the speed of light.
Q8 Energy can be converted into mass and produce particles, if there is enough energy. The mass is always produced in a particle-antiparticle pair.
Q9 Two photons

4. Quarks and Anti-quarks

Page 191 — Application Questions

Q1 a) Charge $= \frac{2}{3} + \left(-\frac{2}{3}\right) = 0$
b) Charge $= \frac{2}{3} + \frac{2}{3} + \left(-\frac{1}{3}\right) = 1$
c) Charge $= \left(-\frac{1}{3}\right) + \left(-\frac{1}{3}\right) + \left(-\frac{1}{3}\right) = -1$
d) Charge $= \left(-\frac{1}{3}\right) + \left(-\frac{1}{3}\right) + \left(-\frac{1}{3}\right) = -1$

Q2 The particle has a charge of –1, and contains 3 quarks, only one of which is a strange quark. The only two quarks which can be added to a strange quark to make –1 are 2 down quarks. Therefore, $\Sigma^- =$ dds.

Q3
$$\begin{matrix} d & s & s & \rightarrow & X & d & s & + & d & \bar{u} \\ -\frac{1}{3} & -\frac{1}{3} & -\frac{1}{3} & & X & -\frac{1}{3} & -\frac{1}{3} & + & -\frac{1}{3} & -\frac{2}{3} \end{matrix}$$
So, $-1 = X + \left(-\frac{2}{3}\right) + (-1) \Rightarrow X - \frac{2}{3} = 0$
$X = +\frac{2}{3}$
The only quark with a charge of $+\frac{2}{3}$ is the up quark. Therefore, X is an up quark.

Page 191 — Fact Recall Questions

Q1 Quarks: up (u), down (d) and strange (s). Anti-quarks: anti-up ($\bar{u}$), anti-down ($\bar{d}$) and anti-strange ($\bar{s}$).
Q2 $\bar{u} = -\frac{2}{3}e$ $\quad \bar{d} = +\frac{1}{3}e$ $\quad \bar{s} = +\frac{1}{3}e$
Q3 proton = uud
neutron = udd
Q4 In beta-minus decay, a down quark decays into an up-quark via the weak interaction, emitting an electron and an antineutrino.
Q5 $u \rightarrow d + {}^{0}_{+1}e + \nu$
Q6 charge

5. Radioactive Decay

Page 195 — Application Question

Q1 a) Since the radiation is blocked by aluminium, but not by paper, it must be beta-minus radiation.
b) The Geiger-Muller tube is still detecting counts from background radiation, so the count rate does not drop to zero (even though no radiation from the source can reach it).

Page 195 — Fact Recall Questions

Q1 When an unstable atomic nucleus breaks down to become more stable, by releasing energy and/or particles.
Q2 a) alpha radiation
b) gamma radiation
c) beta-minus radiation
Q3 Alpha radiation, beta-minus radiation, beta-plus radiation and gamma radiation.

6. Nuclear Decay Equations

Page 197 — Application Questions

Q1 Beta-minus decay involves the conversion of a neutron to a proton when an electron is emitted (along with the release of an antineutrino). So nucleon number A remains constant, and proton number increases by 1:
${}^{137}_{55}Cs \longrightarrow {}^{137}_{56}Ba + {}^{0}_{-1}\beta + {}^{0}_{0}\bar{\nu}$
You can check your answer by making sure that the total charge (proton number) on the left of the equation balances the total charge on the right. Do the same for the nucleon number.

Q2 $^{211}_{85}\text{At} \longrightarrow {}^{207}_{83}\text{Bi} + {}^{4}_{2}\alpha$
Q3 $^{26}_{13}\text{Al} \longrightarrow {}^{26}_{12}\text{Mg} + {}^{0}_{+1}\beta + {}^{0}_{0}\nu$

7. Exponential Law of Decay

Page 201 — Application Questions

Q1 $A = \lambda N = (1.1 \times 10^{-13}) \times (4.5 \times 10^{18})$
$= 4.95 \times 10^5 =$ **5.0×10^5 Bq (to 2 s.f.)**

Q2 Time in seconds $= t = 6.5 \times 3600 = 23\,400$ s
Activity at time t:
$A = A_0 e^{-\lambda t} = (3.2 \times 10^3) \times e^{-(1.3 \times 10^{-4}) \times 23400}$
$= 152.76... =$ **150 Bq (to 2 s.f.)**

Q3 Time in seconds $= t = 35 \times 60 = 2100$ s
Number of unstable nuclei in sample at time t:
$N = N_0 e^{-\lambda t} = (2.5 \times 10^{15}) \times e^{-(9.87 \times 10^{-3}) \times 2100}$
$= 2.49... \times 10^6$
$= 2.5 \times 10^6$ (to 2 s.f.)

Page 201 — Fact Recall Questions

Q1 a) The probability of an atomic nucleus decaying per unit time.
b) The number of atomic nuclei that decay per second.

Q2 a) $\frac{\Delta N}{\Delta t} = -\lambda N$,
where $\frac{\Delta N}{\Delta t}$ is the rate of change of the number of undecayed nuclei, λ is the decay constant and N is the number of undecayed nuclei in sample.
You could also define ΔN and Δt individually as ΔN is the change in number of undecayed nuclei, Δt is the change in time.
b) E.g. spreadsheet modelling

Q3

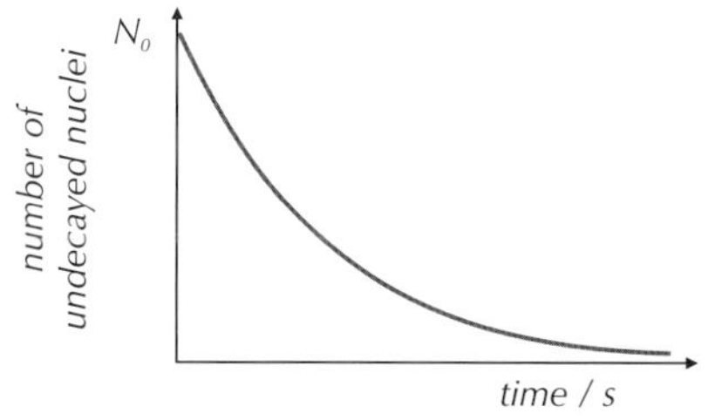

Q4 Rolling a 6 on a dice is a random event with a constant probability, similar to the event of an undecayed nucleus decaying.

8. Half-life and Radioactive Dating

Page 206 — Application Questions

Q1 Find the number of times the activity halved in 24 hours — count the number of times, n, you have to divide 2400 by 2 to reach 75.
$n = 5$, therefore $t_{1/2} = 24 \div 5 =$ **4.8 hours**

Q2 Decay constant $\lambda = \frac{\ln 2}{t_{1/2}} = \frac{0.693...}{483} = 1.435... \times 10^{-3}$
$= 1.44 \times 10^{-3}$ s^{-1} (to 3 s.f.)

Q3 Half-life of carbon-14 in seconds:
$t_{1/2} = 5730 \times (365 \times 24 \times 60 \times 60) = 1.8070... \times 10^{11}$ s
Decay constant $\lambda = \frac{\ln 2}{t_{1/2}} = \frac{\ln 2}{1.8070... \times 10^{11}}$
$= 3.8358... \times 10^{-12}$ s^{-1}
Activity when animal died $= A_0 = 1.2$ Bq
Activity now $= A = 0.45$ Bq
Activity decreases exponentially according to: $A = A_0 e^{-\lambda t}$
Rearranging this for time $t = -\frac{1}{\lambda} \times \ln\left(\frac{A}{A_0}\right)$
$= 2.5569... \times 10^{11}$ s $=$ **8100 years (to 2 s.f.)**

Page 206 — Fact Recall Questions

Q1 The time it takes for the number of undecayed nuclei in a sample of an isotope (or the sample's activity or count rate) to halve.

Q2 Read off the activity at $t = 0$. Halve this value, draw a horizontal line to the curve, then a vertical line down to the x-axis. Read off the time — this the half-life. Find the time taken for the activity to fall to a quarter of the original value and divide by two to check.

Q3 E.g. Set up a Geiger-Müller tube, held steady in a clamp stand, and attached to a Geiger counter. Find the background count by recording the total background counts in e.g. 30 s. Repeat this measurement twice, and calculate an average. Divide this average by the number of seconds to find the background count rate. This value should be subtracted from all subsequent measurements of count rate. Place the sample of Nobelium-255 close to the detector of the Geiger-Müller tube, within the range of an alpha-particle to ensure that the radiation reaches the detector. Record the number of counts in 10 seconds, divide this by 10 and subtract the background count rate to find the count rate due to the Nobelium. Repeat this every 30 s for around 5 minutes. Plot your results on a graph of count rate against time, and draw a line of best fit. Find the half-life from your graph by reading off how long it takes for the count rate to drop to half its initial value.

Q4 E.g. carbon-14.

9. Binding Energy

Page 209 — Application Questions

Q1 Mass defect is equivalent to binding energy.
Convert the mass defect from u to kg:
mass defect $= 0.0989 \times 1.661 \times 10^{-27} = 1.6427... \times 10^{-28}$ kg
$\Delta E = \Delta mc^2 = 1.6427... \times 10^{-28} \times (3.00 \times 10^8)^2$
$= 1.4784... \times 10^{-11}$ J
$= ((1.4784... \times 10^{-11} \div 1.60 \times 10^{-19}) \div 10^6)$ MeV
$= 92.4035...$ MeV
= 92.4 MeV (to 3 s.f.)

Q2 Mass of nucleus $= 15.994915$ u
Number of protons $= 8$, number of neutrons $= 8$
Mass of proton $= 1.00728$ u, Mass of neutron $= 1.00867$ u
Total mass of nucleons $= (8 \times 1.00728\text{ u}) + (8 \times 1.00867\text{ u})$
$= 16.1276$ u
So mass defect = mass of nucleons – mass of nucleus
$= 16.1276\text{ u} - 15.994915\text{ u}$
= 0.132685 u

Q3 Binding energy = binding energy per nucleon × nucleon number $= 8.79\text{ MeV} \times 56 = 492.24$ MeV
Convert from binding energy to mass defect using $\Delta E = \Delta mc^2$
First, convert binding energy from MeV to J
$\Delta E = (492.24 \times 10^6) \times 1.6 \times 10^{-19} = 7.875... \times 10^{-11}$ kg
mass defect $= \Delta m = \frac{\Delta E}{c^2} = \frac{7.875... \times 10^{-11}}{(3.00 \times 10^8)^2}$
$= 8.750... \times 10^{-28}$ kg
$= (8.750... \times 10^{-28} \div 1.661 \times 10^{-27})$ u
$= 0.52684...$ u
= 0.527 u (to 3 s.f.)

Q4 a) 8.6 MeV
Just read the value of the binding energy per nucleon at a nucleon number of 100 from the graph.
b) Multiply the binding energy per nucleon by the nucleon number.
Total binding energy $= 8.6 \times 100 =$ **860 MeV**

Page 209 — Fact Recall Questions

Q1 The energy needed to separate all the nucleons in a nucleus. It is equivalent to the mass defect.

Q2

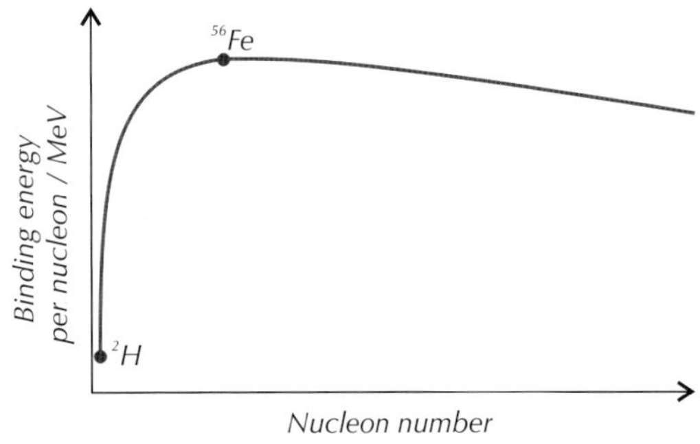

10. Nuclear Fission and Fusion

Page 213 — Application Questions

Q1 a) fusion
b) fission
c) fusion
d) fission

If the nucleon number is less than 56, fusion is more energetically favourable. If it's more than 56, fission is more energetically favourable.

Q2 $^{2}_{1}H + ^{3}_{1}H \rightarrow ^{4}_{2}He + ^{1}_{0}X$

X has a nucleon number of 1 and a proton number of 0. So particle X is a neutron.

Q3 $\Delta m = m_p + m_p - m_{H\text{-}2} - m_e - m_\nu$

$= 1.00728 + 1.00728 - 2.01355 - 0.00055 - 0$

$= 0.00046$ u

$= (0.00046 \times 1.661 \times 10^{-27})$ kg

$= 7.6406 \times 10^{-31}$ kg

$\Delta E = \Delta mc^2$

$= 7.6406 \times 10^{-31} \times (3.00 \times 10^8)^2$

$= 6.87654 \times 10^{-14}$ J

In MeV: $\Delta E = ((6.87654 \times 10^{-14}) \div (1.60 \times 10^{-19})) \div 10^6$ MeV

$= 0.4297... =$ **0.430 MeV (to 3 s.f.)**

Q4 First, calculate the number of neutrons produced by balancing the nucleon number A:

$A_{before} = 1 + 235$, $A_{after} = 94 + 139 + x$

Therefore number of neutrons produced = $x = 3$

So $\Delta m = m_n + m_{U\text{-}235} - m_{Zr\text{-}94} - m_{Te\text{-}139} - 3m_n$

$= 1.00867 + 234.99333 - 93.88431 - 138.90613 - (3 \times 1.00867)$

$= 0.18555$ u

$= (0.18555 \times 1.661 \times 10^{-27})$ kg

$= 3.08198... \times 10^{-28}$ kg

$\Delta E = \Delta mc^2$

$= 3.08198... \times 10^{-28} \times (3.00 \times 10^8)^2$

$= 2.77378... \times 10^{-11}$ J

In MeV: $\Delta E = ((2.77378... \times 10^{-11}) \div (1.60 \times 10^{-19})) \div 10^6$

$= 173.361... =$ **173 MeV (to 3 s.f.)**

Page 213 — Fact Recall Questions

Q1 Nuclear fission is when a larger nucleus splits into two smaller nuclei. It can be induced by firing low energy neutrons at the nucleus. When the nucleus absorbs one of these neutrons, it becomes unstable and will undergo fission.

Q2 The fusing of two smaller nuclei to form one larger nucleus.

Q3 Nuclei are positively charged, so they repel each other via the electrostatic interaction. For fusion to take place, the temperature must be very high so that the nuclei can travel fast enough (and so have enough energy) to overcome this repulsion (so that the strong nuclear interaction can take over and attract them together).

11. Fission Reactors

Page 216 — Fact Recall Questions

Q1 When the neutrons released by nuclear fission cause other nuclei to fission and release more neutrons — and so on.

Q2 The neutrons need to be slowed down in order for uranium nuclei to absorb them, to induce fission.

Q3 The control rods are made of a material that absorbs neutrons (e.g. boron). By inserting them into the reactor, the number of neutrons in the reactor can be limited. The rate of reaction can therefore be controlled by changing how far the control rods are inserted into the reactor.

Q4 The material used for the coolant needs to be a liquid or a gas at room temperature so it can be pumped around the reactor. It also needs to be efficient at transferring heat from the reactor. Water could be used as a coolant.

Q5 Any one of: e.g. facilities to store nuclear waste must be built deep underground, which can permanently change the landscape. / Leaked nuclear waste can contaminate water supplies and damage organisms.

Exam-style Questions — pages 218-220

1 B ***(1 mark)***

2 D ***(1 mark)***

A, B and C cannot occur because charge is not conserved.

3 A ***(1 mark)***

The number of atoms drops to a quarter of its initial value in 10.4 years, so the half-life of cobalt-60 is 5.2 years. Convert this to seconds, and substitute into the equation $\lambda t_{½} = \ln(2)$ to find the decay constant.

4 a) Any one from: e.g. both hadrons and leptons feel and interact with the weak nuclear force / both hadrons and leptons can feel and interact with the electrostatic force (if they are charged) ***(1 mark)***.

Any one from: e.g. hadrons can feel and interact with the strong nuclear force, but leptons cannot / leptons are fundamental particles, but hadrons are not (they are made up of quarks) ***(1 mark)***.

b) Total charge of the Δ^{++} hadron = +2e.

Charge on an up quark = $+\frac{2}{3}e$

So, $2e = \frac{2}{3}e + \frac{2}{3}e + x$

$x = 2e - (\frac{2}{3}e + \frac{2}{3}e) = +\frac{2}{3}e$

The only quark with a charge of $+\frac{2}{3}e$ is the up quark, so quark x is an up quark ***(1 mark)***.

c) mass = 2.19×10^{-27} kg

charge = $-2e$

(1 mark)

d) $\Delta E = \Delta mc^2$

For minimum energy in pair production, Δm = mass of particle + mass of antiparticle = $2m_{strange}$

So, $\Delta E = 2m_{strange}c^2$

$= 2 \times 1.71 \times 10^{-28} \times (3.00 \times 10^8)^2$

$= 3.078 \times 10^{-11}$ J

$= ((3.078 \times 10^{-11}) \div (1.60 \times 10^{-19})) \div 10^6$

$= 192.375$ MeV

$=$ **192 MeV (to 3 s.f.)**

(3 marks for correct answer, otherwise 1 mark for correct method and 1 mark for correct conversion to MeV)

5 a) There are 94 protons.
There are 240 – 94 = 146 neutrons. ***(1 mark)***

b) The strong nuclear force ***(1 mark)***.

c) ${}^{240}_{94}Pu \rightarrow {}^{236}_{92}U + {}^{4}_{2}\alpha$
(2 marks — 1 mark for U numbers correct and 1 mark for α numbers correct)

d) With nine α decays, the proton number will decrease from 94 by 18 to 76. The proton number of Tl is 81, so you need **five β^- decays**.
(2 marks for correct answer, otherwise 1 mark for correct working if answer incorrect)

e) In beta-minus decay, a down quark emits an electron and an antineutrino ***(1 mark)*** and decays into an up quark via the weak interaction ***(1 mark)***.

6 a) Initial activity = 72 Bq.
72 ÷ 2 = 36 Bq
Time when the activity has dropped to 36 Bq = 10 mins.
So, half life = **10 mins *(1 mark)***

b) $\lambda t_{1/2} = \ln(2)$, so $\lambda = \frac{\ln(2)}{t_{1/2}}$
$t_{1/2}$ = 10 mins = 10 × 60 = 600 s
$\lambda = \frac{\ln(2)}{600} = 0.011552...\ \text{s}^{-1}$
Activity at 17 mins = 22 Bq
Rearrange the activity equation, $A = \lambda N$, for number of undecayed nuclei:
$N = \frac{A}{\lambda} = \frac{22}{0.011552...} = 19043.57...$ = **20 000 (to 1 s.f.)**
(3 marks for correct answer, otherwise 1 mark for correct calculation of the decay constant and 1 mark for correct method)

If you got a different answer to part a) and used that answer in this question, you'll still get all the marks, as long as your method is correct.

c) Minimum energy of photon produced by annihilation is given by the mass of one of the annihilated particles converted into energy, $E_\gamma = mc^2$.
The energy of a photon is related to the frequency by $E = hf$. So:
$hf = mc^2$
Rearrange for f:
$f = \frac{mc^2}{h} = \frac{9.11 \times 10^{-31} \times (3.00 \times 10^8)^2}{6.63 \times 10^{-34}}$
$= 1.23665... \times 10^{20}$ Hz
= **1.24×10^{20} Hz (to 3 s.f.)**
(3 marks for correct answer, otherwise 1 mark for correctly equating the two equations and 1 mark for correct substitution)

7 a) $R = r_0 A^{1/3}$
$= 1.4 \times 10^{-15} \times 33^{1/3}$
$= 4.4905... \times 10^{-15}$ m
= **4.5×10^{-15} m (to 2 s.f.)**
(2 marks for correct answer, otherwise 1 mark for correct working if answer incorrect)

b) $\rho = \frac{m}{V}$
Volume of a sphere, $V = \frac{4}{3}\pi r^3$, so:
$\rho = \frac{3m}{4\pi R^3}$, where R is the radius of the nucleus.
Convert mass to kg, $m = (32.97 \times 1.661 \times 10^{-27})$ kg
$= 5.476... \times 10^{-26}$ kg
$\rho = \frac{3 \times 5.476... \times 10^{-26}}{4\pi \times (4.4905... \times 10^{-15})^3}$
$= 1.44378... \times 10^{17}\ \text{kg m}^{-3}$
= **$1.4 \times 10^{17}\ \text{kg m}^{-3}$ (to 2 s.f.)**
(3 marks for correct answer, otherwise 1 mark for correct conversion of mass and 1 mark for correct working)

c) ${}^{33}_{15}P \longrightarrow {}^{33}_{16}S + {}^{0}_{-1}\beta + {}^{0}_{0}\bar{\nu}$
(2 marks — 1 mark for correctly labelled sulfur isotope, 1 mark for correctly labelled beta-minus particle and antineutrino.)

d) Half-life in s = 25.4 × 24 × 3600 = 2 194 560 s
$\lambda t_{1/2} = \ln(2)$
So rearrange:
$\lambda = \ln(2) \div t_{1/2} = \ln(2) \div 2\ 194\ 560$
$= 3.158... \times 10^{-7}$ = **$3.16 \times 10^{-7}\ \text{s}^{-1}$ (to 3 s.f.)**
(2 marks for the correct answer, otherwise 1 mark for correct working if answer incorrect)

e) Number of undecayed atoms given by: $N = N_0 e^{-\lambda t}$
Rearrange for time: $t = -\ln\left(\frac{N}{N_0}\right) \times \frac{1}{\lambda}$
So $t = -\ln\left(\frac{7.0 \times 10^{13}}{1.6 \times 10^{15}}\right) \times \frac{1}{3.158... \times 10^{-7}}$
$= 9.907... \times 10^6$ s
= 114.67... days
= **110 days (to 2 s.f.)**
(2 marks for the correct answer, otherwise 1 mark for correct working if answer incorrect)

8 a) The energy needed to pull all the nucleons in a nucleus apart/the energy released when a nucleus forms ***(1 mark)***.

b)

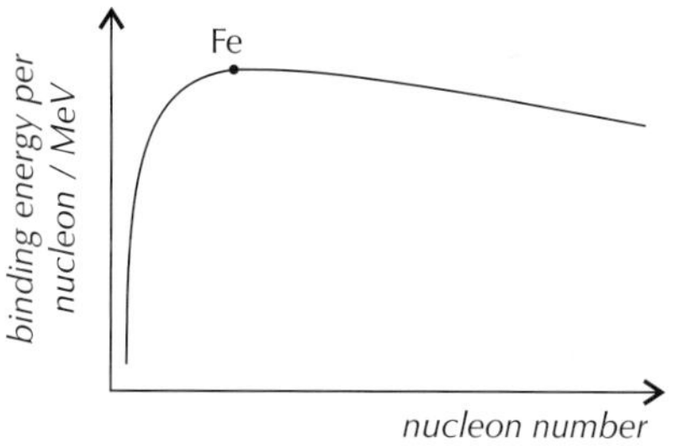

(2 marks — 1 mark for the overall shape, 1 mark for indicating Fe is at the peak of the curve)

c) Binding energy is equivalent to the mass defect.
$\Delta E = \Delta mc^2$
So binding energy per nucleon = $\frac{\Delta mc^2}{\text{nucleon number}}$
Convert mass defect to kg:
$0.62065\ \text{u} = (0.62065 \times 1.661 \times 10^{-27})$ kg
$= 1.0308... \times 10^{-27}$ kg
Binding energy per nucleon
$= \frac{1.0308... \times 10^{-27} \times (3.00 \times 10^8)^2}{66}$
$= 1.4057... \times 10^{-12}$ J
$= ((1.4057... \times 10^{-12}) \div (1.60 \times 10^{-19})) \div 10^6$
= 8.7860... MeV = **8.79 MeV (to 3 s.f.)**
(2 marks for correct answer, otherwise 1 mark for correct method if answer incorrect)

d) Use conservation of charge to find proton number of ${}^{94}Sr$:
a = 92 – 54 = **38**
Use conservation of nucleon number to find number of neutrons: b = (235 + 1) – (140 + 94) = **2**
(2 marks — 1 mark for correctly calculating each of a and b.)

e) The total binding energy of the final nuclei is greater than the binding energy of the initial nucleus ***(1 mark)***. An increase in binding energy means an increase in the total mass defect, so some of the mass from the initial nucleus is converted into energy ***(1 mark)***.

f) $\Delta m = (234.99333 + 1.00867) - (139.89194 + 93.89446 + (2 \times 1.00867))$
$= 0.19826$ u
$= (0.19826 \times 1.661 \times 10^{-27})$ kg $= 3.293... \times 10^{-28}$ kg
$\Delta E = \Delta mc^2 = 3.293... \times 10^{-28} \times (3.00 \times 10^8)^2$
$= 2.9637... \times 10^{-11}$ J
$= ((2.9637... \times 10^{-11}) \div (1.60 \times 10^{-19})) \div 10^6$ MeV
$= 185.23... =$ **185 MeV (to 3 s.f.)**

(3 marks for the correct answer, otherwise 1 mark for correctly calculating the mass defect and 1 mark for attempting to convert it to binding energy.)

If you used your answer from 8d), but got 8d) wrong, you'd still get all the marks for this question if your method is correct.

Section 5 — Medical Imaging

1. X-ray Imaging

Page 224 — Application Questions

Q1 $E_{max} = e \times V = 1.60 \times 10^{-19} \times 85.0 \times 10^3 =$ **1.36×10^{-14} J**
Assume here that all of the kinetic energy of the electrons is converted to X-rays.

Q2 $I = I_0 e^{-\mu x} = 30.0 \times e^{-(2.5 \times 0.69)} = 5.345...$ **5.3 Wm^{-2} (to 2 s.f.)**

Page 224 — Fact Recall Questions

Q1 a)

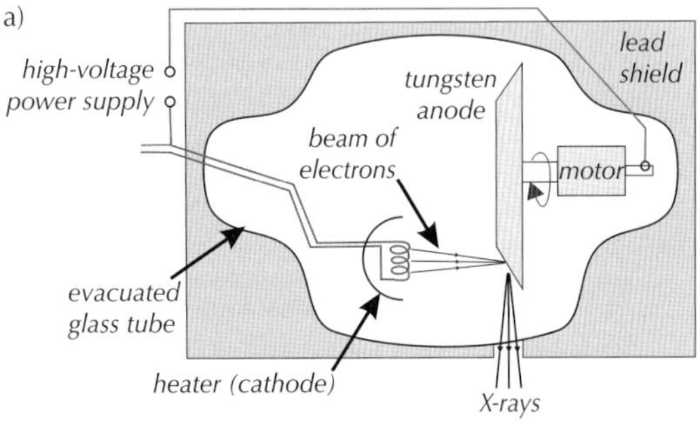

b) A lot of the electrons' energy is converted to heat, so rotating the anode helps spread the heat and prevents it from overheating.

Q2 Simple scattering, the photoelectric effect, Compton effect, pair production.

Q3 A barium meal contains atoms with a high atomic number and so it shows up clearly in X-ray images compared to soft tissue, which is made up of material with lower atomic numbers. A patient can be given a barium meal to swallow. This helps improve the contrast between tissues from the digestive tract and surrounding tissues on an X-ray image.

Q4 A computed axial tomography (CAT) scan is a scan in which an X-ray tube is rotated around the body and the emitted X-ray beam in the shape of a fan is picked up by detectors. These feed a signal to a computer, which works out how much attenuation has been caused by each part of the body and produces a high quality image.

Q5 Any two from: e.g. CAT scans can be processed by a computer to produce more detailed images than regular X-rays / CAT scans are better at imaging soft tissues / data from CAT scanners can be manipulated to generate a 3D image.

2. Medical Uses of Nuclear Radiation

Page 227 — Fact Recall Questions

Q1 The tracer is bound to a compound that is naturally used by the tissue or organ that is being investigated. The tracer is injected into or swallowed by the patient and moves through the body to the region of interest. The radiation emitted due to the isotope in the tracer is then detected by a gamma camera/PET scanner and sent to a computer to produce an image.

Q2 E.g. Technetium-99m. The (gamma) radiation it emits can easily pass outside the body and be detected. It has a half-life of 6 hours, which is long enough for data to be recorded, but short enough to limit the patient's exposure to radiation to an acceptable level. It also decays to a much more stable isotope.

Q3 PET scans.

Q4 Lead shielding — prevents radiation from other sources being detected by the camera.
Lead collimator — only lets gamma rays travelling parallel to the holes in the collimator pass through.
Scintillator — A sodium iodide crystal that produces a flash of light when hit by a gamma ray.
Photomultiplier tubes — detect the flashes of light and turn them into electrical signals.
Electronic circuit — collects the electrical signals and sends the information to a computer which forms an image.

Q5 The patient is injected with glucose that is bound to a radiotracer. The radiotracer emits positrons that annihilate with electrons in the body, producing gamma photons which are detected outside the body. Tissues in the body with a high rate of metabolic activity have a high rate of uptake of glucose, so these areas will have accumulated more radiotracer than areas of low metabolic activity. This means the PET scanner will detect more gamma radiation in areas of high metabolic activity. This information can be sent to a computer to build up an image of metabolic activity in the body.

3. Medical Uses of Ultrasound

Page 232 — Application Questions

Q1 a) $Z = \rho c \Rightarrow c = \frac{Z}{\rho} = \frac{1.5 \times 10^6}{1000} =$ **1500 ms^{-1}**

b) $\frac{I_r}{I_0} = \frac{(Z_2 - Z_1)^2}{(Z_2 + Z_1)^2} = \frac{((1.5 \times 10^6) - (8.0 \times 10^6))^2}{((1.5 \times 10^6) + (8.0 \times 10^6))^2}$
$= 0.4681...$

So the percentage $= 0.4681... \times 100 =$ **47% (to 2 s.f.)**

Q2 $\frac{\Delta f}{f} = \frac{2v\cos\theta}{c} \Rightarrow \Delta f = \frac{2vf\cos\theta}{c}$

$\Delta f = \frac{2 \times 0.31 \times 12 \times 10^6 \times \cos 34^\circ}{1550} = 3979.38...$
$=$ **4.0 kHz (to 2 s.f.)**

Page 232 — Fact Recall Questions

Q1 Ultrasound is a longitudinal wave with a frequency higher than humans can hear (> 20 000 Hz).

Q2 $Z = \rho c$, where Z is the acoustic impedance in kgm^{-2}s^{-1}, ρ is the density of the material in kgm^{-3} and c is the speed of sound in the medium in ms^{-1}.

Q3 The piezoelectric effect is when a material produces a potential difference when it's deformed, or vice versa.

Q4 Ultrasound waves are generated and detected by an ultrasound transducer. Inside the transducer, an alternating potential difference is applied to piezoelectric crystals, which vibrate to create a sound wave. When the wave is reflected and returns, it causes the crystals to vibrate, which in turn creates an alternating potential difference that is detected in an adjoining circuit.

Q5 Ultrasound imaging doesn't work if there are any air gaps between the transducer and body tissue because the difference in acoustic impedance between the air and the soft tissue would cause the majority of an ultrasound signal to be reflected back. Because of this, a coupling medium with similar acoustic impedance to the skin is needed between the transducer and the body, because this allows ultrasound to be transmitted into the body.

Q6 a) Short pulses of ultrasound are sent into the body while an electron beam inside a cathode ray oscilloscope sweeps across its screen. Reflected pulses are detected and show up as vertical deflections on the oscilloscope screen. The difference in time between transmission and detection of the reflected pulse allows a computer to calculate the depth of an object. E.g. A-scans can be used to measure the depth of an eyeball.

b) In a B-scan the amplitude of reflected pulses is shown as the brightness of spots on a screen. A linear array of transducers allows a 2D image to be formed of inside the body. E.g. B-scans can be used to form a 2D image of a fetus.

Exam-style Questions — pages 234-235

1 D ***(1 mark)***

$I_r \div I_0 = (Z_2 - Z_1)^2 \div (Z_2 + Z_1)^2$
$= ((1.70 \times 10^6) - (1.58 \times 10^{-6}))^2$
$\div ((1.70 \times 10^6) + (1.58 \times 10^6))^2$
$= 0.001338... = 0.00134$ (to 3 s.f.)

2 D ***(1 mark)***

Taking the natural logarithm of $(I \div I_0) = e^{-\mu x}$: $\ln (I \div I_0) = -\mu x$
Rearrange to: $\mu = (-\ln (I \div I_0)) \div x$
If the intensity is reduced by 95%, then the intensity of the attenuated beam (I) is $0.05\, I_0$. So $I \div I_0 = 0.05$.
$\mu = (-\ln (0.05)) \div 4.5 = 0.6657... = 0.67 \text{ cm}^{-1}$ (to 2 s.f.)

3 B ***(1 mark)***

PET scanning is the only imaging technique listed which can make use of a medical tracer bound to glucose.

4 a) The gamma rays travel through a lead collimator so that only rays parallel to holes in the lead can get through ***(1 mark)***. The gamma rays hit the scintillator, which emits a flash of light whenever a gamma ray hits it ***(1 mark)***. Photomultiplier tubes detect the flashes of light from the scintillator and turn them into pulses of electricity ***(1 mark)***.

b) How to grade your answer (pick the description that best matches your answer):
0 marks: There is no relevant information.
1-2 marks: *I* points 3 and 4 have been covered. One *S* point has been mentioned. Answer is basic and has lack of structure, with information missing.
3-4 marks: *I* points 1, 3 and 4 have been mentioned. Two *S* points have been mentioned. Answer has some structure, and information is generally relevant.
5-6 marks: All *I* and *S* points have been covered. The answer is well structured, and information is relevant and presented clearly.
Here are some points your answer may include:
Identifying areas of infection (*I*)
1. White blood cells tagged with technetium-99m are injected into the patient.
2. The white blood cells move through the body and accumulate in areas of infection.
3. Gamma radiation is emitted by the technetium-99m attached to the white blood cells.
4. A gamma camera can be used to detect radiation emitted from technetium-99m.
5. Image is formed by mapping the detected radiation, where high concentrations of radiation correspond to areas of infection.

Suitability of technetium-99m (*S*)
1. Technetium-99m emits gamma radiation which readily passes through the body to be detected.
2. The half-life of technetium-99m is long enough for data to be recorded, but short enough to limit the patient's exposure to radiation to an acceptable level.
3. Technetium-99m decays to a much more stable isotope, which limits the patient's exposure to radiation.

5 a) $E = \frac{hc}{\lambda} \Rightarrow \lambda = \frac{hc}{E}$

The X-rays with the smallest wavelength will be produced by the electrons with the highest energy. The maximum energy of the X-ray photons is equal to the potential difference across the X-ray tube multiplied by the charge of an electron, so:

$$\lambda = \frac{(6.63 \times 10^{-34}) \times (3.00 \times 10^8)}{(31 \times 10^3 \times 1.60 \times 10^{-19})} = 4.010... \times 10^{-11} \text{ m}$$
$$= \mathbf{4.0 \times 10^{-11}\ m\ (to\ 2\ s.f.)}$$

(3 marks total, or 1 mark for saying that minimum wavelength X-rays will be produced by maximum energy electrons and 1 mark for correct working)

b) Any one from: the photoelectric effect, the Compton effect, pair production ***(1 mark)***.

c) Iodine has a high atomic number, so it shows up clearly in X-ray images compared to body tissues ***(1 mark)***. Soft tissues have similar attenuation coefficients, so are difficult to distinguish in X-ray images. By injecting iodine into the blood, the contrast between the blood vessels and the surrounding tissue is improved ***(1 mark)***.

6 a) In an A-scan, a short pulse of ultrasound is sent into the eyeball ***(1 mark)***. Part of the pulse is reflected at the boundary at the back of the eye and is detected by an ultrasound transducer in front of the eye ***(1 mark)***. The time it takes for this echo to come back can be used to calculate the depth of the eyeball ***(1 mark)***.

b) $Z = \rho c = 1.23 \times 340 = 418.2 = \mathbf{420\ kgm^{-2}s^{-1}\ (to\ 2\ s.f.)}$ ***(1 mark)***

c) The lungs contain large volumes of air. The acoustic impedance of human tissue is much higher than the acoustic impedance of air ***(1 mark)***. This means most of the ultrasound's energy will be reflected back from the boundary between the lungs and the surrounding tissue because there is a large impedance mismatch, so very little of the ultrasound's energy will be transmitted past the lungs to produce an image ***(1 mark)***.

d) Rearranging $\frac{\Delta f}{f} = \frac{2v \cos\theta}{c}$ into $y = mx + c$ format:

$$\Delta f = \left(\frac{2vf}{c}\right) \cos\theta$$

So the gradient of the graph is equal to $\frac{2vf}{c}$ ***(1 mark)***.

e) $\text{gradient} = \frac{2vf}{c} \Rightarrow v = \frac{\text{gradient} \times c}{2f}$

$$\text{gradient} = \frac{\text{change in y}}{\text{change in x}} = \frac{(2.2 \times 10^3 - 0)}{(0.95 - 0)} = 2.31... \times 10^3$$

$$\text{So } v = \frac{(2.31... \times 10^3) \times 1550}{2 \times (8.1 \times 10^6)} = 0.221...$$
$$= \mathbf{0.22\ ms^{-1}\ (to\ 2\ s.f.)}$$

(3 marks for correct answer, or 1 mark for calculating the gradient of the graph and 1 mark for correct working)

You can calculate the gradient between any two points on the graph, but it's best to choose two that are far apart.

Glossary

A

Absolute scale of temperature
A temperature scale that does not depend on the properties of any particular substance. The unit of this scale is the kelvin (K).

Absolute zero
The lowest possible temperature, 0 K (or –273 °C), that a substance can have.

Absorption line spectrum
A spectrum produced from light passed through a substance, which contains dark lines corresponding to different wavelengths of light that have been absorbed by the substance.

Accurate result
A result that is really close to the true answer.

Acoustic impedance
The density of a medium multiplied by the speed of sound in that medium.

Activity
The number of unstable nuclei in a radioactive sample that decay per second.

Alpha particle
A particle formed of two protons and two neutrons (the same as a helium nucleus).

Alpha (α) radiation
Nuclear radiation made up of alpha particles.

Angular frequency
The equivalent of angular velocity for an object moving with simple harmonic motion.

Angular velocity
The angle a rotating object moves through per unit time.

Annihilation
When a particle and its antiparticle meet and their mass gets converted to energy in the form of a pair of photons.

Anode
A positively-charged electrode.

Antiparticle
A particle with the same mass as its corresponding particle, but equal and opposite charge.

Asteroid
A chunk of rock and mineral that orbits a star.

Astronomical unit
A unit of distance defined as the mean distance between the Earth and the Sun.

Attenuation (absorption) coefficient
A measure of the fraction of an X-ray beam that is attenuated as it passes through a material.

Avogadro's constant
The number of particles contained in one mole of a substance. It is equal to 6.02×10^{23} mol^{-1}.

B

Background radiation
The weak level of nuclear radiation found everywhere.

Beta-minus (β^-) radiation
Nuclear radiation made up of electrons.

Beta-plus (β^+) radiation
Nuclear radiation made up of positrons.

Big Bang theory
The idea that the universe began from a small, very hot and dense region of space, which exploded and has been expanding ever since.

Binding energy
The energy released when a nucleus forms, or the energy required to separate all the nucleons in that nucleus. Equivalent to the mass defect of the nucleus.

Black hole
An object whose escape velocity is greater than the speed of light.

Blue shift
The shift in wavelength and frequency of an EM wave towards (or beyond) the blue end of the electromagnetic spectrum due to the source moving towards the observer.

Boyle's law
At a constant temperature the pressure p and volume V of an ideal gas are inversely proportional.

Brownian motion
The zigzag, random motion of small particles suspended in a fluid.

C

Capacitance
The amount of charge an object is able to store per unit potential difference (p.d.) across it.

Capacitor
An electrical component that can store charge, usually made up of two conducting plates separated by a dielectric.

Cathode
A negatively-charged electrode.

Centripetal acceleration
The acceleration of an object moving with circular motion. It's directed towards the centre of the circle.

Centripetal force
The force on an object moving with circular motion. It's directed towards the centre of the circle, and is responsible for the object's curved path.

Chain reaction (nuclear)
When the neutrons released by a nuclear fission cause other nuclei to fission and release more neutrons — and so on.

Chandrasekhar limit
The maximum mass of a star core for which the electron degeneracy pressure can counteract the gravitational force.

Citation
A citation is included in the main text of a report to show that a particular bit of information was taken from a source.

Comet
A ball of ice, dust and rock which orbits a star in a highly elliptical orbit.

Contrast medium
Substance with a high atomic number that shows up clearly in X-ray images and can be followed as it moves through a patient's body e.g. a barium meal or iodine.

Control rod
A rod inserted into a nuclear reactor to control the rate of fission by absorbing neutrons.

Control variable
A variable that is kept constant in an experiment.

Correlation
A relationship between two variables.

Cosmic microwave background radiation (CMBR)
A continuous spectrum of microwave radiation, corresponding to a temperature of 2.7 K, that exists throughout the universe.

Cosmological principle
On a large scale, the universe is homogeneous and isotropic, and the laws of physics are the same everywhere.

Coupling medium
A substance applied during ultrasound scans to avoid air gaps between the ultrasound transducer and the body.

Critical damping
Damping such that the amplitude of an oscillation is reduced in the shortest possible time.

Critical mass
The amount of fuel needed for a fission chain reaction to continue at a steady rate on its own.

D

Damping
A force which causes an oscillating object to lose energy and so causes the amplitude of the object's oscillation to decrease.

Dark energy
A type of energy that fills the whole of space, and might explain the accelerating expansion of the universe.

Dark matter
A source of mass in the universe that cannot be seen, but makes up a large amount of the known of the universe.

Decay constant
The probability of an unstable nucleus decaying per unit time. A measure of how quickly an isotope will decay.

Dependent variable
The variable you measure in an experiment.

Diffraction grating
A slide or other thin object that contains lots of equally spaced slits very close together, used to show diffraction patterns of waves.

Doppler effect
The change in the frequency and wavelength of a wave emitted (or reflected) from a source moving towards or away from an observer.

Driving frequency
The frequency of a periodic external driving force which causes an object to oscillate.

Dwarf planet
A planet-like object that orbits a star, but does not meet all the conditions to be considered a planet.

E

Electric field strength
The force per unit positive charge experienced by a body in an electric field.

Electric potential
The work done in bringing a unit positive charge from a point infinitely far away to a specific point in the electric field. / The electric potential energy that a unit charge would have at a specific point.

Electric potential energy
The energy stored by a charge due to its position in an electric field.

Electromagnetic induction
The induction of an electromotive force (e.m.f.) across the ends of a conductor due to a changing external magnetic field.

Electromotive force (e.m.f.)
The amount of electrical energy a power supply transfers to each coulomb of charge.

Electron degeneracy pressure
The pressure that stops electrons being forced into the atomic nucleus, e.g. in the dense core of a star.

Emission line spectrum
A spectrum of bright lines on a dark background corresponding to different wavelengths of light that have been emitted from a light source.

Equation of state of an ideal gas
A combination of the three gas laws, given by $pV = nRT$ or $pV = NkT$.

Escape velocity
The velocity that an object would need to travel at to have just enough kinetic energy to escape a gravitational field.

Event horizon
The boundary of the region around an object inside which the escape velocity is greater than c.

Exponential relationship
A relationship in which two variables are related by the exponential function, e.g. $y = Ae^{kx}$.

F

Faraday's law
The induced e.m.f. is directly proportional to the rate of change of flux linkage.

Field lines
A way of representing a force field.

Force field
A region where an object will experience a non-contact force.

Forced vibration
The oscillation of an object due to an external driving force.

Free vibration
The oscillation of an object with no transfer of energy to or from the surroundings.

Frequency
The number of complete revolutions or cycles that a rotating or oscillating object makes per second.

Fundamental particle
A particle which cannot be split up into smaller particles.

G

Gamma (γ) radiation
Nuclear radiation made up of high-frequency electromagnetic waves/photons (known as gamma rays).

Generator
A device that converts kinetic energy into electrical energy by rotating a coil in a magnetic field.

Geostationary satellite
A satellite that orbits directly over the equator and is always above the same point on Earth. Its orbit takes exactly one day.

Gravitational field
A force field generated by any object with mass which causes any other object with mass within the field to experience an attractive force.

Gravitational field line
Arrows showing the direction of the force that masses would feel in a gravitational field.

Gravitational field strength
The force per unit mass, $\boldsymbol{g}$, experienced by a body in a gravitational field.

Gravitational potential
The gravitational potential at a point is the work done in moving a unit mass from infinity to that point.

Gravitational potential energy
The energy an object has due to its position in a gravitational field.

H

Hadron
A particle made up of quarks that is subject to the strong nuclear force.

Half-life
The average time it takes for the number of undecayed nuclei (or the activity or count rate) in a sample of a radioactive isotope to halve.

Hertzsprung-Russell (HR) diagram
A graph showing a luminosity-temperature plot of many stars. It shows distinct areas corresponding to main sequence stars, red giant stars, super red giant stars and white dwarf stars.

Homogenous
Every part of something is the same as every other part.

Hubble's law
The recessional velocity of a distant object in space is proportional to its distance away from an observer, $\boldsymbol{v} \approx H_0 d$.

Hypothesis
A specific testable statement, based on a theory, about what will happen in a test situation.

I

Ideal gas
A (theoretical) gas that obeys the three gas laws at all temperatures.

Independent variable
The variable you change in an experiment.

Internal energy
The sum of the random distribution of kinetic and potential energies associated with the molecules of a system.

Inverse square law
Any law in which a physical quantity is inversely proportional to the square of the distance.

Ionising radiation
Radiation which, when it hits an atom, can cause the atom to lose an electron.

Isotope
One of two or more forms of an element with the same number of protons but a different number of neutrons.

Isotropic
Everything looks the same in every direction.

K

Kepler's first law
Each planet moves in an ellipse around the Sun, with the Sun at one focus.

Kepler's second law
A line joining the Sun to a planet will sweep out equal areas in equal times.

Kepler's third law
The period of a planet's orbit and the mean distance between the Sun and the planet are related by $T^2 \propto r^3$.

Kinetic energy
The energy of an object due to its motion.

Kinetic model of matter
The idea that solids, liquids and gases are made up of tiny moving or vibrating particles.

Kinetic theory
See kinetic model of matter.

L

Lenz's law
The induced e.m.f. is always in such a direction as to oppose the change that caused it.

Lepton
A fundamental particle that is not subject to the strong nuclear force.

Light year
The distance that electromagnetic waves travel through a vacuum in one year.

Luminosity
The total amount of energy emitted by an object in the form of electromagnetic radiation each second. Also known as power output.

M

Magnetic field
Force field in which a force is exerted on magnetic materials.

Magnetic flux
The total magnetic flux, ϕ, passing through an area, A, perpendicular to a magnetic field, $\boldsymbol{B}$, is equal to $\boldsymbol{B}A$.

Magnetic flux density
The force on one metre of wire carrying a current of one amp at right angles to the magnetic field.

Magnetic flux linkage
The product of the magnetic flux passing through a coil and the number of turns on the coil cutting the flux.

Main sequence
A phase of a star's evolution in which the star is fusing hydrogen in its core.

Mass defect
The difference between the mass of a nucleus and the sum of the individual masses of the nucleons. Equivalent to the binding energy of the nucleus.

Mean square speed
Represents the mean of the squared speeds of all the particles. It has units of m^2s^{-2}.

Medical tracer
A radioactive substance that is used in medicine for imaging inside the body.

Model
A simplified picture of what's physically going on.

Moderator
A material (often water) in a nuclear reactor that slows down neutrons so they can be captured by uranium nuclei (or other fissionable nuclei).

Mole
An amount of substance containing N_A particles, all of which are identical. N_A is the Avogadro constant.

N

Natural frequency
The frequency of an object oscillating freely.

Neutrino
A lepton with zero charge and (almost) zero mass.

Neutron
A neutral hadron with a relative mass of 1, and quark composition udd.

Neutron star
A star mainly made up of neutrons, formed by the collapse of a red giant with a high core mass.

Newton's law of gravitation
The force acting between two point masses is proportional to the product of their masses and inversely proportional to the square of the distance between their centres of mass, or $\mathbf{F} = -\frac{GMm}{r^2}$.

Nuclear fission
The spontaneous or induced splitting of a larger nucleus into two smaller nuclei.

Nuclear fusion
The fusing of two smaller nuclei to form one larger nucleus.

Nuclear model of the atom
A model of the structure of the atom that states that the atom consists of a small, positively charged nucleus containing protons and neutrons, orbited by electrons. The atom is mostly empty space, and the nucleus contains most of the mass of the atom.

Nuclear radiation
Particles or energy released by an unstable nucleus as it decays, e.g. alpha, beta-minus or beta-plus particles, or gamma rays.

Nucleon
A particle in the nucleus of an atom (which can be a proton or a neutron).

Nucleon number
The number of nucleons in the nucleus of an atom.

O

Orbital period
The time taken for a satellite to complete a full orbit.

Orbital speed
The speed at which a satellite travels.

Overdamping
Heavy damping such that the system takes longer to return to equilibrium than a critically damped system.

P

Pair production
The process of converting energy to mass to produce a particle-antiparticle pair.

Parallax
A measure of how much a nearby object (e.g. a star) appears to move in relation to a distant background due to the observer's motion (e.g. as the Earth orbits the Sun). Measured as an angle of parallax.

Parsec
The distance of an object if its angle of parallax, as measured from Earth, is equal to 1 arcsecond = $(1/3600)^\circ$.

Peer review
Where a scientific report is sent out to peers (other scientists) who examine the data and results, and if they think that the conclusion is reasonable it's published.

Period
The time taken for a rotating or oscillating object to complete one revolution or cycle.

Permittivity
The permittivity of a material is a measure of how difficult it is to create an electric field in that material.

Phase difference (of two waves)
A measure of how much one wave lags behind another wave. It can be measured in degrees, radians or fractions of a cycle.

Piezoelectric effect
The production of potential difference across piezoelectric crystals when they are deformed (and vice versa).

Planet
A natural object in space which orbits a star and satisfies certain conditions.

Planetary nebula
The remnants of the outer layers of a red giant that are ejected as the star becomes a white dwarf.

Planetary satellite
Objects which orbit a planet, such as moons and artificial satellites.

Positron
The antiparticle of an electron. Sometimes called a β^+ particle.

Potential energy
Energy that is stored (e.g. elastic potential energy is energy stored in something that has been stretched or compressed, like a spring).

Precise result
A result that is really close to the mean.

Prediction
See hypothesis.

Pressure law
At constant volume, the pressure p of an ideal gas is directly proportional to its absolute temperature T.

Proton
A positively charged hadron with a relative mass of +1, and quark composition uud.

Proton number
The number of protons in the nucleus of an atom.

Protostar
The earliest stage in the life cycle of a star. Protostars are formed when the force of gravity causes clouds of dust and gas to spiral together.

Q

Quark
A fundamental particle that makes up hadrons.

R

Radial electric field
An electric field where the field lines all point towards or away from the central point.

Radian
A unit of measurement for angles. There are 2π radians in a complete circle.

Radioactive dating
A means of dating objects by analysing the activity of an isotope with a known half-life within the object.

Radioactive decay
When an unstable nucleus breaks down to become more stable, by releasing energy and/or particles.

Recessional velocity
The speed at which an object is receding from Earth.

Red giant
A phase of a star's evolution in which the star is fusing larger elements than hydrogen in its core.

Red shift
The shift in wavelength and frequency of EM radiation towards (or beyond) the red end of the electromagnetic spectrum due to the source moving away from the observer.

Red super giant (or super red giant)
A phase of a massive star's evolution during which fusion of large nuclei (up to iron) takes place.

Reference
A detailed description of where a source of information can be found, usually at the end of a document.

Resonance
When an object, driven by a periodic external force at a frequency close to its natural frequency, begins to oscillate with a rapidly increasing amplitude.

Resonant frequency
The frequency of an object oscillating freely.

Root mean square speed
The square root of the mean square speed.

S

Satellite
Any smaller mass which orbits a much larger mass.

Simple harmonic motion
The oscillation of an object where the object's acceleration is directly proportional to its displacement from the midpoint, and is always directed towards the midpoint.

Solenoid
A wire wound into a long coil.

Specific heat capacity
The amount of energy needed to raise the temperature of 1 kg of a substance by 1 K (or 1 °C).

Specific latent heat
The quantity of thermal energy required to change the state of 1 kg of a substance (e.g. to melt or vaporise it) without changing its temperature.

Standard form
A number written in the form $A \times 10^n$, where A is a number between 1 and 10 and n is a whole number.

Standard notation
Summarises all the information about the nucleus of an element in the form ^A_ZX, where A is the nucleon number, Z is the proton number and X is the element's chemical symbol.

Stefan's law
The luminosity of a star is directly proportional to the fourth power of the star's surface temperature and to its surface area: $L = 4\pi r^2\sigma T^4$.

Strong nuclear force
A force with a short range which is attractive at small separations and repulsive at very small separations.

Supernova
The explosion of a high-mass star after its red super giant phase, caused by the core collapsing and the outer layers of the star falling in and rebounding, creating huge shockwaves.

T

Theory
A possible explanation for an observation.

Thermal equilibrium
When two objects are at the same temperature with no net flow of thermal energy between them.

Thermal neutron
A neutron in a nuclear reactor that has been slowed down enough by a moderator that it can be captured by uranium nuclei (or other fissionable nuclei).

Time constant
The time taken for the charge on a discharging capacitor to fall to $\frac{1}{e}$ (about 37%) of its initial value, or for the charge of a charging capacitor to rise to about 63% of the full charge.

Transformer
A device that makes use of electromagnetic induction to change the size of the voltage for an alternating current.

U

Ultrasound waves
Longitudinal waves with higher frequencies than humans can hear (> 20 kHz).

Uniform electric field
An electric field with the same electric field strength everywhere.

V

Valid conclusion
A conclusion that answers the original question.

Validation (of a theory)
The process of testing a theory by repeating experiments, or by using the theory to make new predictions and then testing these new predictions with new experiments.

Variable
A quantity that has the potential to change, e.g. temperature, time, mass and volume.

Velocity selector
A device used to separate out charged particles of a certain velocity from a stream of accelerated charged particles moving at a range of speeds.

W

Weak nuclear force
A force that has a short range and can change the flavour of a quark.

White dwarf
A star with a low luminosity and a high temperature, left behind when a low-mass star stops fusing elements and its core contracts.

Wien's displacement law
The wavelength corresponding to the peak intensity from the continuous emission spectrum of a star is inversely proportional to its surface temperature: $\lambda_{max} \propto \frac{1}{T}$

X

X-ray tube
An electrical circuit contained in an evacuated tube that is used to produce X-rays.

Acknowledgements

OCR Specification statements throughout the book are adapted and reproduced by permission of OCR. OCR, A level Specification Physics A, H556, 2017

Image on p 8 contains public sector information published by the Health and Safety Executive and licensed under the Open Government Licence. http://www.nationalarchives.gov.uk/doc/open-government-licence/version/3/

Photograph acknowledgements

Cover Image © **Paul Fleet** - stock.adobe.com, p 1 Science Photo Library, p 8 (top) **GIPhotoStock**/Science Photo Library, p 14 **Natural History Museum, London**/Science Photo Library, p 15 **Martin Dohrn**/Science Photo Library, p 20 **Science Stock Photography**/Science Photo Library, p 22 Science Photo Library, p 23 **Martin Shields**/Science Photo Library, p 28 Science Photo Library, p 30 **GIPhotoStock**/Science Photo Library, p 39 **Lawrence Berkeley Laboratory**/Science Photo Library, p 44 **Steve Allen**/Science Photo Library, p 47 **GIPhotoStock**/Science Photo Library, p 62 **Victor De Schwanberg**/Science Photo Library, p 66 **Henn Photography**/Science Photo Library, p 73 **European Southern Observatory**/Science Photo Library, p 74 **NASA**/ Science Photo Library, p 76 **Detlev Van Ravenswaay**/Science Photo Library, p 80 **Damian Peach**/Science Photo Library, p 81 **John Chumack**/Science Photo Library, p 86 (top) **Royal Observatory, Edinburgh**/Science Photo Library, p 86 (bottom) **Kim Gordon**/Science Photo Library, p 87 (top) **Royal Observatory, Edinburgh**/ Science Photo Library, p 87 (bottom) **CXC/SAO/F. Seward et al/NASA**/Science Photo Library, p 88 **NASA/ ESA/STSCI/J. Bahcall, Princeton IAS**/Science Photo Library, p 89 **National Library of Congress**/Science Photo Library, p 96 **Carlos Clarivan**/Science Photo Library, p 101 **NASA/WMAP Science Team**/Science Photo Library, p 104 **Emilio Segre Visual Archives/American Institute of Physics**/Science Photo Library, p 105 **Massimo Brega, The Lighthouse**/Science Photo Library, p 112 **GIPhotoStock**/Science Photo Library, p 114 **Sheila Terry**/ Science Photo Library, p 119 **GIPhotoStock**/Science Photo Library, p 129 **Chris Knapton**/Science Photo Library, p 137 **Ted Kinsman**/Science Photo Library, p 140 **Charles D. Winters**/Science Photo Library, p 146 **GIPhotoStock**/Science Photo Library, p 147 **New York Public Library**/Science Photo Library, p 148 **Emilio Segre Visual Archives/American Institute of Physics**/Science Photo Library, p 150 **Nikola Tesla Museum**/ Science Photo Library, p 154 **Lawrence Berkeley Laboratory**/Science Photo Library, p 156 **Lawrence Berkeley Laboratory**/Science Photo Library, p 159 Science Photo Library, p 164 **Chemical Heritage Foundation**/ Science Photo Library, p 166 **Trevor Clifford Photography**/Science Photo Library, p 169 (top) Science Photo Library, p 169 (bottom) **National Physical Laboratory © Crown Copyright**/Science Photo Library, p 172 **Trevor Clifford Photography**/Science Photo Library, p 176 **Prof. Peter Fowler**/Science Photo Library, p 179 **Klaus Guldbrandsen**/Science Photo Library, p 182 **David Parker & Julian Baum**/Science Photo Library, p 185 **Science Source**/Science Photo Library, p 186 **Lawrence Berkeley National Laboratory**/Science Photo Library, p 193 (top) **Andrew Lambert Photography**/Science Photo Library, p 195 **Hank Morgan**/Science Photo Library, p 200 **Trevor Clifford Photography**/Science Photo Library, p 203 **http//woelen.homescience.net/ science/index.html**, this image is licensed under the Creative Commons Attribution-Share Alike 3.0 Unported License, p 205 **James King-Holmes**/Science Photo Library, p 208 Science Photo Library, p 214 **Patrick Landmann**/Science Photo Library, p 215 **Sputnik**/Science Photo Library, p 222 **General Electric Research And Development/Emilio Segre Visual Archives/American Institute of Physics**/Science Photo Library, p 223 **Alfred Pasieka**/Science Photo Library, p 224 **Maximilian Stock Ltd**/Science Photo Library, p 227 **National Institute On Aging**/Science Photo Library, p 229 **Doncaster And Bassetlaw Hospitals**/Science Photo Library, p 231 **Zephyr**/Science Photo Library, p 236 (top) © **David Maliphant**, p 236 (bottom) © **David Maliphant**, p 243 **Wladimir Bulgar**/Science Photo Library

Index

Data Tables

This page summarises some of the constants and values that you might need to refer to when answering questions in this book. Everything here will be provided in your exam data and formulae booklet somewhere... so you need to get used to looking them up and using them correctly. If a number isn't given on this sheet — unlucky... you'll need to remember it as it won't be given to you in the exam.

Physical constants

Quantity	*Value*
acceleration of free fall, g	$9.81\ ms^{-2}$
elementary charge, e	$1.60 \times 10^{-19}\ C$
speed of light in a vacuum, c	$3.00 \times 10^{8}\ ms^{-1}$
Planck constant, h	$6.63 \times 10^{-34}\ Js$
Avogadro constant, N_A	$6.02 \times 10^{23}\ mol^{-1}$
molar gas constant, R	$8.31\ Jmol^{-1}K^{-1}$
Boltzmann constant, k	$1.38 \times 10^{-23}\ JK^{-1}$
gravitational constant, G	$6.67 \times 10^{-11}\ Nm^{2}kg^{-2}$
permittivity of free space, ε_0	$8.85 \times 10^{-12}\ C^{2}N^{-1}m^{-2}\ (Fm^{-1})$
electron rest mass, m_e	$9.11 \times 10^{-31}\ kg$
proton rest mass, m_p	$1.673 \times 10^{-27}\ kg$
neutron rest mass, m_n	$1.675 \times 10^{-27}\ kg$
alpha particle rest mass, m_α	$6.646 \times 10^{-27}\ kg$
Stefan constant, σ	$5.67 \times 10^{-8}\ Wm^{-2}K^{-4}$

Quark charges

Quark	*Charge*
up	$+\frac{2}{3}e$
down	$-\frac{1}{3}e$
strange	$-\frac{1}{3}e$

Conversion factors

Unit	*Conversion*
unified atomic mass unit	$1\ u = 1.661 \times 10^{-27}\ kg$
electronvolt	$1\ eV = 1.60 \times 10^{-19}\ J$
day	$1\ day = 8.64 \times 10^{4}\ s$
year	$1\ year \approx 3.16 \times 10^{7}\ s$
light year	$1\ light\ year \approx 9.5 \times 10^{15}\ m$
parsec	$1\ parsec \approx 3.1 \times 10^{16}\ m$

PRATB61